THE HISTORY OF
THE WORLD
IN
100 ANIMALS

The Animals Entering the Ark *by Jacob Savery (1593–1627)*.

THE HISTORY OF
THE WORLD
IN
100 ANIMALS

SIMON BARNES

SIMON &
SCHUSTER

London · New York · Sydney · Toronto · New Delhi

MMXX

CONTENTS

Detail from The Entry of the Animals into Noah's Ark *by Jan Brueghel the Elder (1568–1625)*.

FOREWORD

'The difference between man and the higher animals,
great as it is, certainly is one of degree and not of kind.'

Charles Darwin, *The Descent of Man*

We are not alone.

We are not alone in the universe. We are not alone on the planet. We are not alone in the wilderness. We are not alone in the farmed countryside. We are not alone in cities. We are not alone in our homes. We are not even alone in the bath or the shower: *Demodex* mites live on our facial skin.

We are humans and we love the idea of our uniqueness. Our thoughts, philosophy, religion, art and even a good deal of science are all based on the assumption of human uniqueness... so much so that we divide the world into animals and humans. The word 'inhuman' is the worst insult in human culture, and we reserve it for Adolf Hitler and Pol Pot, overlooking the fact that a Labrador puppy, a kitten with a ball of wool and the horse the queen rides are all equally inhuman.

The fact is that we humans are as much members of the animal kingdom as the cats and dogs we surround ourselves with, the cows and the fish we eat, the bees who pollinate so many of our food plants and those mites on our faces. We are vertebrates, we are mammals, we are primates, we are apes and we share more than 98 per cent of our DNA with chimpanzees and bonobos.

Our lives, our history and our thoughts are inextricably intertwined with our fellow animals. Non-human animals shaped human lives when our ancestors first walked on the savannahs of Africa 3 million years ago and they have done so ever since.

We have domesticated animals for food and for transport. Animals powered agriculture and so made civilization possible. Animals drove warfare right up to the twentieth century; my grandfather was a sergeant in the Royal Garrison Artillery in Salonika in Greece during the First World War and worked with the horses that pulled the big guns.

'Our lives, our history and our
thoughts are inextricably intertwined
with our fellow animals.'

'There are many estimates of the number of species in the animal kingdom: let's choose 10 million – which is round about the middle – to be going on with. It can be argued that every one of those 10 million has affected humanity in some way or other, even if we don't know about it. And it can also be argued that humanity has affected every single non-human species.'

A species of flea came close to destroying human civilization in Europe. The slaughter of a species of bovines was used to create one civilization and destroy another. Rats have been our despised fellow travellers across the centuries and yet they have provided some of our greatest medical discoveries. Pigeons made possible the biggest single breakthrough in the history of human thought.

We have filled our minds with animals and made them symbols of good and evil. In many religions, including Christianity, God is frequently represented in the form of an animal. We have doves of peace and eagles of war. We have turned to the sea, found a series of ideal foods and hunted the relevant species close to extinction. We humans have looked at the slaughter on the seas and vowed to reform as a species and to make peace with the world and our fellow animals.

We have taken animals into our homes to love and to be comforted by. We have created myths of unimaginable ferocity from the creatures of the wild; we have also used them to create myths of peace-loving nobility. We have tried to understand the world and our place within it by means of non-human animals, and in doing so we have led ourselves through revolutions in the way we understand our lives and the way that we run the planet that we live on.

There are many estimates of the number of species in the animal kingdom: let's choose 10 million – which is round about the middle – to be going on with. It can be argued that every one of those 10 million has affected humanity in some way or other, even if we don't know about it. And it can also be argued that humanity has affected every single non-human species.

It follows that choosing my century of animals – selecting 100 from 10,000,000 – has been a difficult business. Some are obvious: cattle and rats have always been with us. Others are about a more recent awareness: like gorillas, like the species found only on the Galápagos Islands. Some have a profound but less than obvious relationship with our species, like earthworms and wolves. Some species have timeless myths attached to them; others have inspired more modern myths, often subverting the old. Some have changed the human worldview.

I write here in lean unlovely English, but what I write about is not, even remotely, confined to England, to the English or to the English-speaking world. It is a global thing. My subject is the relationship between our own species and the other 9,999,999 – give or take – that make up the kingdom Animalia to which we humans belong.

Zoologists talk of symbiosis: the way that two different species interact. A classic example is the relationship between large African mammals and the two species of oxpecker. These birds relieve buffaloes, hippos and others of external parasites. The process feeds the one and brings relief and better health to the other. Both parties benefit from the fact that they are not alone.

We humans can claim to have a symbiotic relationship with the rest of the kingdom Animalia (and, by extension, to the rest of life on Earth in the other kingdoms in the domain of Eukaryota, these being Plantae, Fungi, Chromista and Protista, and the life in the other domains of Bacteria and Archaea). I hope that these pages can bring this network of relationship into the forefront of our minds and allow us to understand it better. The heresy of human uniqueness has led us across the millennia along the path of destruction. If we were to understand our place in the world better, we might do a better job of looking after it. That might save the whales. That might save the polar bears – the modern emblem of impending loss and destruction. It might even save ourselves.

> *'If we were to understand our place in the world better, we might do a better job of looking after it. That might save the whales. That might save the polar bears – the modern emblem of impending loss and destruction. It might even save ourselves.'*

Note: *When I refer to the conservation status of many of the species in this book, I use the categories and conclusions of the International Union for the Conservation of Nature (IUCN), who run the Red List. The categories are: Extinct; Extinct in the Wild; Critically Endangered; Endangered; Vulnerable; Near Threatened; Least Concern; Data Deficient.*

Is it not brave to be a king?

ONE
LION

'*Wrong will be right, when Aslan comes in sight,*
At the sound of his roar, sorrow will be no more,
When he bares his teeth, winter meets its death,
And when he shakes his mane, we will have spring again.'

C. S. Lewis, *The Lion, the Witch and the Wardrobe*

If you pay a visit to the museum at the Olduvai Gorge in Tanzania, you will see a cast of some footprints. They come from the nearby Laetoli Gorge and they're perhaps the most moving set of footprints on the planet, at any rate as far as humans are concerned. The footprints are 3.6 million years old and they are quite obviously human. That's what makes them interesting: it's not what makes them moving.

Atavistic terror: Daniel in the Lions' Den by Peter Paul Rubens (c.1614–16).

Look closer, then. Some of the footprints are half the size of the others. There seem to be two, perhaps three creatures who made these prints. *People* who made these prints. One, maybe two adults. And their child.

The prints march close together, those of the child alongside that of a grown-up, but not overlapping. Surely – surely this ancient pair are walking hand-in-hand. And with a rush, the twenty-first-century parent and the twenty-first-century child feel a surging time-travelling rush of empathy with those long-vanished walkers.

A parent walks hand-in-hand with a child for restraint, reassurance, protection, love. 'Hold my hand crossing Lupus Street,' my father would demand during my early childhood in Pimlico in London. Not much traffic to worry about in the Laetoli Gorge 3.6 million years ago, but there were other dangers that demanded protection and reassurance. And they're still there.

Lions. Perhaps the most ancient enemy of humankind. Humans first walked upright on the savannahs of Africa, and there they walked with lions. Humans grew and developed and evolved within the senses of lions: lions saw us, heard us, smelt us, felt us and tasted us. Humans were not the dominant animals of the ecosystem: they were prey. Part of us still knows this. I know it from personal experience.

I have worn my best lion story to shreds but not, thank God, to death. Here's the edited version. Me walking. Unarmed. In the Luangwa Valley in Zambia. Surprising a male lion from his sleep. He stood up in his anger perhaps twenty paces away from me. I did exactly the right thing. Nothing. Every muscle locked. Had I turned to run, I would have triggered the chase-reflex and been caught in half-a-dozen strides. But I didn't turn. I stood. And – because he wasn't hungry, because he was the one taken by surprise – he was the one who backed down.

The point of the story is my wholly appropriate response – one that came to me from the very dawn of our species. Part of us still knows what it's like to be prey; what's more – what is a very great deal more to me, because I owe my life to it – is that part of us still knows how to deal with it.

Not that there is much a pre-firearm human can do against a lion who really wants to kill him. In 1898–9, two lions preyed on the humans building the Kenya–Uganda railways. The project was run by Lt-Col. John Henry Patterson, who published an account of this in 1907 called *The Man-Eaters of Tsavo*, in which he claimed the lions managed to kill 135 people before he shot them. I read an account of this unfortunate railway delay in a newspaper under the headline 'The Wrong Kind of Lions'.

We like to believe that the eating of humans is aberrant behaviour: that it will only happen to a lion who is lame or weak-minded or gone in the tooth. Man-eating goes against the natural order: only depraved and decadent lions go in for it. But that's nonsense. Picking off humans in the bush is natural to a lion: and always has been.

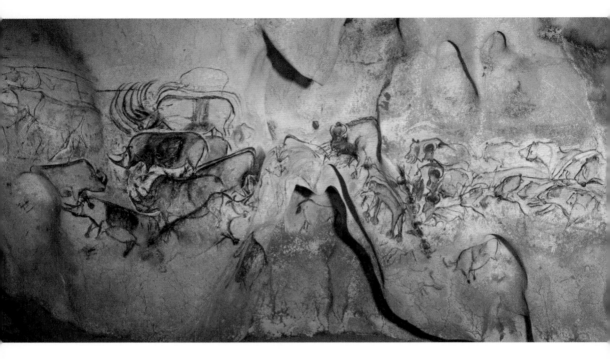

Admired enemy: painting in the Chauvet Cave, France, 32,000–30,000 BC.

We tend to assume that lions are only significant in the deep past of our species: the times before agriculture, before settlement, before civilization. The historical range of lions contradicts that view. Lions were once European beasts: found in Spain, France, Italy and Greece. There were Euro-lions on the Caucasus as late as the tenth century: lions were found in Turkey, across Asia and down to the foot of India. The retreat of lions is the story of the advance of humanity.

Lions have always preyed on humans; and yet humans have always venerated them above all other creatures. The main reason is their drastic sexual dimorphism: the way that a fully grown male is so different from a female that it looks like a different species. The male lion is associated not just with masculinity but also with kingship: king of the jungle, crowned and garlanded with fur, a monarch of all he surveys. Medieval rulers named for lions include: Richard the Lionheart of England; Henry the Lion, Duke of Saxony; William the Lion, King of Scotland; and Robert III, the Lion of Flanders.

Lions are the dominant beasts in heraldry: the monarch of England is represented by three lions passant guardant, and that of Scotland by a lion rampant. Lions abound in Aesop's fables: a mouse rescued the lion that once spared its life, and the notion of the lion's share can also be found there. The first of the twelve labours of Hercules was to slay the Nemean lion.

Lions became double-edged symbols, representing not only courage, manhood and kingship but also human power over nature. As the human victory eventually became a rout, so the dominion of humankind – Man, as people once preferred to say – was travestied in the circus, where lion-tamers walked unafraid in a cage full of beasts and made them sit up on their bottoms and wave their paws in the air: symbols of courage meeting a still greater courage: the most fearsome weaponry of nature meeting a weaponry still more powerful: and that nothing less than the human mind.

In the early part of the twentieth century, big-game hunting was a way of showing how rich and powerful you were, and a lion, of course, was the ultimate bag. There was a feeling that the people who killed them were not only fearfully brave, they were also doing a good deed for humankind. But as the human population grew and grew, the lions were increasingly squeezed out. And then came the backlash. As the Environment Movement began to gather momentum in the 1960s it slowly became clear that the resources of nature and the wild world were not, after all, infinite. And catching this new wave came Elsa.

Elsa showed, as it were, the human side of lions. She was a cub adopted by George and Joy Adamson in Kenya. Joy wrote a book, *Born Free*, that caught the imagination of the world. Camera crews flocked to the bush to see the lion living in harmony with humans: a scene from Eden in which, if the lion didn't lie down with the lamb, she certainly lay down with Joy; the two frequently shared a bed. Elsa played her part in the change of the world's attitudes to the environment and to non-human life. She became a symbol of a new dream: of the new way of looking at the world: one of kindness and tolerance and decency and gentleness: of life respecting life: of love, peace, joy, harmony and understanding. The story was filmed, starring Virginia McKenna as Joy.

The reality was more complex. Elsa came to the Adamsons because George had shot her mother. George later shot one of Elsa's cubs after it had killed his assistant and attacked a child. Both Joy and George were murdered in separate incidents. This was never an idyll: always a tale of violence. But in print-the-legend terms, the story of happy coexistence with lions survived. Unlike anyone else involved.

A more genuine understanding of lions came with prolonged ethological studies that began with George Schaller. He showed that lions scavenge kills from hyenas, a classic example of myth-busting; lions are no more noble than any other species. The social life of lions was revealed as intense, but rather ad hoc and informal, compared to that of wild dogs or, for that matter, hyenas. At its heart is the pride – and that, busting another myth, is not an extension of the glory of the male lion. Every pride is held together by the lionesses: mothers, daughters, aunties, sisters and cousins, tied to each other by blood, upbringing and shared experience: by affection, we might call it, if we weren't so terrified of anthropomorphism. Love, you will be even more inclined to say if you have seen a pride sleeping it off after a feast, all in one great big furry huddle, a big-pawed, warm, lazy camaraderie that everyone who gets close to them half wishes to join, jumping in to roll and roll and roll with the lions.

Lions are classified as Vulnerable. Their population dropped 43 per cent between 1993 and 2014, or in about three lion generations. There are still around 300 lions in India, in the Gir forest. The decline of lions worldwide is put down to indiscriminate killing by humans to protect livestock and human populations, by depletion of their wild prey and by continuing destruction of wild habitat.

But inside the great national parks of Africa lions survive. They do so because people want them to. They are the big prize for the annual influx of tourists, and they have great meaning for the people who live in Africa: why else would Cameroon call their football team the Indomitable Lions? There was global outcry when a lion wandered outside Hwange National Park in Zimbabwe and was killed by an American dentist armed with a bow and arrow. The lion bore the unexpected name of Cecil: proving that, if you want widespread sympathy for the killing of a non-human creature, you must give it a name. The United States Fish and Wildlife Service subsequently put lions on their Endangered species list, making it more difficult for US citizens to kill lions.

Lions matter. They stand on guard in Trafalgar Square to protect Nelson's Column; they snarl on national badges; they stand for any number of mostly male virtues. The Emperor Haile Selassie was the lion of Judah. Aslan dominates C. S. Lewis's *Chronicles of Narnia* as a messiah must. An encounter with lions is on everybody's bucket list. Lions were part of our lives from the dawn of our species: and they haunt our imaginations to this day.

TWO

DOMESTIC CAT

'I am the cat that walks by himself, and all places are alike to me.'

Rudyard Kipling, *Just So Stories*

More or less as soon as we humans began to invent civilization – and thereby began the process of separating ourselves from our fellow animals – we started to take non-human species into our homes. Was the cat the pioneer? The first pet? One theory is that we brought cats into our lives deliberately, in order to control rodents. In other words, the first miracle of human civilization was the herding of cats. I'm inclined to doubt this.

The first and the greatest revolution in human history took place about 12,000 years ago. It was probably a more or less simultaneous event that took place in different parts of the world, give or take the odd millennium, but so far as European civilization is concerned the great advance took place in what's called the Fertile Crescent: that is to say, along the Lower Nile, the Tigris and the Euphrates rivers, an arc of land where humans invented agriculture and irrigation – along with writing and the wheel. It was nothing less than the invention of civilization.

So let us turn to feline history: and as we do so we see that this dramatic change had two simultaneous effects. The first is that humans established stores of grain, to tide them over the less fertile seasons and give them something to plant the following year when the time was right. These stores were inevitably a magnet for mice and rats: and therefore a magnet for their predators. These included the wild cat *Felis silvestris*. Genetic research has shown that our modern domestic cats are descended from the wild cats of West Asia.

In the beginning, these cats were surely tolerated rather than purposefully introduced. But, as we know, a cat will always push its luck. Being bold and curious creatures, with great faith in their own powers of flight, as in running away, they inevitably entered human habitations in search of shelter and an easy meal. Here they would have met with a mixed welcome. It was at this point they were able to unleash their secret weapon.

Purring.

Cats purr for a number of reasons; one of them is the expression of contentment. It's generally suggested that this function of the purr is part of the mother–kitten

The Cat that Walked by Himself: *illustration by the author for* Just So Stories *by Rudyard Kipling (first published in 1902).*

bond: a kitten will purr when getting a good lick-over. But it's a plain fact that humans find the purr beguiling. As the human scratches the cat between the ears, recreating the mother's rough-tongued wash, so the cat purrs – and conveys a sense of meditative calm to the scratcher, bringing down pulse rate, blood pressure and rate of respiration: a pleasure shared across the barrier of species. When cats invaded human homes they brought something with them.

A cat skeleton was found buried with a human in a Neolithic tomb in Cyprus; it is 9500 years old. If a cat was important enough to share a tomb with a human, it must have been a seriously significant animal, either as a species or as an individual. This cat went to its grave more than 4000 years before the first depiction of cats in Egyptian art, busting the theory that the ancient Egyptians invented the domestic cat.

The Egyptians were certainly very keen on cats, and held them in considerable reverence. Bastet was a goddess who originally took the form of lioness, but as Egyptian civilization progressed she became a cat. She was the goddess of the home, women's secrets, childbirth and, of course, cats. Cats were embalmed and buried with humans: it was as if the great advances made by the Egyptians in a thousand different forms were all centred on the domestic god of the cat. You can find the famous figure of what's known as the Gayer-Anderson cat in the British Museum: an elegant Egyptian cat of immense dignity, clearly taking the worship of humanity in its stride, perhaps as a basic feline right.

Thus human civilization advanced to the sound of the purring cat. Cats were not the drivers of civilization, but perhaps they were something of civilization's reward. As humans domesticated themselves, so they brought cats for company. By abandoning the hunter–gatherer life, by moving on from nomadic pastoralism, and by taking on agriculture, humans bound themselves to a lifetime of hard labour. The payback was an increased certainty about existence, along with a permanent dwelling: humans could lay their heads in the same place every night, with the same family or extended family around them, and the same tribe within calling distance, while the domestic cats kept the worst of the rodents clear of the seed corn and purred their songs of contentment when times were good.

Cats cannot help but remind humans of our most ancient foe, the lion. The two species are members of the same family of Felidae: and much of their body language is the same. They are both supple and strong and prone to long bouts of sleepfulness. But here's a fact: lions can't purr. Nor can the other members of the *Panthera* genus: tiger, leopard, jaguar and snow leopard. But they can all roar, the only cat species that can. Perhaps the purring of the cat is the ultimate antidote to the sound of the lion roaring in the night. Roaring is associated with the hyoid bone in the throat: it is incompletely ossified in the *Panthera* cats, and so permits the roaring that is so important a part of their social and territorial behaviour.

One obvious reason for the early domestication of cats is that it was so easy. Cats, for all their social instincts, are at the same time strongly independent. This was the trait celebrated by Rudyard Kipling in what might be the greatest of all cat tales, *The Cat that Walked by Himself*, in the *Just So Stories*. 'I am the cat that walks by himself, and all places are alike to me.' The story is about the tension of the cat's dual nature – half wild, half tame. The cat will amuse the baby and be a comforting presence about the house, but will also take itself off and attend to its own needs.

This mixture of wild and tame is part of the attraction of the domestic cat: and it also makes for a low-maintenance pet. The provision of food and shelter is all that is required: the cat does the rest by himself, on his wild(ish) lone.

Thus the domestic cat is very different to the domestic dog. Unlike most domestic dogs, most domestic cats are, given a reasonably viable environment, perfectly capable of operating as wild creatures, without reference to humankind. Unlike dogs, they have not been bred into a fantastic variety of shapes and forms and behaviours: broadly speaking, a cat is a cat.

And they have indeed gone feral in enormous numbers, all across the world. It has been estimated that there are 25–60 million feral cats in the United States: the colossal variability in those numbers makes it clear that we have no idea at all.

Here they have established a fascinating social life. Most wild cats of the ancestral species *Felis silvestris* are largely solitary, of necessity, because food resources are slight. But, when there is ample food, feral cats live in large colonies, a society based, like that of elephants, around a matriarchy. These dominant females, or queens, will help each other out, suckling each other's kittens and even helping each other to give birth.

Cat as god: image of the Egyptian deity Bastet, c.500 BC, from the British Museum.

It's been estimated that in the United Kingdom cats, both pets and feral animals, kill 64.8 million birds a year: which makes cats controversial creatures in wildlife conservation. Filling the world with obligate carnivores, many of them given strength by human feeding and by scavenging around humans, is not the best way to conserve wild populations of birds, small mammals, reptiles and amphibians, but a campaign against cats would alienate many of the people who happily pay their subs to conservation organizations. The Royal Society for the Protection of Birds (RSPB) in the UK said that cats had 'no significant impact' on bird numbers. However, it has been claimed that cats played a major part in the extinction of eighty-seven bird species, and twenty small mammal species in Australia alone. In 1894, Lyall's wren (sometimes called Stephen's Island wren) was by then found only on Stephen's Island in the Cook Strait, between North and South Island in New Zealand. Shortly after it was first described for science the last one was killed by the lighthouse keeper's cat. The cat was called Tibbles.

The domestic cat, with his wild streak and his tame streak, remains an archetype of modern life. The Cheshire Cat, with his enduring grin, appears in *Alice's Adventures in Wonderland*; Tobermory, the cat that learnt to speak and knew far too much for anyone's comfort, comes in the eponymous short story by Saki; T. S. Eliot took refuge from the existential crises of modernism and the complex consolations of religion by turning to cats, and writing *Old Possum's Book of Practical Cats*; and it was from this work that Andrew Lloyd Webber created the musical *Cats*.

I have no wish to have a cat in my house today, because of the harm it would do to the wildlife of wet wild marshland that surrounds the place I live. But I have occasional pangs of nostalgia for the cats of the past: particularly for Beauty (my wife, then a professional actress, was in a production of *Beauty and the Beast* when we acquired the cat). This cat was indeed comely, also large, ginger and uncompromising. She slept the day away – cats like to sleep for 12–16 hours out of 24 – in my in-basket, making an incongruous oblong of her round ginger body and making a good deal of work morally impossible. When I summoned up the courage to disturb her, rummaging for some essential piece of paper, she purred.

THREE

GORILLA

'The more you learn about the dignity of the gorilla,
the more you want to avoid people.'

Dian Fossey, as quoted in *Los Angeles Times*

The film *King Kong* was made in 1933 and depicts 'the fiercest, most brutal monstrous damned thing that has ever been seen', according to his creator, Merian C. Cooper. In 1979 the thirteen-part wildlife documentary *Life on Earth* was shown on British television; it was broadcast in a hundred territories and viewed by 500 million people.

The elapsed time between the two filmed sequences is almost incomprehensibly brief: no more than forty-six years, a tiny amount even in terms of merely human history. But the moral gap between them is light-years wide. Here is a classic example of species revisionism: the way we have reimagined some of the creatures we share the planet with and completely reversed our previous attitude.

During the final part of the *Life on Earth* series, inevitably dealing with primates, the presenter, David Attenborough, talks about gorillas while sitting arrestingly close to a gorilla family group in Rwanda – and then, in a few brief seconds of broadcast footage, the young male gorillas seek him out and play with him – to his obvious delight – as if they were broadcasting live from Eden.

In the film, King Kong is a giant gorilla. He is captured on Skull Island (where else?) and brought (where else?) to New York. He gets loose, causes havoc and climbs the Empire State Building: yes, a deeply familiar image even to the millions who have never seen the film, one of the most famous images the cinema has ever created. The image of King Kong running amok in Manhattan – nature in a head-on collision with civilization – has become a primal cultural notion. There he still is, and always will be: raging at the marauding aeroplanes with 'a face half-beast, half-human', as Cooper said. King Kong captures the actress Ann Darrow, played by Fay Wray, and his moment of tenderness towards her is what makes him vulnerable. So the planes and their clattering machine guns finish him off. 'Oh no – it was not the aeroplanes. It was beauty killed the beast', as the closing lines of the film explain.

King Kong was released eight years after the notorious Scopes Monkey Trial (more later in Chapter 30), in which a schoolteacher, John T. Scopes, was tried

Enemies no longer: David Attenborough with mountain gorillas during filming for the BBC Life on Earth *series, Rwanda, 1979.*

for teaching evolution. It was a major story all over the United States. Are humans apes or angels? The film showed the giant ape as the savage side of humanity, mixed with just the tiniest drop of not-quite-humanizing tenderness: a brute groping hopelessly for his own forever-inaccessible humanity. He is close to the salvation that comes from being human – but alas it's a little way beyond him.

Attenborough's documentary told another story, for all that his subject was also evolution. He is explaining the advantages of the primate's opposable thumb when the gorillas interrupt. The sequence was, almost inadvertently, committed to film. It was not intended for public consumption, and there were major arguments at the BBC as to whether or not the images of Attenborough and the gorillas at play should be included in the film: many considered it too frivolous. But the sequence – no more than thirty-three seconds of running time – stayed in, bringing to life Attenborough's earlier speech to camera: 'If ever there was a possibility of escaping the human condition and living imaginatively in another creature's world, it must be with the gorilla.' It is a sequence of extraordinary tranquillity, and Attenborough emphasizes that point. 'It seems really very unfair that man should have chosen the gorilla to symbolize everything that is aggressive and violent – all that a gorilla is not and we are.'

The gorilla was once seen as the unchecked, uncontrolled, uncontrollable, bestial side of humanity: all that we have risen above, all that we might sink back into if we let ourselves go. In 1859 – the same year that Charles Darwin published *On the Origin of Species* – Emmanuel Frémiet had a *succès de scandale* at the Paris Salon with his sculpture *Gorilla Carrying Off a Woman*. To complicate matters, it was a female gorilla doing the carrying-off, though a revised 1887 version showed a male. Both were pure fantasy pieces. The gorilla was a symbol of sexual violence and sexual incontinence: human reduced to beast. Objective fact had nothing to do with it. I wonder what those who thrilled to this sculpture would make of the zoological fact that a gorilla's erect penis is 1½in (4cm) long; among the apes by far the largest penis – in both relative and absolute terms – is that of the human.

Gorillas were discovered and described for science comparatively recently, in 1847. But myths and rumours of a great ape, or a bestial human or a human-like beast, had existed for a great deal longer: other primate species had been familiar to humans for centuries, and to postulate a larger and more human-like version was no great imaginative step. The Carthaginian explorer Hanno the Navigator found or was otherwise aware of gorillas in 500 BC; the name gorilla comes from Ancient Greek and means a tribe of hairy women.

The explorer Paul du Chaillu brought the first dead specimens of gorilla to Europe in 1861: two years after the publication of Darwin's *Origin*. There is an illustration of du Chaillu in action: shooting a gorilla standing before him in a pose of apparent supplication. It is captioned 'My first gorilla'.

It was deep in the twentieth century before gorillas were seen as something other than symbols of natural ferocity. The process began with the first serious investigation of the way they actually live. The science of animal behaviour is ethology, and the ethologist George Schaller, already met in these pages as the pioneer observer of lions (see Chapter 1), worked with gorillas and in 1964 produced an excellent popular-science book, *The Year of the Gorilla*. The non-violent gorilla was now in the public domain.

Schaller was followed by Dian Fossey, who studied mountain gorillas in Rwanda from 1966 till her death in 1985. Her extraordinary intimacy with the gorillas, and their hard-worked-for tolerance of her presence, allowed her to understand the gorilla's way of life as no one ever had before. (Attenborough's gorillas were members of the habituated groups that Fossey was studying.) She discovered the way females transfer from one troop to another; the wild world is full of behavioural devices that prevent inbreeding. She also recorded the range and meaning of gorillas' vocalizations, their hierarchies and their social relationships. The most frequent of these vocalizations is a rumbling belch, a sign of contentment. You could, if you like, call that a gorilla's purr. She summed up gorillas as: 'dignified, highly social, gentle giants with individual personalities and strong family relationships'.

Her work is celebrated in her book *Gorillas in the Mist*, which was made into a film in 1988 starring Sigourney Weaver as Fossey. Fossey campaigned against poaching and made enemies. She was said to keep suspected poachers captive, and to beat them. She was murdered.

There are two recognized species of gorilla: eastern and western. Gorillas have often been described as the largest living primate, though extreme examples of humanity can beat them fairly comprehensively: a big male gorilla can reach 430lb (195kg); human records go up to 1400lb (635kg).

Gorillas and humans had a common ancestor – well, so did everything that lives on Earth. It's the point of divergence that is significant, and humans and gorillas split about 7 million years ago. That is to say, pretty recently. Our two species have 95–99 per cent of DNA in common. Gorillas are exclusively vegetarian and live in troops of females and young, with one dominant male, the renowned silverback. Troops with multiple males also exist; subdominant males will defer to the silverback, and will be in pole position to take over when the silverback dies.

Gorillas make and use tools: one gorilla was observed using a stick to measure the depth of a river before crossing, another making a bridge from a tree stump. Gorillas have cultures that vary from place to place, they laugh, grieve, think about the past and the future, and even, it's been claimed, possess what seem to be religious or spiritual feelings.

A gorilla named Koko was taught sign language and used this to communicate with her handlers at the Gorilla Foundation in California. Philosophers and philologists have weighed in, claiming that what Koko did is not language. Anyway not *real* language… and as they explained this they changed the definition of language in an effort to keep the language club exclusive to a single species. In other words, moving the goalposts. What is certain is that gorillas are highly social and great communicators, and so are humans. And it's also been claimed that gorillas have a sense of humour: at one point Koko tied her handler's laces together and then signed 'chase'.

FOUR

GALÁPAGOS MOCKINGBIRDS

*'It never occurred to me, that the production of islands
only a few miles apart, and placed under the same
physical conditions, would be dissimilar.'*

Charles Darwin

There's an odd kind of intimacy in visiting an exhibition while it's still being put together: dustsheets, nameless objects still enclosed in corrugated cardboard and bubble-wrap, priceless treasures on the floor so be careful where you put your feet, and a thrilling opportunity to catch things of immense significance off their guard. I have had this experience a couple of times at the Natural History Museum in London. You don't get the full sweep of what the curators are trying to achieve, but you can sometimes strike up a relationship of unexpected closeness with one or other of the exhibits.

It was like that for me in 2008. Amid the clutter and the bones and furniture there were two birds lying side by side on a purple cushion. To be more accurate, they were bird skins: mere feathers stuffed with cotton waste. They had labels tied to their feet.

But here were two birds – these very birds, these actual specimens, every one of these confirmed and authenticated feathers – that changed the way we humans think about ourselves and about our place in the world. They were mockingbirds: two different species, both found in nearby islands of the Galápagos archipelago.

They were found and shot by Charles Darwin; he also wrote the labels – those very labels – in ink, using a dip pen with a scratchy old nib. And as he completed his five-year journey on HMS *Beagle*, he puzzled about these birds and made notes. Darwin's mind was like a rock crusher: and he thought through his pen. His notebooks are a slow-motion replay of his mind in action. He wrote about the similarities and the differences between the specimens of mockingbirds he had acquired in the Galápagos, and then added: 'If there is the slightest foundation for these remarks, the zoology of the archipelago will be worth examining, for such facts undermine the stability of species.' These last five words seem now to be written in letters of fire.

The birds that changed the world: Galápagos mockingbirds by John Gould, who recognized Darwin's specimens as separate species.

We look back at Darwin's time on the *Beagle* with the glorious frustration of hindsight: look, Charlie, look: it's so *obvious*! But of course it wasn't. The unique nature of the Galápagos wildlife gives a hundred clues to Darwin's great idea, but while he was there he hadn't had it yet – so he didn't see the clues for what they were. He might have got there from the marine iguanas, which he described as 'disgusting clumsy lizards'. But he had seen a museum specimen labelled with the false information that it came from the South American mainland, so the uniqueness of the creature was hidden from him.

Darwin might have got there from the giant tortoises, and in a sense he did. He was intrigued by the fact that the prisoners who then inhabited the islands claimed to be able to tell which island any given tortoise had come from. Darwin took notes: but not action. They took plenty of tortoises onto the *Beagle*, but not one adult made it back to London. They ate them all. The carapaces, which were the most hefty clues, were all thrown overboard.

And he might have got there from the group now known as Darwin's finches, and they were, indeed, important to his thinking: a considerable evolutionary radiation from a single ancestral species. But they weren't proper evidence because he didn't trouble to put the name of the individual island on the labels of the finches he shot. It didn't occur to him then that it mattered.

But with the mockingbirds, he was more meticulous. Darwin liked them, writing that they 'are lively, inquisitive, active, run fast and frequent houses to pick the meat'. There are seventeen species of mockingbird normally recognized today; the northern mockingbird is the one found in the United States and the one in Harper Lee's novel *To Kill a Mockingbird*. Darwin noticed that the mockingbirds of the Galápagos were different to those on the mainland. Crucially, he also noticed that they differed from one island to another. Not that this proved anything. Rather, it raised a number of questions: and the more he thought about the birds, the more questions there were to answer. 'Each variety is constant to its own island,' he wrote. 'This is a parallel fact to the one mentioned about tortoises.'

A 'variety' is a term not much used in zoology now; it means a subgroup within a single species, one with a small degree of difference. When Darwin got back to England, his mockingbirds were formally described for science by the great bird painter and ornithologist John Gould in 1837. Gould was unequivocal. These were not varieties. They were distinct species.

Why did each island have a different species of mockingbird? And then a still more pertinent question: how?

Darwin did not invent the idea of what was then called 'the transmutation of species': that is to say, the idea that new species can arise. His own grandfather Erasmus Darwin had written on the subject in *Zoonomia*. The idea had also been raised by Jean-Baptiste Lamarck among others.

It was not original. But it was still speculation, and pretty unpopular speculation at that. It's an exaggeration to say that biblical literalism was a universal orthodoxy, that everyone believed that the Earth was 6000 years old and that God created light and then, three days later, the sun. But the idea that God was behind it all – in some fairly committed hands-on role – was central to the worldview of Western civilization. This was best expressed by William Paley, who, a generation or so earlier, had famously speculated on what a logical person would think on finding first a stone and then a watch. You'd have no option but to conclude – surely – that 'there must have existed, at some time and at some place or other, an artificer'.

The idea that anything other than purposeful creation lay behind every creature on Earth was anathema. If creation is not purposeful, how can you explain us humans? That's why Darwin's big idea met with such hostility: not because it dethroned God but because it dethroned humanity. We are not God's

Darwin before the prophet's beard: watercolour by George Richmond (1840).

special creation, we are just one more species in the animal kingdom.

Darwin returned from the voyage of the *Beagle* in 1836. It was 'by far the most important event in my life', he said. It 'determined my whole career'. But it was still twenty-three years before the ideas he scratched into his notebook on board ship were available to the world. *On the Origin of Species* was finally published in 1859, and Darwin said that it was 'like confessing to a murder'.

It wasn't just that he suggested that evolution took place. He also explained, with a series of small steps that followed each other with impeccable and remorseless logic, exactly *how* it happened. We shall look at that later on in this book, most notably when we move on to pigeons (see Chapter 22): but, with the publication of *The Origin*, evolution and transmutation were no longer speculation. They demanded acceptance as irrefragable facts. Humans now knew how life worked: and they didn't like it a bit. Still don't.

The Galápagos was not Darwin's eureka experience. The idea that he saw the finches and was instantly enlightened is not, alas, true. Rather, his time on the Galápagos was the crucial event that made the eventual eureka moment possible. And it was the mockingbirds, with their accurate labelling and scrupulous identification, that played the most significant part in the subtle, almost furtive development of Darwin's big idea.

The eureka experience took place a couple of months after his return. Before he set out on the *Beagle* he had decided to become a parson; that idea no longer made sense. He was reading widely, trying to put together a coherent view of life that would hold good after his travels and the life-changing things they had shown him. In 1838 he read Thomas Malthus's *An Essay on the Principle of Population*.

Not all humans born grow up to adulthood. Why? Darwin took this a giant step farther: not all non-human animals grow up to become ancestors. Why? And what does that say about the ones that do survive and become ancestors? Could they possibly have some small advantage that those that failed did not? And over time – in the sense of the relatively recent concept of Deep Time (for more on Deep Times see Chapter 13 on *Tyrannosaurus rex*) – might not these advantages add up and change the nature of the species in question?

Why were there four species of mockingbirds in the Galápagos archipelago? How did they get there? How did they get to be the way they are?

Sigmund Freud said:

Humanity has in the course of time had to endure from the hand of science two great outrages on its naive self-love. The first was when it realised that our earth was not the centre of the universe but only a speck in a world-system of a magnitude hardly conceivable… The second was when biological research robbed man of his particular privilege of having been specially created and relegated him to a descent from the animal world.

(Freud added that he had made the third great outrage: that humans can't even console themselves with the thought that we are rational animals.)

We still haven't really got over Darwin. And it all began with those mockingbirds: with the man who wrote the labels, noted the island of provenance, tied each label with careful inky fingers and then began to think about what he had done. He lifted up the pen again and once more wrote in his notebook.

FIVE

AMERICAN BISON

*'My great forte in killing buffaloes was to get them circling by
riding my horse at the head of the herd and shooting their leaders.
Thus the brutes behind were crowded to the left, so that they were
soon going round and round.'*

William Cody aka Buffalo Bill

Just as no one sees the skyline of New York for the first time, no one sees an
American bison – a buffalo if you prefer – for the first time either. We have
seen images of both too many times already: almost as a shared human archetype,
an image that is now part of the mythology of humankind.

We have watched a thousand chases through those canyon streets of
Manhattan, we have witnessed a thousand kisses in the parks and avenues, we
have heard a thousand songs. In the same way, we have all seen a thousand images
of the buffalo of North America: a towering hump, its tip 6ft (1.8m) from the
ground and, some way below it, that huge head with its neat, economical and
purposeful horns.

The Wild West has become everybody's romantic past: cowboys and Indians,
lawlessness, the people who went west to grow up with the country, the films that
turned all this into a mythology of hard, dangerous men, virgins and whores, good
and evil and the most desperate violence set in a landscape to die for or, of course, in.

The essential emblem of this myth is the buffalo: a beast that stands for a life
and a nation that was once wild, and whose loss is about what America became.
At the turn of the nineteenth century the United States of America had existed
as a nation for only twenty-four years; at the beginning of the twentieth it was
well on its way to becoming the most powerful nation on Earth. At the beginning
of the nineteenth century there were 60 million buffaloes in North America; at
the beginning of the twentieth there were 300 left in the USA. These facts are
not unrelated.

Scientists mostly prefer the term bison, to distinguish the species from
the Cape buffalo of Africa and the water buffalo of Asia. The scientific name is
Bison bison, which is pretty unequivocal. Vernacular English tends to prefer
buffalo; it's even been suggested that the bison is a white-man's term, unacceptable
to Native Americans.

They are large herbivores with an intense social life and a profound need to be together in herds: maternal herds and bachelor groups often combining. They are by nature migratory, like the wildebeest on the African savannahs: they follow the seasons and the flushes of grass on the Great Plains.

There are two subspecies usually recognized, the plains and the woodland buffalo; the woodland being bigger, top males weighing in at 2200lb (1000kg). They are the world's third largest living bovid (cattle relative), after the gaur and the water buffalo, both Asian species. They are more closely related to the European bison or wisent, which still hangs on in small numbers.

American buffaloes evolved to live in large numbers: in a people-scarce world there were vast areas in which they were the dominant large mammals. Their only natural predators were wolves, who would never attempt to tackle a fit adult. The buffaloes lived in a great triangle of land with the Great Bear Lake in Canada at the top, going down as far south as Durango and Nuevo León provinces in Mexico, and east to the Atlantic... so, ironically, the symbol of the Wild West never reached out to the far west.

They were exploited by the Native Americans, the Plains Indians in particular. Buffaloes were food, clothing and shelter: the hides and muscles – that is to say, meat – of buffaloes formed the basis of their civilization, and so it is hardly surprising that buffaloes were regarded as sacred animals.

There weren't domesticated by Native Americans, partly because there was no need. Buffaloes are also hard animals to contain, being huge and lacking the placidity required to make domestication easy. Even today they don't like being fenced in: they can jump 6ft (1.8m), nearly 7ft (2m). They can also trash most kinds of fencing, including razor wire.

It would be overdoing the romanticism to say that, before the white man came, humans and buffaloes lived in harmony. The early humans in North America were responsible for a number of extinctions. The most important technique of slaughter was to drive the target animals over a cliff, and buffaloes were hunted in this way. Fire was used to drive the animals in dry seasons. But there were many buffaloes and not so very many humans: the situation was stable until the European settlers came in.

I first set eyes on a buffalo in Badlands National Park in South Dakota: and it was an unexpectedly powerful experience. I felt a deep personal connection: which doesn't make obvious sense as I am English. But American mythology has become everyone's mythology: I too have ridden with the outlaw Josey Wales and with Ten Bears, and I too dreamt of a land in which both our races could live together in peace, and in harmony with nature. Their dream was my dream. As with Eden, I didn't have to believe in its literal truth; like Eden it has an inescapable meaning for us all.

I had travelled by road to the Badlands park: through Nebraska, the great open-air food factory of the USA. To English eyes it is an abomination: no attempt to retain any form of natural or even semi-natural beauty. It is an uncompromisingly functional landscape. And it was here, rather than on the poor soils of the Badlands, that the buffaloes had their heartland.

No longer. They were first slaughtered by European settlers in the ancient belief that nature is a bottomless well, endlessly self-replenishing. Then, when it became clear that this was no longer so, they were slaughtered as a matter of ad-hoc policy.

William Cody got a contract to supply the Kansas Pacific Railroad with meat; in eighteen months, during 1867–8, he killed 4282 buffaloes. He later had a match with Bill Comstock, the winner to have the rights to the name of Buffalo Bill. In eight hours Cody killed sixty-eight buffaloes to Comstock's forty-eight.

Cody played his part in the destruction of wild America. But he played a still greater part in the creation of its legend. First he was the hero of a novel: *Buffalo Bill, The King of the Border Men* by Ned Buntline, first published in 1869. This was serialized (front page) in the *Chicago Tribune* and had many sequels. After that success Cody went on to found *Buffalo Bill's Wild West* in 1883. It was a touring show that in 1887 travelled to Europe, including Britain: making the western myth an international event.

Killing buffalo was seen by many as essential to the future of the nation. A US army colonel, unnamed, told a wealthy hunter: 'Kill every buffalo you can! Every buffalo dead is an Indian gone!' It was a war on the Indian food supply and on their way of life. Railway companies would provide opportunities to shoot buffalo

The horizon-filler: Buffalo Coming to Water *by E. J. Sawyer, already nostalgic in c.1925.*

from train windows. At the 1870s peak, buffaloes were killed at a rate of 5000 a day. The hide was taken for the leather industry, the hump and the tongue for meat. The rest was mostly left to rot. Some of the Dante-esque scale of the industry is captured in the otherwise unmemorable Kevin Costner film *Wyatt Earp*.

This slaughter was never official government policy. It just happened. But it was certainly encouraged. You could make money by killing buffaloes in large numbers. The buffalo hunters weren't supposed to go onto reservation land, but hell that's where the buffaloes were... Major General Philip Sheridan, who played a major part in claiming the Great Plains for the invaders, said of the buffalo hunters: 'These men have done more in the last two years to settle the vexed Indian question than the entire regular army has done in the last 30 years.' Scorched earth policy – the denial of resources to the enemy – has always been a part of warfare, but the near-extinction of a wild species drove this to a new level.

These are the facts. It is easy to greet them with phoney toughness or with revisionist sentimentality. Either way, the truth is undeniable: the USA would not be the nation it is today without the near-extirpation of the buffalo.

And with this devastating truth has come a modern backlash. Almost a need to apologize. Buffaloes have come back. The relict herd of twenty-five that were left in Yellowstone National Park in Montana is now up to 5000: the project to bring them back in numbers was an early example of species-directed conservation. Buffaloes have been reintroduced into many places where they once roamed, Badlands National Park included.

Many are kept on private land, where they are managed for the meat trade: buffalo meat is less fatty and has lower cholesterol than beef from domestic cattle.

Plenty more where that came from: photo from 1870 showing a pile of buffalo skulls to be ground down and used as fertilizer.

There are around half a million buffaloes in the world today; of which the IUCN reckons that 15,000 are genuinely wild. 'If we bring our herds back to life, we bring our people back to life,' said Fred Dubray of the Cheyenne River Sioux.

Well, we're all on that side these days, in our romantic hearts. Now that the buffaloes have mostly gone, it's an easy thing to believe. All of which left me sitting on the sparse, well-cropped grass in Badlands National Park with a scattered herd of buffaloes not too far away. Not too near either. The beauty of the present was caught up with the imperishable beauty of the past: and in the beautiful myths of wild America. In the shaggy fur, the immense head and the impossibly lofty hump, there seemed to be all the good past, all the great memories of a time I never knew and which in some ways never really was.

In 2016 President Barack Obama declared that the buffalo was now the national animal of the United States. I have no doubts that he was aware of the eternal ironies of that decision.

SIX

ORIENTAL RAT FLEA

*'So, naturalists observe, a flea
Has smaller fleas that on him prey;
And these have smaller still to bite 'em,
And so proceed ad infinitum.'*

Jonathan Swift, 'On Poetry: A Rhapsody'

A lot has been claimed for this species of flea – that it caused the Reformation, perhaps the Renaissance as well, that it killed 200 million humans in a few short years, that it was responsible for wholesale social and cultural changes besides wiping out one-third of the population of Europe all in one go.

Not that the fleas got anything out of it. Neither the deaths nor the consequent changes in human history came from the fleas' own wishes and needs: in defiance of them, if anything. The fleas were a kind of innocent bystander, caught in the crossfire of biology and of history… though they did an awful lot of biting and drank a fair amount of blood in the course of this story.

The oriental rat flea is the vector for the pathogen *Yersinia pestis*, which causes the plague. The plague comes in three forms: bubonic, pneumonic and septicaemic. Bubonic is the least lethal of the three; untreated, its death rate is around 50 per cent, and accounts for three-quarters of plague cases. Bubonic plague can be passed on only by means of a flea bite; pneumonic can also pass from person to person.

Fleas are insects. Like most insects they pass through four life forms: egg, larva – in the fleas' cases, tiny, worm-like things that feed in the dust on dried skin and droppings. The larva then pupates and hatches out as an adult: a small (0.1in/2.5mm long), wingless, brownish – the better to hide in fur – flat-bodied – the better to live in dense fur – creature with long hind legs. These enable it to do the fleas' second most famous trick: they can jump 20in (50cm), about 200 times their own body length, so in proportional terms they are probably the finest leapers that ever evolved. About 2500 species of flea have been described.

It's crucial at this point to understand that most fleas can feed on a variety of warm-blooded creatures – birds and mammals – but tend to breed only on one species, or small group of related species. This association makes it possible for them to meet others of their kind and, therefore, to reproduce. Oriental rat fleas

A black jest: William Blake, The Ghost of a Flea *(1820).*

are, unsurprisingly, associated with rodents, especially the black rat and the great gerbil. It's been speculated that humans evolved hairlessness to outcompete fleas: deprived of a thick pelt to live in, fleas can't exploit humans as semipermanent hosts. But, as anyone who has owned a dog can attest, fleas associated with non-human species of mammal can be trying companions for humans as well as dogs. They bite humans on an opportunistic basis, but prefer dogs as a permanent home.

The mouth of the flea is an admirable and elegant construction, which squirts saliva and part-digested blood into a wound and also sucks blood. It is the first half of this operation that allows the pathogens to be passed on, though the flea gains nothing from the process. When the plague pathogen infects a rat flea, it starts to reproduce in the flea's gut, and the increasing population forms a blockage. As a result, the flea, maddened by hunger, will start to feed aggressively, regurgitating the pathogen into the host as it does so.

> Mark but this flea, and mark in this,
> How little that which thou deniest me is;
> It first sucked me, and now sucks thee…

We could, I suppose, choose this moment to discuss sixteenth-century typography and how the similarity between S and F gives the poet John Donne (quoted here) the opportunity for a dirty joke within a dirty joke, but rather let us understand the poem as society's casual acceptance of fleas, and the inevitability of their biting. No one was aware that plague came from pathogens and that fleas carried them: back then it seemed infinitely more reasonable to assume that the plague was an expression of the wrath of God.

The pathogen, the fleas and the rodents have been present in and around human society for millennia and are still around today. But in times past and in favourable circumstances, the pathogen could get on a roll, and episodes of great killing punctuate history. Signs of plague have been found in the DNA (taken from teeth) of Bronze Age humans. The Justinian Plague took place in the sixth century AD, in the time of the Byzantine emperor Justinian I, reaching Constantinople in 542, spreading east to Asia and west to infect Mediterranean ports. The historic patterns of the plague suggest that it used the Silk Road. Shipboard rats also took the fleas and their lethally charged guts across the known world. (I should add here that the theory of the flea as vector for the plague has been challenged. Some suggest that the plague passed from human to human in the manner of ebola and Covid-19.) The Black Death of 1346–53 is remembered as the most fearsome plague event of them all. Boccaccio wrote in *The Decameron*: 'It was the common practice of most of the neighbours, moved no less by fear of contamination by the putrefying bodies than by charity towards the deceased, to drag the corpses out of the houses with their own hands… and to lay them in front of the doors, where anyone who made the rounds might have seen, especially in the morning, more than he could count.'

It begins with shivering, vomiting, headache, giddiness and intolerance of light, and goes on into limb pain, sleeplessness, apathy and/or delirium. The bubonic plague is marked by swollen lymph nodes – buboes – on the armpit and groin.

We yearn for a big number, to come to terms with the horror: all across Europe and Asia, combined, perhaps 75 million people died; though there are claims of far more, even 200 million. It was world-changing: or seemed so. It took a century

The Black Death: Dance of Death *by Michael Wolgemut (1434–1519).*

or two to repopulate, and half a millennium to reach the modern point of overpopulation: an event that can be seen as an equal and opposite disaster to the Black Death. The Black Birth, perhaps.

It is only in human terms that the Black Death is seen as one of the greatest disasters of all time. It can't be compared to, say, the Permian extinction, in which 96 per cent of all forms of life on Earth was wiped out (see Chapter 100 on polar bears). But for stricken humanity it seemed like the end of the world and a destruction of all certainties.

Religion offers an answer to the biggest question of all: what happens to us when we die? It followed that the Black Death was a religious event. It was accompanied by religious fervour, and – common factor in all times of crisis in human history – persecution of out-groups. It was believed in some places that Jews caused the plague by poisoning the wells. Accordingly, Jews were persecuted and murdered: wiped out in Strasbourg, Mainz and Cologne. This didn't happen in England, because the Jews had already been expelled in 1290. Foreigners, beggars and Romany people were also persecuted.

The church taught that the plague was the result of the sinfulness of humankind, but clergy died as often as the laity: what conclusions were people to draw from such a thing? Still more importantly, the plague showed that the church was powerless to intervene between God and humankind. The priests could do nothing. Live priests fled dying communities, fuelling anticlericalism. Their absence opened the way for the laity to take on the task of devotion for themselves: the beginnings of Protestantism – rejection of the Catholic Church – can be found here.

The Black Death also led to increased social mobility. As the plague receded, a smaller population faced a world with larger opportunities. This was a major blow to the feudal system, for peasants could find work for good wages and escape their traditional ties to the land and its lord. It was, in a way, a chance for society to throw a double six and start again – at least in some areas.

Other outbreaks of plague followed, but in the west there was nothing to beat the Black Death. The Great Plague of London was a comparatively small affair in 1665. The diarist Samuel Pepys wrote on 31 August of that year:

> Up, and after putting several things in order to my removal to Woolwich, the plague having a great increase this week beyond all expectation... Thus the month ends, with great sadness upon the public through the greatness of the plague, everywhere through the Kingdom almost. Every day sadder and sadder news of its increase. In the City died this week 7496, and all of them 6102 of the plague. But it is feared that the true number of dead this week is nearer 10,000, partly from the poor that cannot be taken notice of through the greatness of their number, and partly from the Quakers and others that will not have any bell rung for them.

There were subsequent outbreaks of plague in North Africa, Turkey, Poland, Austria and Germany from 1675 to 1684. Later came what is usually termed the Third Plague Pandemic. It began in Yunnan in southwest China in the 1850s, and hit Hong Kong in 1894. It killed 12 million people in India and China, reached every inhabited continent and was considered active until 1960.

But by the beginning of the twentieth century the disease was on the retreat as a major killing event among the human population. This was probably down to the increased separation of humans and rats: a greater efficiency at keeping rats out of human dwellings, along with more efficient rat-proofing of ships.

The cause of the plague was established; Alexandre Yersin isolated the pathogen in 1894 and Charles Rothschild described the flea in 1903. By the 1930s, antibiotics had been invented and they provide an effective treatment. The plague still occasionally infects individual humans, but the conditions that caused the great dying no longer exist.

However, during the Cold War in the second half of the twentieth century, both sides attempted to develop the plague as a biological weapon (so never doubt which is the most lethal species in this book). In modern life the idea of the plague is more potent than its reality, so this was a weapon designed for panic. Both sides stockpiled countermeasures.

Meanwhile, across the world oriental rat fleas go on biting rodents whenever they get the opportunity: heedless of the role they have played in human history.

SEVEN

CATTLE

'And God said: *"Let us make man in our image, after our likeness: and let them have dominion over the fish of the sea, and over the fowl of the air, and over the cattle, and over all the earth and over every creeping thing that creepeth upon the earth".'*

Genesis 1:26

We humans have chosen a non-human species to define our own and it is cattle. We are what we eat, or so we like to think: and the human food of choice across the millennia has mostly been the muscles of cattle. We have sought to attain the might of the bull – the power of the bull – the bulk of the bull – the prestige of the bull – the masculinity of the bull – by consuming the flesh of the bull. (Bulls, like gorillas and rhinos, get their strength and virility from a strictly vegetarian diet, so if they are what they eat it is because they are grass. Logic plays a relatively small part in the affairs of humankind.) Cattle have shaped human lives throughout history and they direct the way we manage the planet we live on today. Cattle drove the construction of our civilization; they are now making a significant contribution to its destruction.

Cattle. The singular doesn't exist: you can't say, I saw a cattle the other day; most of us would say a cow, even it wasn't female. Cattle originally meant property: like goods and chattels. It's been speculated that cattle were the first form of wealth, and therefore the first form of theft… though that's probably just another layer of bovine mythology.

Entire male cattle are bulls, but you don't see all that many of them, because they are hard to manage. (A neighbour of mine was killed by a bull.) A mature female is a cow. Unweaned cattle are calves; once weaned they become stirks or, sometimes less colourfully, yearlings. Before her first calf, a young female is a heifer; a castrated male – much easier for humans to deal with than an entire bull – is a bullock in most English-speaking countries, but in America he's a steer. To Americans a bullock is a young (entire) bull. A cow or heifer close to calving is a springer; a castrated adult male used for draught work is an ox.

This richness of language surrounding cattle reflects years of human closeness to them. We domesticated them from the wild auroch, which was found in Europe, Asia and North Africa. We were so successful that wild aurochs –

Father of thousands: A Prize Bull *by Edmund Bristow (1787–1876).*

aurochsen if you prefer a funkier plural – were pushed out to the margins, away from human habitation. They went extinct in the seventeenth century.

But there was an intense relationship between humans and cattle long before domestication. Aurochs and the hunting of aurochs are depicted in cave paintings, most famously at Lascaux, where the images are 17,000 years old. A successful hunt resulted in a bonanza of highly nutritious food – but food that didn't keep. There was no option but a feast: and, to this day, the great annual occasions of the church are called feast days (see also Chapter 83 on turkeys). We celebrate with festivals, a word with the same root, and we celebrate carnivals: the Italian word *carnevale* literally means farewell to meat: it marked the last permitted feast day before a season of abstinence, of which the principal one is Lent. We like to separate the ordinary from the special and we have traditionally marked special days (Sundays in the Christian calendar) with meat, and very special days with a very great deal of meat. Feasting is an essential part of human culture: and cattle have played a part in this from the very beginning.

How did we domesticate them? Aurochs were fearsome beasts, the biggest 6ft (1.8m) tall at the shoulder, with colossal and lethal horns. There are various theories, but the most likely is a natural extension of the hunter–gatherer lifestyle.

Following the herd creates an intimacy: an understanding of the herd as a collection of individuals. If you live close to a herd you know each member, and how it differs from the others, in size, in markings, in horn size and shape, in behaviour. You also get to learn the different natures of the individuals. It makes obvious sense to kill (and eat) the most belligerent, when it is safe to do so. That way the herd becomes safer for humans next time around.

What's more – what's a great deal more – is that over time – over considerable periods of time – this process allows the easier-natured beasts to do the breeding. Understanding the powerful effects of time – what to us are unimaginable extents of time – is crucial to understanding not only geology and evolution, but also the development of early human culture and the alterations in the animals we choose to live with (see Chapter 13 on Tyrannosaurus rex). Without needing to think the matter through, humans found themselves selectively breeding for docility. Century by century they were faced with a herd of smaller, more modestly armed animals, all possessed of a gentler nature than their ancestors. They had tamed the auroch: they had created the cow.

The exact process remains a mystery, if not a very profound one. Where and when it happened is fact. All our domestic cattle are descended from a herd of eighty individuals who lived in West Asia 10,500 years ago; this has been demonstrated by DNA analysis. Thus it was that Those Cattle became Our Cows. Or My Cows. Now we could feast whenever we wanted to. Now we could – or some of us could – be rich.

Other species of cattle have been domesticated: the gaur in the Indian subcontinent; the swamp and the river buffalo in East Asia; and in Tibet, the yak. But it is the species descended from the auroch – *Bos taurus* – that rules the world.

Once domesticated, cattle gave humans more than just meat and their hides. They could also be milked. They could be ridden, they could pull stuff: so cattle also advanced the cultivation of crops. An acre (0.4 hectares), traditionally 22 × 220 yards (metres), is the area of land that a pair of oxen could plough in a single day on medium soil. Even the dung of cattle is useful, as fertilizer and, when dried, as fuel.

Advancing technologies allowed humans to keep milk fresh for longer, with yoghurt, which is created by natural fermentation, and cheese. Cheese is made by coagulation of the milk protein casein, and the process creates a product of protein and fat with the addition of rennet, which is a collection of enzymes found in the stomach of ruminants – including cattle. Cheese has the advantage of being highly portable. This meant that people could now travel substantial distances without fear of starving. How big a role did cheese play in the advance of civilization and the opening up of the world? It's a less glamorous product than meat, so that question is not often asked.

Hindus traditionally venerate cows; they are not worshipped as gods, but they are symbols of the bountiful giving of the gods. In the *Mahabharata* it is suggested that you give cows 'the same respect as your mother'. The beloved god Krishna was brought up among cowherds and has the title Govinda, protector of cows; Shiva rides on a buck called Nandi. McDonald's outlets in India sell no beefburgers.

The main asset of domestic cattle is their ability to make grass palatable for humans. Cattle turn grass into protein, and do so with remarkable efficiency. They have stomachs with four compartments, and that allows them to get far more nutrition from grass than animals with simpler digestive systems, like horses and elephants. Cows, like all ruminants, regurgitate semi-digested food and chew it all over again; and this second crack yields double value from what they have eaten.

The French called English *les rosbifs*: a derisive term aimed at their brutish appetites: just bloody great hunks of meat. The English took it as a compliment: no effeminate ragouts and cutlets for us.

This aspirational nature of beef has become a global phenomenon: what was once a treat, a feast, is now widely seen as a daily right applicable to all. As a result, beef production has soared: it's reckoned that there are US 1.4 billion domestic cattle in the world. Meat and dairy herds have become separate. An average English dairy cow can supply 40 pints (22 litres) of milk every day; the problem is that she needs to produce a calf annually to stay in milk. Female calves go on to make milk in their turn; males are mostly slaughtered for veal.

The animal that mattered: detail of a cave painting from Lascaux in France, 15,000–10,000 BC.

The commercial demand for beef means that great areas of land are dedicated to pasture: land that would produce more food if it was devoted to arable farming. Huge tracts of forest have been cut down to make pasture. In Britain ancient woodlands have been reduced across the centuries to a few small relict patches; in more recent times Brazil has destroyed a great deal of the Amazon rainforest and now has more cattle than any other country on Earth.

In other places, notably the United States, beef cattle are fed not on grass but on grain, so it is arable crops rather than grass that get processed into meat. This is efficient in terms of commerce, less so in terms of optimizing land use for food production. Increasingly intense methods of farming require increasingly intense management: the cattle live their lives in barns, in very close proximity, eating, drinking and, when the time comes, dying. This has stimulated a continuing debate on ethics: are we prepared to accept that sentient mammals like ourselves should live drastically impoverished lives? (The global consensus is probably this: yes, so long as we don't have to watch.)

Intense cattle farming of every kind – but especially factory farming – involves antibiotics. About 40 per cent of all the antibiotics produced go into animal feed products, largely on a pre-emptive basis. This speeds up the process of weight gain, which is essential for commercial beef in huge quantities. There are increasing fears that profligate use of antibiotics will encourage strains of pathogens immune to antibiotics: the World Health Organization (WHO) has warned of the danger of a post-antibiotic world in which common infections can no longer be cured. Farm use of antibiotics is rising as medical use is diminishing, though there is now greater awareness of the potential dangers. The use of antibiotics has actually dropped by 53 per cent since 2014 in the UK.

There is a further problem associated with cattle farming and that is methane. Cattle produce it in great quantities: gases expelled at the mouth and the anus. Methane is twenty-eight times more efficient as a greenhouse gas than carbon dioxide. We are, it seems, eating our way to our doom: we have set in motion the process that could make us the first species to fart itself to extinction.

In recent years there has been a backlash against beef. It has been driven more by ideas about health than by ethics. It's been claimed that eating a great deal of red meat gives you a greater risk of heart disease and cardiovascular problems, also cancer, diabetes, obesity and – amusingly, given the traditions involved – erectile dysfunction. There are, needless to say, vigorous counterclaims from the beef industry.

But as the human population rises, so does the population of cattle. We are still in thrall to the prestige, the myth, the magic of beef: the tradition that beef is the food of all food: the food that makes us special, the food that lifted us up from the rest and made us truly human. Our love of beef might yet play a part in the destruction of the planet's ecosystem and, therefore, of ourselves.

EIGHT

BLUE WHALE

'*... for there is no folly of the beast on Earth which is not
infinitely outdone by the madness of men.*'

Herman Melville, *Moby Dick*

You think you're ready for the size of the damn thing, but of course you're not. You couldn't possibly be. It arrives with a hugeness that runs against all your intuitive understanding of the natural world. But at least you know you're going to be amazed by sheer size: you're just amazed at how amazing it really is. What comes without any anticipation whatsoever is the grace, the slenderness of the blue whale. This is not only the largest living animal on the planet; it is also the largest animal that has lived on this planet. It makes dwarves of dinosaurs and minnows of even its nearest relatives.

I have walked under and all around the great models of blue whales in the Natural History Museums in London and New York, but their stillness and glassy-eyed expression cannot prepare you for the real thing: the sound and the movement of the beast: that colossal and simultaneous exhalation and inhalation, the endless rolling past of its body as it prepares to submerge again – more and more and more of it, so you think it will never end. By the time you reach the last yard (metre) or two, it is so slim that its perfect streamlining is quite obvious and, when it slicks up the great tail flukes to dive, quite astonishingly graceful.

Very large animals have haunted human minds from the very beginning, not least because our early human ancestors first walked among them. Naturally, great monsters that come from the unknowable oceans have played a part in our nightmares and myths for centuries: and no doubt many more centuries before we began to write them down. Lord, what must it have been like for hungry humans when some great whale stranded itself on the shore? A miraculous warm Everest of protein: the cruel ocean's capricious gift.

Monsters, ship-eating sea serpents, the terrors and the bottomless bounty of the sea: such things have been part of human understanding for as long as we have been aware of the ocean. And of all the creatures that live beneath the waves, none has captured human imagination more completely than the blue whale, the biggest of them all.

Here be monsters: a blue whale skeleton at the Natural History Museum in London, 2017.

We love stats about hugeness, but when we are confronted with the real thing we don't need them to make the experience count. But let's settle on a length of 100ft (30m), which is a rough-and-ready – and only slightly exaggerated – guide to the biggest ones. The maximum recorded weight is 173 tonnes: but weight is hard to calculate with precision. You can't get a blue whale onto your bathroom scales, after all, so they've never been weighed whole. Some more stats, then: heart as big as a small car, tongue as heavy as an elephant – a small one, around 2.7 tonnes – inside a mouth that can hold 90 tonnes of food and water. A blue whale calf drinks 84 gallons (380 litres) of whale milk a day. A blue whale's penis can measure up to 10ft (3m); unsurprisingly the biggest on record.

We humans tend to equate size with clumsiness, confusing the notion with obesity in our own species. Blue whales make other large whales look stocky, lumpen: their streamlining is extravagant and effective, for blue whales can shift: up to 30mph (48km/h) in bursts (usually in excited social encounters) and have a sustained cruising speed of 12mph (19km/h). At the lower end it seems that they can go on for ever, travelling huge distances with sweeps of that great tail. Up and down: as a cheetah runs, by bending its spine from an N-shape to straight – or even into a shallow U – and back again, so a whale swims.

Blue whales lack teeth: instead they have great hairy sieves in their mouths called baleen plates, around 300 of them. It has become a classic wonder of nature that the biggest animal lives almost entirely on krill, a crustacean the size of your little finger. An adult can consume 40 million in a day: so on an individual basis the blue whale probably causes more daily animal deaths than any other species in the animal kingdom, though we humans can no doubt outcompete them as a species, once we factor in insecticides and so forth. The whales feed by taking an immense mouthful of water and swarming krill. They then push out the water with their elephant-sized tongues, while their sieves retain the krill to be swallowed.

The blue whale was first described by Robert Sibbald in 1694 after an individual got stranded in the Firth of Forth in Scotland; he measured it at 78ft (24m). In 1735 Linnaeus – Carl von Linné – then described it in his great work *Systema Naturae* (see Chapter 11 on platypus, for more on Linnaeus). He gave the blue whale the scientific name *Balaenoptera musculus*. That is also a Latin pun; the second or specific name means both muscled and mouse. No doubt the great Swedish count chuckled into his wig about that one (see Chapter 62 on head louse, for more on wigs, and see Chapter 93, for more on mice).

Whales were not only exploited as a source of meat. They were also a source of light. The blubber, after it had been boiled down in the process called flensing, rendered an oil that was used to fuel lamps. The availability of a genuinely bright source of illumination changed the rhythm of human lives and made night less a time to be feared and endured, and more a time that could be used for enjoyment and profit. In that way, whales changed human possibilities.

A superior oil came from sperm whales. Sperm whales are the largest whales to have teeth, and their huge head allows them to echolocate with fantastic precision. They do this by means of an organ called the melon, and it contains a substance called sperm oil – though it's actually a wax that looks a bit like sperm. The sold stuff was odourless and was a premium product – and that explains why Captain Ahab and the crew of the *Pequod* set sail in search of Moby Dick, who was a sperm whale.

'Be it known that, waiving all argument, I take the good old-fashioned ground that a whale is a fish, and call upon holy Jonah to back me,' declares Ishmael, the narrator of *Moby Dick*. But of course it's not. It's a mammal like us and it breathes air. This was the key point of a trial in New York in 1818. Fish oil, which could also light lamps, was subject to inspection by the authorities, who could levy a fine on those who refused. But it's not fish oil because a whale is not a fish, claimed the owner of three barrels of whale oil. After a lengthy show trial, the jury took fifteen minutes to decide in favour of... fish. Taxation is more important than scientific truth.

An industry in action: Harpooning a Sperm Whale *from William Jardine* The Naturalist's Library *(hand-coloured engraving, 1837).*

Despite this conclusion, fish were on the run. The old biblical categories of life were being challenged by science, and in many places, the Bible was quite obviously wrong. And if the Bible was wrong about whales, its absolute authority was again called into question. Whales played a small but significant part in the secularization of society.

With advancing technology, blue whales became accessible to the whaling industry. As the twentieth century advanced, explosive harpoons were available, fired from factory ships that could drag whales on board through vast doors in the stern, giving the whale no chance to sink. Now blue whales were a target species: it's been estimated that between 1900 and the mid-1960s 360,000 blue whales were killed. Things slowed down after that, because blue whales were a lot harder to find. They were killed for meat and for blubber. Margarine (invented in 1860) is a blend of animal fat, milk and salt, and whale oil was widely used in its manufacture. If you were alive during the 1950s and 1960s, the chances are that you've eaten blue whale, a thought that makes me feel a trifle queasy.

But the 1960s was a time of profound change. In those years the way we saw the world and the way we chose to live our lives were under question. What could be more revolutionary than the idea that humans are not the only animal species that matters? What could be more shocking than the idea of looking after the animals we had until then, to immense profit and satisfaction, been killing without heed?

Save the whales!

It was a radical cultural shift. It was widely agreed that whales were a good thing, that whales needed to be saved, that human lives were richer for the knowledge that whales still swam across the globe. This challenged the view that the wild world was a bottomless pit from which we could forage throughout eternity. Whales became not just a finite resource but also creatures to admire and love.

Greenpeace, founded in 1971, embodied the new ideas about whales and the need for a humane relationship with them. The organization was founded as an anti-nuclear campaigner, but it rapidly widened its ideas. Soon it was filling news bulletins with footage of committed young people recklessly propelling inflatable boats between whales and whaling ships: ram us, harpoon us, do what you will but we're not leaving the whales.

This was vivid stuff and the world warmed to the spectacle, the courage and the new-minted philosophy. It was the first widely understood direct action of what became known as the Environment Movement: and it is probably still its most successful campaign. On 23 July 1982, the International Whaling Commission agreed to a moratorium on whaling, and this came into effect in 1986. Japan, Norway and Iceland rejected this and continue to kill whales.

It is still a remarkable turn around. We refuse to kill wildlife – wildlife whose deaths make a useful or at least profitable contribution to human lives – because we'd sooner have the wildlife. Whale watching is now a recognized business, and seeing a whale is a bucket list staple. I have seen blue whales off Sri Lanka; this is the subspecies humorously known as the pygmy blue whale. I have visited San Ignacio Lagoon in Baja California, Mexico, a place that was once a killing-field, and the grey whales that mass there were known as devil-fish for the way they would attack small boats and deliberately overturn them. I have been out in that lagoon in a small boat and whales have deliberately approached – to be patted and tickled and made much of. Have whales also changed their culture, then?

The current world population of blue whales is estimated by the IUCN at 5000–15,000. That's a fair old margin, but they're not easy animals to census. The numbers are increasing but the species is still classified as Endangered. The Antarctic population was reckoned to be 230,000 in the nineteenth century; it has been estimated that by 2100, assuming the whaling ban is still in place, the Antarctic population will still be less than half its nineteenth-century peak. This is a long-lived, slow-breeding species – a female raises a calf about once every three years – and recovery can only be slow. Continuing threats include ship strike and the effect of undersea human-generated noises on the whales' ability to navigate and to find food and each other by means of sound.

NINE

CORAL

'I can mention many moments that were unforgettable and revelatory, but the single most revelatory three minutes was the first time I put on scuba gear and dived on a coral reef.'

David Attenborough

You can't get greater dependence on an animal than by making the animal your home. The nation of Kiribati (unexpectedly pronounced 'Kiribass') is essentially an animal. Or many animals. It had a population of a little over 118,000 humans in 2018; its landmass is 313sq. miles (811sq. km), all of this dispersed over 1.3 million sq. miles (3.5 million sq. km) of the Pacific Ocean. The nation comprises thirty-two coral atolls and one raised coral island.

Coral is the greatest constructor in the animal kingdom; its scope and range outstrip humans and do so by a considerable margin. The biggest construction on the planet is the Great Barrier Reef, off the coast of Queensland in Australia: 1400 miles (2300km) long and famously visible from space. It comprises 2900 individual reefs over an area of 133,000sq. miles (344,400sq. km). As nations and plutocrats race each other for the brief prestige of the world's tallest building, a notion that Freud would have explained pithily enough, so humankind's best efforts at boastful construction are shamed into pitiful smallness by the work of creatures a good deal smaller than your smallest finger wide and a few centimetres long.

Corals belong to the phylum of cnidarians and so are related to sea anemones. In the same way, they are essentially a soft bag of life with tentacles surrounding a central mouth. They tend to live colonially. They can reproduce both with and without sex; they can clone themselves into a series of genetically identical individuals acting more or less uniformly. Some corals are soft. Some cold-water species can live at considerable depths; they have been found at 10,800ft (3,300m).

But the corals that have had the greatest impact on human lives are the reef-building species: those with individual polyps that are able to secrete calcium carbonate to make a hard shell – the shell that we call coral. As they form colonies, so they create extraordinary structures, including reefs.

Most coral species feed by catching tiny flecks of animal life – loosely called zooplankton – with stinging cells on that central tentacle-surrounded mouth. But the reef-builders have another way of gaining nourishment. They have evolved

The land that lives: Millennium Atoll, sometimes called Carolina Atoll, from the coral island nation of Kiribati.

– co-evolved – a complex symbiosis with algae that live within the tissues of the coral polyps. Being plants, the algae live by photosynthesis – converting light into life. This process also gives the coral polyp 90 per cent of its nutritional needs, while the algae benefit from the waste products of the coral.

The simple brilliance of this strategy has given corals the energy they need to make those impossible structures. There are, of course, limits to the effectiveness of this ploy: and these limits are defined by the availability of the light. In order for the algae to photosynthesize, corals need warm, clear, bright shallow seas. Which explains why they are mostly tropical.

There are certain species or groups of species that create a habitat, create an environment. The Great Plains of North America were kept open by the grazing of millions of bison (see Chapter 5); beavers, as noted later (see Chapter 91 on beavers), manage the hydrodynamics of the places where they live; sea otters maintain kelp forests by preying on the sea urchins that eat kelp. These are

Environment of complexities: octopus in coral reef, off Christmas Island in the Indian Ocean.

keystone species: species without which the entire ecosystem would collapse. Corals create a habitat that strikes human observers with a sense of wonder and disbelief that feels like a physical blow. Rainforests are famous for their diversity, but it can be hard to grasp this as a human walking 100ft (30m) below the great life-filled canopy, hearing only enigmatic sounds and aware only of plaguing insects. If you want to get something of an idea of what biodiversity really means, plunge your face – preferably with a mask – into the waters above a coral reef.

Nothing tells us more vividly of nature's plenty than a coral reef: an explosion of colours, of great numbers, of immense variety. Here in a single splash is a sermon on biodiversity and its conjoined twin bio abundance. We have been shown such things from the 1950s by adventurous film-makers, most notably Jacques Cousteau, whose film *The Silent World* won the Palme d'Or at the Cannes Film Festival in 1956. More recently we have seen the same phenomenon fictionalized in the animated *Finding Nemo* films: a thin plot accompanied by stunning visuals of coral-reef communities.

A single second is all it takes to appreciate the extraordinary nature of coral-reef habitat. I remember my first attempt to snorkel in the Red Sea. It lasted all that length of time – before I emerged spluttering and gasping in wonder, unable to control my breathing in the face of this unfeasible extravagance.

The great abundance of life made possible by the existence of coral reefs has created a food resource long exploited by humans. Here, in a setting that was almost literally like shooting fish in a barrel, was nature's bounty: a swimming buffet. Coral reefs have also helped humans by protecting the shores behind them from wave action and from storms.

Coral reefs create and make possible a long-term, rich and stable ecosystem. They are slow by the works of humans, but quick by the works of geology: it's reckoned that the Great Barrier Reef, though it has been in existence in one form or another for more than half a million years, has existed in its current structure for only 6000–8000 years.

But for how much longer? That, alas, is the question that must be asked. Coral reefs are dying all over the world. The phenomenon is called bleaching, but it has nothing to do with bleach: that just describes the appearance of dead coral, which loses its colour. The process is mostly the result of the rise in sea temperatures. When this occurs, the coral polyps can no longer provide for the needs of the algae – the zooxanthellae – and so the algae can no longer provide for the needs of the coral. This creates a death spiral and reefs are dying. It's been calculated that, in 2016, between 29 per cent and 50 per cent of the Great Barrier Reef was already dead.

This is not the only effect that climate change has had on coral. The Maldives are 80 per cent coral, with more than 400,000 human inhabitants; there are 1192 coral islands in the archipelago. The average height above sea level is 5ft (1.5m); the high point is nearly 8ft (2.4m). Rising sea levels – the result of rising sea temperatures and the consequent melting of the polar icecaps – makes life on such places, as it does for life on Kiribati, increasingly precarious. It has been estimated that the Maldives will be uninhabitable by 2100.

TEN
EAGLE

*'The eagle never lost as much time, as when
he submitted to learn of the crow.'*

William Blake

There came a point in human history when we stopped looking at other species of animals in purely practical terms. We ceased to divide them into those you can eat and those that can eat you; those that can kill you and those that run away; those that sting and those that don't. We started to look at animals with admiration, affection, envy, dislike. We found animals we could identify with; animals to which we could assign arbitrary personalities; animals who could bring us luck; animals who made us laugh. Some of them were painted onto the walls of caves: the lions of Chauvet Cave are 30,000 years old and, if you too have walked with lions and got too close, the images will give you a chill. These painters knew lions all right.

Eagles and military might: The Distribution of the Eagle Standards *by Jacques-Louis David (1810).*

Perhaps they were created in a sort of religious admiration; perhaps there was a kind of magic involved, one intended to bring abundant prey for humans. Perhaps they were supposed to give humans the powers these animals represented: there are plenty of theories to choose from. What is unambiguously clear is that there was an important shift in thinking. We no longer looked at our fellow animals just for what they did. We were increasingly concerned with what they meant. We were creating symbols.

Humans have never sought to add eagles to their diet. Eagles have seldom eaten humans, though you hear occasional tales of a martial eagle taking a baby that had been left to sleep at the side of a field. Before the invention of the shotgun there was little direct interaction between humans and eagles – though no doubt shepherds killed eagles and destroyed nests when they got the opportunity – but eagles still played a major role in human life. We chose them to represent power and might, both personal and national. We even chose them to represent God.

Even today, few people are indifferent to eagles. They command the attention. They are huge, first of all, and they fly with such nonchalance, looking down on the earth as if they owned the place. Who could fail to envy that? To aspire to that? And should you see an eagle perched, the ferocious nature of the enormous hooked beak inspires instant respect. There is also a classic piece of anthropomorphism going on here: in the eagle's face, in the two very large, bright, forward-facing eyes – eyes that can see a great distance and which give stereoscopic vision for homing in on prey – we seem to see an expression of gorgeous ferocity. It is in our human nature to see human faces and human expressions in the branches of trees and in the passing clouds: naturally we do the same thing in the faces of non-human animals. We see cheekiness in mice, wry grins on dogs, sweet gentleness in cows – and ferocious majesty in eagles.

Should we define an eagle? It's not a precise term ornithologically. You could confine it to the genus *Aquila*, sometimes called 'true' eagles, but that misses out bald eagles and African fish eagles, so Americans, Zambians, Zimbabweans and Namibians, for whom these are national birds, would be greatly put out. The birds we loosely call eagles are not all closely related: they just do roughly the same thing: flying extremely well and taking large(ish) vertebrates as prey. In the King James Bible, some of the eagles are certainly vultures and I suspect that others are black kites.

Just about every human civilization has adopted eagles for symbolic and religious purposes, for eagles are found on every continent except Antarctica. Vishnu's chosen mount is Garuda, who has a man's body and the wings of an eagle; the Aztec sun god Huitzilopochtli takes the form of an eagle; so does the Nordic storm giant Thiassi.

Eagles have their being between the ground and the sky, between earth and heaven if you prefer. Zeus took the form of an eagle to abduct the boy Ganymede,

A bird to glory in: bald eagle based on an illustration by John James Audubon, first published in his book The Birds of America *in the 1830s.*

described in the *Iliad* as the most beautiful of mortals, and carried him up to Mount Olympus, where he received the gifts of immortality and eternal youth, and with them the job of acting as Zeus's cup-bearer. This represented a socially acceptable erotic relationship, but not many people see the eagle as a symbol of homosexuality.

The eagle was probably the world's first logo. Because it was Zeus's bird – for the Romans, Jupiter's bird – it was adopted as the symbol of the Roman army. It was portrayed on each legion's standard, above the initials SPQR (Senatus Populusque Romanus, meaning the senate and people of Rome). To lose such a standard was the most terrible disgrace.

Eagles have a powerful dual meaning. They were birds representing human military might; they were also the birds of God. In many a church in Britain you will find the Bible resting on a lectern that takes the shape of an eagle with outspread wings: an eagle is the symbol of St John, for the eagle, caught between heaven and earth, is an intermediary between humans and God. That is precisely what the Bible is supposed to be, particularly the Gospels, and, for many, most particularly the Gospel of St John.

The bird of God is also the bird of the armed human. Eagles appeared on the shields of the great just as lions did. Charlemagne predated heraldry but was given the retrospective coat of arms of an eagle, which became the symbol of the Holy Roman Empire. In 1211, the eagle acquired a second head, looking both ways to show that the empire spanned both Europe and Asia.

Napoleon adopted the eagle as his personal symbol, borrowing the idea from the Romans he admired so much. In 1782 the bald eagle was adopted on the

Great Seal of the United States, carrying thirteen arrows, for the original thirteen states, in one claw and an olive branch in the other; see Chapter 22 on pigeons for the significance of the olive branch.

Prussia adopted the eagle as its symbol; this was taken on by all Germany under the Weimar Republic between 1919 and 1933. It was subsequently claimed and modified by Adolf Hitler, and it became the secondary symbol of Nazi Germany; the eagle was often depicted with a swastika in its claws. It was modified again to symbolize post-war Germany, becoming a less belligerent beast altogether, colloquially referred to as *die fette Henne* (the fat chicken).

As eagles proliferated on coins and flags and banknotes and badges, so they declined in the wild. Many were shot, partly to protect shooting and fishing interests, and partly for reasons of atavistic fear: fear that they would carry off livestock and young humans. It was claimed that, in twelve years to 1930, 70,000 bald eagles were shot in Alaska. Worse was to follow. As the use of pesticides proliferated, most notably dichlorodiphenyltrichloroethane (DDT), eagles were killed more than insects. There will be more on this later (see Chapter 23 on mosquitoes), but suffice it to say now that the poisons got into the food chain. Birds of prey, exposed by the nature of their diet to the highest concentrations of DDT (by eating the insect-eaters), could no longer lay viable eggs. It's been estimated that in the eighteenth century there were up to half a million bald eagles in the contiguous United States; by 1950 there were a reported 412 nest sites. The same pattern was repeated across the world with eagles and all other birds of prey.

The recovery – the partial recovery – from this disaster will be told later (see Chapter 59 on falcons). What is true now is that, in many places, eagles have made a comeback and once established they attract not only admiration but also visitors and the money of visitors. The white-tailed eagle went extinct in Britain at the start of the twentieth century and was reintroduced in the 1990s. The island of Mull, off the coast of Scotland, now has a thriving tourist industry based around eagles: you can see two species of eagle on the island (and also taste two locally distilled single malt whiskies).

In this way eagles have shifted their symbolic ground a little. They are still fierce, but we are now aware of their vulnerability. Where once they stood purely for might they now stand also for fragility. Where once they were feathered conquerors they are now birds that require cherishing. As such they are symbols of humanity's ability and willingness to change. Whether humans will be able to change enough remains to be seen… but if you stand on Mull with a crowd of tourists, many of whom have never looked at birds before with any seriousness, and certainly never through binoculars, you will see people responding to eagles with the atavistic awe of our distant ancestors. At some impossibly deep level, eagles matter to us humans.

ELEVEN

PLATYPUS

*'... perfect resemblance of the beak of a duck... grafted
on the head of a quadruped... naturally excites the idea
of some deceptive preparation by artificial means.'*

George Shaw, *The Naturalist's Miscellany*

'**M**y own suspicion is that the universe is not only queerer than we suppose, but queerer than we can suppose.' These much-quoted lines of the scientist J. B. S. Haldane were written in 1927 but they are essentially timeless: relevant now, no doubt relevant one hundred years on, and equally relevant at the end of the eighteenth century.

The eighteenth century is traditionally seen as the century of the Enlightenment. It was a revolution in the way we suppose. It was also a continuous revelation of the queerness of life. It was the century of Carl von Linné, or Linnaeus, who attempted to name the entire planet, organizing animals and plants into separate kingdoms, and placing each species into different categories according to their degrees of relatedness: anticipating the discovery of genetics and paving the way for the revelations of Charles Darwin.

Where does it fit in? How does it work? Who are its cousins? Is a secretary bird a stork with a hooked bill or a bird of prey with long legs? It was all in the details: small details that ask some of the great questions about life, how it works and what it means. You can find these details in Linnaeus's great work *Systema Naturae*, which was published in a series of thirteen editions in the eighteenth century: first edition 1735, tenth and most important 1758.

The Enlightenment was about many things – perhaps most especially liberty, progress, tolerance, empirical knowledge and rational thought. This led inexorably away from the automatic assumptions of the past. The authority of the church and of the monarch had, under the new way of thinking, to be questioned.

New discoveries brought constant doubting of the old certainties. Ships could travel the length and breadth of the Earth and come back with creatures never mentioned in the Bible, queerer than we had imagined, queerer than we were capable of imagining. So when the first specimen of platypus came to London from Australia in 1798, the scientific establishment was at first unable to deal with it, and diagnosed it as 'a high frolic practised on the scientific community by some colonial prankster'.

Searching for food with electricity: platypus in Tasmania (photograph by David Watts).

Certainly it was a creature hard to believe: the beak of a duck, the fur of a mole, the limbs and feet of an otter, the tail of a beaver: it was against nature, and such a thing could never be. That original skin still exists and can be found at the Natural History Museum in London; apparently if you look closely you can see the marks of the scissors which the scientist George Shaw used in an attempt to reveal the stitches that he believed attached the beak to the head.

But he looked in vain, and this outlying curiosity had to be first accepted and then classified according to the methods and practice of Linnaeus. But it was not until 1884 that we discovered how bewilderingly queer the platypus really is. The males bear venomous spurs on their hind limbs – venom is unusual among mammals, though it's common in reptiles. It was also discovered that the females lay eggs, like many reptiles, but like only four other species of mammals; the other three are all of them echidnas, spiny-covered insect-eaters also found in Australia.

Had Darwin known about the eggs he would have been overjoyed: here was a transition species, halfway between reptiles and mammals. *The Origin* is full of apologies for the lack of transitional creatures, which consistently failed to turn up in the fossil record of life (but see Chapter 31 on Archaeopteryx). Here was a living example: and, what's more, Darwin knew about it. He had even been involved in the killing of one during his travels on the *Beagle*. He wrote: 'I consider it a great feat to be in at the death of so wonderful an animal.' It was more wonderful than even he was capable of imagining; at least at the time.

The discovery of the reptilian links, the betwixt-and-between status of the platypus, led to the notion that the platypus was 'primitive': that it was somehow incomplete and had been superseded by more sophisticated creatures in less forgiving parts of the world than eastern Australia. This reveals a widespread error about evolution: that it is essentially progressive: that it has a goal of increasing proximity to perfection: that evolution reveals a hierarchy of creatures, starting at the bottom with the 'lowliest' creatures, meaning those most unlike humans, and ending up, of course, with our own glorious gorgeous fantastic selves: the goal, the end result, what all this evolution is *for*.

There'll be more on this later, but that is the pleasant, reassuring and smug message that was read from the platypus. Further study showed it was quite wrong. They are remarkable creatures, able to live a semiaquatic life in the tropical rainforests of Queensland and the chilly heights of the Tasmanian Mountains. Their thick fur, capable of trapping air, keeps them warm in water, where they feed mostly on the bottom. They can exploit fast-flowing riffles and still ponds with equal facility. They dig burrows above the water, and in these the female lays her famous eggs. It's worth noting that the Aboriginal people of Australia knew that platypuses lay eggs and carry poisonous spurs, so maybe it would have paid early investigators to ask them. There are ancient Aboriginal tales about the platypus and its refusal to

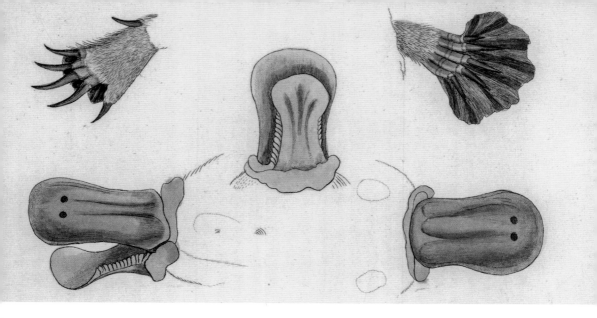

Platypus anatomy: from George Shaw's The Naturalist's Miscellany *of 1799 (artwork by Frederick Polydore Nodder).*

join one of the more conventional groups of creatures: finding dignity and meaning in its singularity, not needing to be part of a crowd for self-justification.

When they dive beneath the surface they close their eyes, ears and nose: posing the question: how do they find the worms, larvae and shrimps they feed on? The answer is perhaps queerer than we can suppose: by electricity. Platypuses detect electrical fields generated by the muscular contractions of their intended prey. This means that they can operate in complete darkness, and so they hunt mostly by night, when, of course, they are a great deal safer from predators. The possession of this highly sophisticated system disqualifies the animals from any disobliging notion of primitiveness. This is a fully valid modern creature perfectly capable of making its living in a twenty-first-century world, in all places where humans haven't destroyed their habitat or too drastically affected the quality of the water.

These days humans study the platypus for knowledge of genetics, and for its degree of relatedness to humans and to the rest of us mammals. The platypus genome, decoded in 2008, shows that the species has its beginnings at the earliest mammalian offshoot from reptiles, a branching that took place 166 million years ago. The platypus contains genes not found in other mammals: but it feeds its young – once they're hatched – on milk and so bears the genes for lactation. That defines it for all time as a mammal. In other words, mammals fed their young on milk before they began to bear live young.

For two centuries and more the platypus has been teaching humans about the meaning of life: and showing us that there are creatures queerer than we are capable of supposing, and they provide the first step on the way to greater knowledge and deeper understanding.

TWELVE

HONEYBEE

'A land flowing with milk and honey; unto the place
of the Canaanites, and the Hittites, and the Amorites,
and the Perizzites, and the Hivites… and the Jebusites.'

Exodus 3:8

I was sitting on the trunk of a fallen tree, taking my ease and drinking a cup of tea. Seeing this, a bird dropped down from the topmost branches and began to sing to me: a long, unbroken and insistent chattering call. Follow me, the bird was saying. Follow me! So I did – and the bird led me straight to the sweat of heaven, the saliva of the stars, the food of the gods…

Honey! For this bird was a greater honeyguide. The human affinity for bees is so intense that it has modified the behaviour of a third kind of animal. And no, this is not folklore: this is hard fact. The bird lives on the savannahs of Africa, and over the course of endless millennia of sharing this habitat with humans it has evolved the trick of honey guiding. The bird will lead you, step by step, to a hive of wild bees. A human honey-hunter will bring down the nest, and the honeyguide will feast on the larvae inside the wild combs. The legend says you must always leave the bird some of your trove because, if you don't, next time you meet him he will lead you to a lion. I neglected to chop down the honey-filled tree; I hope the bird doesn't bear grudges.

For our ancestors out there on the savannah, honey was not just a little bonus, a fun extra in a diet. Honey is the most energy-rich food in nature. Calculations on the diet of a modern hunter–gatherer society in Africa estimated that the people get about 15 per cent of all calories from honey – and that's not including the grubs ingested with the honey. To our ancestors, hunting honey was as important as hunting for meat, and it required just as much cooperation – as well as a considerable willingness to endure pain. Honey was essential to the way the first humans lived. It provided high-calorie nutrition: in short, brain food. We are what we are because of honey; because of the bees.

As our species developed and invented civilization, we came up with ways of making honey easier to come by. No longer were you required to climb a tree, or chop it down: if you caught a swarm and provided it with a place to live, the bees would constantly return to it – and within it they would make honey.

Bringing sweetness to ancient civilizations: hieroglyphs from Amun Temple, Egypt.

There are 20,000 species of bees worldwide. Many of them are solitary. Some are semi-social. But a few are what is termed eusocial, which means benignly social (see Chapter 99 on ants, for more on this). This state of being requires a stratified and highly evolved society that is defined by self-sacrifice and the delegation of labour. The bees' strategy for existence is based around honey: honey gets the members of the hive through the lean periods when there are no flowers around: the winter or, in hotter places, the dry season.

The perils of domesticating bees are obvious: bees are dangerous and defend themselves vigorously. A hive comprises a single queen, who lays the eggs; thousands of worker bees, all female, daughters of the queen; and a smaller number of males, drones, who do nothing except fertilize the queen. The females will guard the hive and attack intruders: an intervention that is usually fatal to the bee, who must leave part of her body behind with the inserted sting; the sting is a modified ovipositor, or egg-laying device, and that is why only females sting.

Bees are basically vegetarian wasps. This change of diet gave them a great advantage over wasps: a wasp looks for food that tries to get away; a bee looks for food that is more or less literally offered to her on a plate. Bees co-evolved with flowers: they carry pollen from plant to plant, allowing them to reproduce sexually. Bees gather pollen as high-protein food to feed the larvae; as they collect

Genius of the bees: Lorenzo Langstroth (1810–95). (Photo courtesy of the American Philosophical Society)

it they scatter some of it into the different flowers they visit, inadvertently creating sexual opportunities as they do so. The flowers respond by offering nectar as a bribe, which makes possible the honey-making and the close-season life of the bees. Flower and bee, bee and flower: inextricable.

Most of the long history of the domestication of bees was about capturing a swarm and providing it with suitable accommodation. The bees establish themselves and get on with the job of filling the place with honey – but then, at the moment of harvesting, the beekeeper had no option but to destroy the hive and with it the colony. If you wanted more honey, you had to start all over again.

Keeping bees was difficult and dangerous, but the rewards were considerable. Sweetness: how many words and phrases in our language equate sweetness with love? Sugar was scarcely known outside Asia until the fifteenth century, and was still considered a luxury in the eighteenth century. If you wanted sweetness, you had to look for honey. 'Thy lips, oh my spouse, drop as the honeycomb: honey and milk are under thy tongue.' This from the Song of Solomon, written it is claimed around 1000 BC and presented here in the Jacobean English of the King James Bible. Both the original author and the translator find the notion of honey inspiring: as sweet as the act of love should be.

There is a strong connection between bees and religion. Beekeeping was a traditional art of the monastery, and the variety of bees bred in Glastonbury Abbey is still universally admired for its combination of productivity and docility. The wax of bees provided good candles: light in the darkness, nothing less. At the Easter Mass, the triumph of life over death is symbolized by the light of a single candle in the darkened place of worship: 'On this night of grace, oh holy father, accept this candle, a solemn offering, the work of bees and of your servants' hands... Never dimmed... for it is fed by melting wax drawn out by mother bees to build a torch so precious.' (From the Exsultet or Eater Proclamation.) Part of the bees' appeal to religion was their apparently perfect society, a model for our own. The idea has been taken up with varying degrees of enthusiasm by Aristotle, Plato, Virgil, Seneca, Erasmus, Shakespeare, Tolstoy and Marx. In Shakespeare's *Henry V*, the Archbishop of Canterbury, taking on the idea with great enthusiasm and a certain confusion of gender – but emphatically grasping the point about division of labour – explains how honeybees:

Teach
The act of order to a peopled kingdom.
They have a king and officers of sorts,
Where some like magistrates correct at home,
Others like merchants venture trade abroad,
Others like soldiers armed with the stings
Make boot upon the summer's velvet buds.

But bees did even more than bring sweetness, light and a moral lesson to our ancestors. They also brought booze. It is possible that mead, made from honey, is the oldest form of alcohol enjoyed by humans. It is a glorious fact of life that practically everything will turn into alcohol if left to its own devices, and honey exposed to natural yeast will ferment. When you water this down you have a drink that has the usual sort of effect, with the percentage of alcohol varying from 3 to 20 per cent. In northern latitudes without grapes, mead was for centuries the drink of choice. It was relegated to quaintness by successive waves of fashion, but is now having something of a revival among young people, not least because of the popularity of the medieval fantasy *Game of Thrones*.

Beekeeping moved very slowly towards the concept of the sustainable hive, the hive from which you can remove combs of honey without destroying the colony. Advances were made in the eighteenth and nineteenth centuries; the American parson Lorenzo Langstroth exploited the discovery that bees create pathways in their hives by inventing the moveable frame, which can be lifted out from the hive and the honey extracted. I remember my grandfather – he who fought in Salonika (see Introduction) – performing this task in his bee-veil and gauntlets: the honey from his hives was one of the great tastes of my childhood.

Modern beekeeping has been greatly troubled by the phenomenon called colony collapse disorder. This happens when worker bees desert the hive, leaving behind a queen and larvae, and often plenty of food. The precise causes are unknown: suggestions include pesticides, mites, fungi, antibiotics and malnutrition, perhaps a combination of some or all of these. It was noticed and named in the United States in 2006. There have been problems in Europe, with low success among overwintering hives in some years, but most experts agree that this is not an identical situation. Perhaps the ultimate problem is the fact that we are, step by step, building a countryside less suitable for bees: fewer flowers, fewer wild plants, fewer opportunities for bees to forage and feed.

This creates a further problem, because pollination is important to humans. It is estimated that every third mouthful we eat is the result of pollination. This subject will be discussed at greater length later (see Chapter 65 on bumblebees), but let us note in passing that the shortage of wild pollinators has created an industry in which honeybees are bussed in at considerable expense to perform the pollination services that, in earlier times, were done for free. The problems of colony collapse disorders have put the costs of pollination up by 20 per cent and more.

The honeybee remains central to human civilization. We lose bees at our very great peril: lose bees and we lose ourselves. And we are losing bees.

THIRTEEN

TYRANNOSAURUS REX

*'Tyrannosaurus and Allosaurus are two of the dinosaurs
that really deserve to be called terrible.'*

Bertha Morris Parker, *The Golden Treasury of Natural History*

We barely understand time at all. A week is a long time in politics; a half-term is an eternity for a young child; Christmas has come around again only twelve months since the last time and it still takes us all by surprise: where *did* the time go?

We have a certain rough-and-ready grasp of time as it affects the life of an individual human: a decade, a generation, two or even three generations… but after that it starts to go fuzzy. Beyond the century – our own best-expectation lifespan – we lose our instinctive understanding of time and it becomes an abstraction. Two or three centuries is like a journey to the moon: the 2000-odd years since the start of Christianity is beyond easy imagining.

It's possible – just possible – to understand multiple thousands. No doubt that's why in former times we decided that the Earth was 6000 years old: it was a number with plenty of mystery, but still just about within our imaginative scope. The invention of agriculture 12,000 years ago is beyond our reach.

And then came the discovery of Deep Time: and with it the shattering, impossible realization that the Earth had been around for a great deal longer than our own species. In fact, just about the entire history of the planet is non-human. But how can anyone imagine 4.5 billion years? Most people get millions and billions confused: our minds are not built for such numbers. As the scientist Richard Dawkins wrote: 'Our brains have

Active hunter: the T. rex known as Sue, mounted in dynamic pose at the Field Museum, Chicago.

Haunter of imaginations: T. rex and Triceratops (twentieth-century painting by Charles R. Knight).

evolved to deal with the problems within the orders of size and speed which our bodies operate at.'

So we need an image: something we can get our minds around. The writer John McPhee, in a much-quoted example, wrote: 'Consider the Earth's history as the old measure of the English yard, the distance from the king's nose to the top of his outstretched hand. One stroke of a nail file on his middle finger erases human history.'

The discovery of Deep Time was initially a matter for the geologists. James Hutton came up with the notion in the eighteenth century; and the idea was further examined and developed by Charles Lyell in his *Principles of Geology*, which was published in three volumes between 1830 and 1833. Darwin read Lyell on the *Beagle*: without the concept of Deep Time he could never have developed his own ideas.

But since the beginning of the twentieth century a single creature has given us all some kind of understanding of what Deep Time really means, and allowed us to get hold of the idea that humans haven't always dominated the planet. It's a lesson we learn as children and one we take on avidly: for who could resist the most fearful predator that ever walked on the surface of the Earth?

All dinosaurs have their thrilling side, but *Tyrannosaurus rex* is the most thrilling of them all. It was not only big and fierce but, to add to the glamour and the mystery, it is also dead, extinct, gone, never to be seen again. I remember discovering dinosaurs as I read *The Golden Treasury of Natural History* by Bertha Morris Parker, and I gazed at a painting that showed a battle to the death between Tyrannosaurus and the horned, helmeted dinosaur Triceratops: so big! So fierce! So dead! And so long ago...

The first Tyrannosaurus teeth were found in North America in 1874; vertebrae were found in 1892. In 1900 Barnum Brown found a partial skeleton in East Wyoming; he found another in the Hell Creek Formation in Montana in 1902. The species was named in 1905, and the naming was a piece of creative genius. The man responsible was Henry Fairfield Osborn, president of the American Museum of Natural History. The mystery of their names is part of every dinosaur's glamour: they are all known by their scientific name: no common names for dinosaurs. Scientists call a house sparrow *Passer domesticus* but nobody else does. But everyone calls *Tyrannosaurus rex* by his scientific name, one that rolls off the tongue. It means king of the tyrant lizards.

Linnaeus (see Chapter 8 on blue whales and Chapter 11 on the platypus) established science's double-name system, or binomials. *Tyrannosaurus rex* is perhaps the most widely known binomial in the world, with the possible exception of *Homo sapiens*. The basic working principle of the way we understand and categorize life on Earth is taught to us by the king of the tyrant lizards, and it comes with an understanding of Deep Time.

Our vivid grasp of the nature of *Tyrannosaurus rex* is largely the gift of Charles R. Knight, who made a speciality of painting extinct animals. In 1926 he began a massive project of twenty-eight murals for the Field Museum of Natural History in Chicago. It took him four years and the most famous shows – yes – Tyrannosaurus in battle with Triceratops. It was an image that, in today's terms, went viral: and, in altered form, it made it into my great childhood text book.

These early discoveries were the only Tyrannosaurus fossils known until the beginning of the 1960s; since then another forty-two have been found. One of the most complete was discovered by the amateur palaeontologist Sue

Hendrickson, who found it in the same Hell Creek Formation; it was purchased by the Field Museum for US$7.6 million and has been on display since 2000; it has been nicknamed Sue, though no one knows which sex she really is. She is 40ft (12m) long, stands more than 12ft (3.6m) tall at the hip and is the biggest known specimen of Tyrannosaurus.

Knight portrayed the animal with its body at 45 degrees to the ground, dragging a heavy tail. The specimen in the American Museum of Natural History in New York was mounted in that position in 1915 and displayed that way for the next seventy-seven years. There is, in Knight's work, always a hint of the redundancy of the dinosaurs: as if they were always doomed to be superseded. He was of course correct in this, but dinosaurs were the dominant megafauna on Earth for about 150 million years, a figure that certainly gives us humans something to aim for. It pleases human vanity to depict dinosaurs as creatures that paved the way for their betters – so much so that we refer to a person with old-fashioned views as a dinosaur. In the film *GoldenEye*, M, played by Dame Judi Dench, tells James Bond: 'You're a sexist misogynist dinosaur.'

The James Bond franchise was going through a bit of revisionism at the time. We have also gone through a process of revision in the way we portray dinosaurs. In the film *Jurassic Park* the Velociraptors were shown as fast, active, intelligent and cooperative. There has also been some extensive revisionism around Tyrannosaurus. It is now accepted that he carried his body parallel to the ground, big tail in the air as a counterweight to the vast head; the New York specimen has been rearranged in this position. The 5ft (1.5m) skull with its 6in (15cm) teeth would have delivered the largest bite of any terrestrial animal in history.

There has been debate about whether Tyrannosaurus was a scavenger rather than an apex predator; current thinking suggests both. Certainly Tyrannosaurus has obvious two-eyed – stereoscopic – vision, like predators across all taxa. There have been debates about whether the species was warm-blooded, and whether it had feathers, how fast it was able to travel: there have been suggestions that it was able to run at 43mph (70km/h), while more recent studies say it couldn't run at all – no moment of suspension being possible – but could walk briskly at 11mph (18km/h). Subsequent studies say that *Tyrannosaurus rex* was a decent runner after all; no doubt this one will run and run.

Tyrannosaurus remains a perpetual hunter of the human imagination: first encountered in childhood and never forgotten. It tells us about the fact of evolution and about the long history of life on Earth that made it possible. It remains a problem for biblical fundamentalists who, in some moods, insist that in the 6000 years of Earth's history Tyrannosaurus somehow failed to make Noah's Ark. Tyrannosaurus is also our first lesson about extinction, and that's a topic that, alas, we cannot help but revisit more than once later in this book.

FOURTEEN
SHARK

'Just when you thought it was safe to go back in the water.'

Peter Benchley, *Jaws*

Of all the claims that we humans make about ourselves, perhaps the most absurd is that we are rational, and that our rationality is the quality that sets us apart from the beasts. The fact of the matter is that we don't live rational lives. Perhaps we are incapable of doing so; perhaps we don't even try very hard.

Take monsters. It seems that we cannot live without them: that we have no wish to live without them. Monsters are part of every human culture: some terrible mindless force of nature that aims to destroy human civilization. In every age monsters have a new face – yesterday a dragon, today the Incredible Hulk – but they are all related to each other: the terror that lurks unseen just below the horizon, ready to come from the darkness, from the deep, from the unknown, from within ourselves. Rationally, it would make more sense to fear bad drivers and speeding vehicles.

Perhaps it all stems from lions: the fact that our ancestors lived and walked with lions every day of their lives (see Chapter 1 on lions). Our minds evolved over the course of Deep Time, and, although there have been many drastic changes to human life over the past few thousand years, our minds have not changed at the same rate. We still fear monsters: and part of us still relishes that fear.

We fear sharks. We fear sharks perhaps more than any other non-human creature on the planet. They rise from the depths and kill us. They make rather a point of killing us while we are enjoying ourselves, on holiday, and that makes things even worse. They stand for the motiveless malevolence of the non-human world: they don't destroy us because they want to or because they need to: they destroy just because they can. Their mindlessness is part of the threat: brute engines of destruction.

We have savoured monsters from the deep since we first put our toes in the water. In Greek legend Lamia was bedded by Zeus and Zeus's wife Hera killed Lamia's children in revenge. Zeus responded by turning Lamia into a monster shark, so that she could revenge herself on the world by eating the innocent children of others. In the eponymous Old English epic, Beowulf dives into the

depths of the lake to fight Grendel's mother: not specifically a shark but one of the legion of water monsters that have haunted human imagination for uncountable years.

Seafaring people have always cherished legends about sea serpents and other monsters of the deep that destroy ships and eat sailors for the fun of it. Water is not our element: we enter as intruders: when we take to the waves the advantages are all on the side of the enemy. The fear of what might strike from the sea is deeply embedded in us: and in modern times, as we learnt more about the nature of life under the ocean, this fear came to centre on the shark.

The shark became a monster for us all, no longer confined to ancient mariners. As sea bathing became a recognized form of human pleasure, we found the pleasure mixed with an atavistic tinge of fear. Perhaps that's a small but important part of the experience: like the Tabasco in a Bloody Mary.

Everybody who has ever swum in the sea has swum with sharks. They're everywhere: in every sea and at every depth to about 6500ft (2000m). There are about 500 different species of shark, of which four have been involved in attacks on humans. Sharks range in size from the lantern shark, which is about 7in (17cm) long, to the whale shark, at 40ft (12m). Neither is a danger to humans: one is too tiny, the other filter-feeds on plankton. A couple of species, the river shark and the bull shark, can live in fresh water as well as salt water. Sharks were known as sea dogs till the sixteenth century. You can still find echoes of that; a small shark found in British waters is called the dogfish.

One of the reasons that sharks seem so instantly menacing is that they are not like most other fish you see. The fact is that, in zoological terms, the word 'fish' means nothing. You can't put all 'fish' into one big related group, as you can birds, otherwise known as the class of Aves. There are four different groups of 'fish' – each one representing a different class in the phylum of vertebrates. The largest class is the ray-finned fish, sometimes called the bony fish: like goldfish, salmon and cod (see Chapter 17 on cod), like most of the fish you see swimming over a coral reef.

The second largest order is cartilaginous fish: sharks and rays. Their skeletons are not made of hard bone like ours, like most of the other creatures in our home phylum. They are made of cartilage, like your ears and your nose. Cartilage is lightweight and flexible, and allows the sharks their exaggerated side-to-side swimming action.

Ray-finned fish can hang perfectly still in the water, because they possess a swim bladder: it is their most obvious characteristic, as your goldfish will show you. Cartilaginous fish do not: so they must either sink to the bottom or carry on

Coming to get you: poster for the film Jaws (1975).

moving. The related rays take the first option; so do some sharks, like the nurse shark. But most sharks go in for constant swimming. They need to keep a flow of water over their gill slits to take oxygen from the water. Nurse sharks can push water past while remaining stationary; most sharks must keep moving, even in their sleep. Their skeletons are lighter than bone, and their vast livers aid buoyancy. But they need to maintain dynamic lift to survive, and so they must swim every moment of their lives.

The jaws that fascinate us so much are filled with teeth that move on a conveyor belt system: it's been estimated that a shark may lose as many as 30,000 teeth in the course of its lifetime. They are embedded in the gums rather than attached to the jaw like our own, and they are constantly discarded and replaced. The jaws are not attached to the cranium like our own, and that gives sharks their extraordinary alien expression, as we humans perceive it.

We have never quite decided whether we want our monster shark to be an evil genius of the hunt or driven by blind idiot rage. Is intelligence or stupidity more frightening? Take your choice, but the fact is that sharks are remarkable and effective predators. Their sense of smell is as acute as legend has it: sensitive to blood in the water at parts of one to 1 million. This sense is also directional, like our own hearing. That's because the distance between the shark's nostrils gives them a crossbearing; it's been speculated that this ability in heightened form explains the head-shape of the hammerhead sharks. Sharks have good underwater vision and acute hearing. They also possess a sense of electroreception, like the platypus already met (see Chapter 11); hammerheads are particularly adept here. The brain-to-body proportions of all sharks are much the same as most mammals and birds; learning ability has been observed, and sharks have been seen to play. Some species are highly social; social life is a potent driver of intelligence.

Sharks won a sharp and dizzying promotion in the human hierarchy of monsters during the summer of 1916. Between 1 July and 12 July in New Jersey, a shark (or sharks) killed four people and injured another. It was a time when the combination of a heatwave and a polio epidemic encouraged people to seek refuge by the sea. As the deaths began, the journalists rolled into town and the shark attacks became a huge story. Polio was horrifying, but these deaths were thrilling.

Aspects of this story were recreated in Peter Benchley's 1974 novel *Jaws*, which a year later became a classic monster film, directed by Steven Spielberg. Following the physiological truth that it's always better to have two jaws than one, *Jaws 2*, directed by Jeannot Szwarc was released with a tagline that became part of the English language: 'just when you thought it was safe to go back in the water'.

This is all about fear – but it is certainly not a rational fear. The International Shark Attack File lists 2785 unprovoked shark attacks on humans between 1958 and 2016, of which 439 were fatal. That averages at a little under eight shark deaths per year. Only three of the 500-odd species have been heavily involved in these deaths: great white, tiger and bull sharks. These figures don't take into account unreported deaths; records from developing nations are scanty.

On the other hand, humans kill approximately 100 million sharks a year. A lot of these we eat. Sharks were once a staple of British fish and chips, under the euphemism of rock salmon, until they became scarce through overfishing. Shark is widely served under many other aliases, including flake, huss, steak fish, lemon fish and even tofu fish.

In Chinese culture, shark's fin soup is a delicacy, and one that people will pay a great deal of money for: US$150 a bowl has been quoted. The fish are caught, the fins removed and the rest is chucked back into the sea to die. The fin has no taste: the texture is chewy. It is said to boost sexual potency, to restore energy and rejuvenate, and to prevent cancer. It is a highly prestigious dish to serve. In recent years the dish has somewhat fallen out of favour: the basketball star Yao Ming, who played in the National Basketball Association of the United States, headed a campaign against the consumption of shark's fin soup. As a result consumption has dropped.

Sharks are also routinely culled in many places, to protect bathers, for example in Australia, South Africa and Réunion. This is not so much to protect bathers from sharks as to protect bathers from the fear of sharks: and so encourage visitors and boost profits. The plot of *Jaws* highlighted the same notion: that sharks – or rather, the fear of sharks – is bad for local business. Between 2001 and 2017, 10,480 sharks were killed in Queensland, Australia, including around the Great Barrier Reef. Since then courts have ruled that this indiscriminate killing can't go on.

Humans have reason to fear sharks, but sharks have far more reason to fear humans. The annual score in the match between humans and sharks is 100 million to us and eight to them.

FIFTEEN

COCKROACH

*'You can name a cockroach after your ex in time
for Valentine's Day and it only costs $2.'*

Internet stunt

Cockroaches are the most hated animals on the planet: hated, at least, by the species that's best adapted for the task of hating. We don't fear them as we fear sharks; we don't even fear them as some of us fear spiders. Cockroaches inspire disgust more than fear. The fear lurks deeper, well behind the disgust: and it is the atavistic dread of losing control: the idea that non-human life is winning the fight against humanity.

There might be cockroaches in your own block of flats, in your workplace, in your favourite restaurant; worse still, in your nearest hospital: and that's certainly a thought to inspire fear and loathing. Pictures, especially moving pictures of cockroaches assembled round some source of food, water or warmth, bring to the viewers a feeling close to panic.

We spend a great deal of time and money trying to kill them, and yet they survive. Their resilience is itself a form of horror: they are rare, almost unique creatures that have slipped beyond our powers of destruction. The cockroach's tenacity for life has created one of our favourite modern myths: that the only creatures who will survive the nuclear winter will be cockroaches: that despite all the richness of human civilization – *Ulysses*, the Mass in B Minor, *Hamlet*, *Starry Night* – it is the cockroaches who will inherit the Earth. A lethal dose of radiation for cockroaches is about ten times higher than it is for a human, though some other insects are even more resistant. But the idea of cockroaches – the night crawlers, the dirt dwellers, the eaters of the most disgusting things we can imagine – taking over our world excites a deep level of fascinated disgust.

Cockroaches belong to the order of Blattodea as we belong to the order of Primates. They are not beetles: their resemblance to some members in the vast order of Coleoptera is superficial. There are around 4500 species of cockroaches and only twenty or thirty of them are associated with humans, and, of these, only four can be categorized as 'pests'. Most species are prodigious recyclers, playing a subtle but important part in the wild ecosystems they inhabit; there are species living within the Arctic Circle, but they are most numerous in the tropics.

The night crawlers (artwork by Steve Roberts).

The four species problematic for humans are the American, oriental, German and brown-banded cockroaches. These have spread widely by hitching rides on human transport, and even in difficult climates, like that of Britain, they can survive and thrive in sympathetic places.

Perhaps their greatest advantage is that they can eat almost anything and take sustenance from it. There is a lesson here. In a world radically changed by humans, the generalists have the best chance of surviving: those that can adapt, improvise, live in different ways and take on many kinds of food. A woodpecker or a penguin can't make a living in a built-up environment but pigeons can. There are many highly specialized species of beetles; the order Coleoptera contains perhaps a million or even more species – but few of them could thrive in a hospital.

What's more, cockroaches can survive a month without any food at all, and can keep going on very little: the gum in a book of stamps, leather, flakes of skin, hair, soiled clothing, faecal matter. They're not what you'd call fussy.

It's as if their evolution had predesigned them for life among humans. That's an illusion, but a compelling one. They are creatures of the night, most active when we are least active, and despite their chunky appearance they are adept at squeezing through impossibly narrow cracks, flattening their bodies and splaying out their legs to do so. They are very good at keeping out of sight: the fact that you can't see any cockroaches doesn't mean that there are no cockroaches there. They are also swift and efficient breeders, in a manner that routinely dismays humans: a cockroach can mate at two months old and, given helpful circumstances, a few pioneers can become a thriving population in less than a year.

They have large compound eyes that make them comfortable in low light, and they have long antennae – described by the great Gerald Durrell as 'like moustaches of a Mandarin' – to feel their way in the dark. They have a great turn of speed, able to cover 11in (28cm) in a second. They have intense social lives and are known to take decisions about food collectively: so they are complex and sophisticated creatures, not just blind mouths. Some species have been observed offering parental care to live young, a rare thing in insects.

How much threat do they bring to humans? Certainly some: they can carry pathogens on the surface of their bodies, which can be transferred to humans. Their droppings are associated with some human allergies, notably those that cause asthma. They also leave an unpleasant smell.

But there are other and greater threats to human health. In restaurants, it is more important to worry about good practice in the storage, preparation and cooking of meat, and about the threat of human faecal contamination by human hands, than about cockroaches.

Perhaps the reason humans have such a violent hatred of cockroaches is the fact that a cockroach infestation is a glaring sign of failure. They are living proof that we have failed to maintain sufficiently high standards of hygiene: that we have left the door open a crack and stinking nature has crept back in. Professional pest control organizations are always keen to stress the reputational damage that is caused by the presence of cockroaches: they don't just harm you because they might make people ill, they could close your business through the hysteria they generate. The more people hate cockroaches, the more people *need* to hate cockroaches: it's economic common sense.

Though there are some benign facts about cockroaches. In China twice-fried cockroach is considered a good medicinal food; there are even cockroach farms to supply the market. Cockroaches are also eaten in Thailand and Mexico. (Why not? They are arthropods, like shrimps and crabs and lobsters.)

The English-language name of cockroach is a corruption of the Spanish: and, of course, 'La Cucaracha' is a song everybody has heard. It has a million versions: in one it is associated with the Mexican Revolution of 1910–20 and the *cucaracha* in question was President Victoriano Huerta.

And here's a joyous fact: the only terrestrial animal to have mated in the weightless conditions in space and returned to Earth to produce offspring is a cockroach. Her name was Nadezhda and she was fired into space by the Russians in 2007. Her offspring thrived and so, for that matter, did her grandchildren. But then thriving is what cockroaches do best.

SIXTEEN

PANDA

'Pandas are bad about sex and picky about food.
These genetic misfits might have died out years ago,
had they not been so adorable.'

The Economist, 2014; quoted by Lucy Cooke,
Wall Street Journal, 2018

We don't expect pandas to survive the nuclear holocaust. We're not sure if they have what it takes to survive the next decade. They have become an emblem of the fragility of non-human life: the exact opposite of the cockroach: cuddly, lovable and in desperate need of cherishing – the sort of cherishing humans are pretty keen on handing out.

Pandas were not known to the West until 1869, when Armand David, missionary and zoologist (for whom Père David's deer was named), acquired a skin. In 1916 Hugo Weigold, a German zoologist, purchased a panda cub, becoming the first Westerner to see a living panda. In 1929, either Theodore or Kermit Roosevelt, sons of President Theodore Roosevelt, became the first Westerner to shoot one; they took joint credit; you can find what's left in the Field Museum in Chicago. In 1936 Ruth Harkness, an American fashion designer, brought one back to the United States, stepping off the boat with the panda in her arms. It was displayed at Brookfield Zoo in Chicago.

At the beginning of the 1950s, as part of the revolution in global thought that followed the Second World War, the extinction of large mammals was becoming accepted as a real possibility. The fact that nature is neither bottomless nor endlessly forgiving is one of the greatest and most traumatic revelations in human history: and the panda played a significant role in this new understanding of the possibilities of life.

There are a lot of good reasons for this. The panda is one of the most recognizable species on the planet: nothing else looks even remotely like a panda. It also excites the cuteness response in humans. This phenomenon was identified by the ethologist Konrad Lorenz in the 1930s; he concluded that a creature with a large round head, especially with large eyes – in other words, like a human baby – draws a cuteness response in humans: a need to cherish and nurture something that so obviously needs hugging. (You will notice that Disney

Pandas as seldom seen: in the wild in Sichuan, China.

The ultimate fashion accessory: American fashion designer Ruth Harkness with her panda.

figures, human or anthropomorphic animals, follow the same principle. Mickey Mouse has a large round head and large eyes – he doesn't look much like a murid rodent. This phenomenon is described by the great palaeontologist and writer Stephen Jay Gould.)

The fact is that the panda's head shape is related to the immense muscles it needs to power its remarkable bite, and the black eye patches that emphasize the eyes so dramatically are used for signalling to other pandas and for recognizing each other as individuals. But when we humans see a panda we see something we want to make much of: in 2006 a drunk jumped into the enclosure at Beijing Zoo to embrace a panda and was horribly bitten.

The first pandas in the West excited human compassion. And our response was seriously wrong. We thought the answer was zoos and captive populations. The prevailing orthodoxy was that pandas were too fragile – too useless – to survive in the wild unaided. It was up to us humans to rescue them from the wild and look after them properly.

This was tied up with the emergence of China from willed seclusion, back into the social whirl of global politics; pandas are only found in China. China began to practise what was called Panda Diplomacy: lending captive pandas to zoos in major Western cities. The pandas were madly popular: no one can resist a panda. Their presence in Western cities showed China as the good guys: looking after a fragile cuddly creature when the country might be building nuclear missiles. At the same time the pandas gave the people of the West a newfound sense of the fragile nature of nature.

But again, we understood this the wrong way. Pandas were popularly seen as a charming but terrible mistake. They were bears – members of the order Carnivora – who had chosen to become vegetarian, with their diet 99 per cent bamboo. To a Westerner that sounded unfeasibly picky, even though bamboo is just another kind of grass and there were once endless tracts of the stuff.

And to make them seem even more helpless, it was impossible to make the damn things have sex. It was as if they didn't want to make more pandas; as if they were self-doomed to extinction, despite our most heroic efforts. Chi Chi, a female from London Zoo, was sent to Moscow to meet a male called An An: but they reacted so aggressively to each other that they were promptly separated. Nowhere in the world could these captive pandas be persuaded to mate. They were shown images of pandas mating – referred to in newspapers as 'panda porn' – but no result.

But the answer wasn't about zoos. It wasn't about what humans could do. It was about what humans could stop doing. The reason there were fewer pandas in the wild was because humans had encroached onto their bamboo forests and destroyed them. Sometimes there are many and complex reasons for the dwindling populations of wild animals, but the one you find time and again is habitat destruction. You destroy the places where animals live, and so they die. Sometimes it really is as simple as that. Pandas were now restricted to remaining bamboo forests on higher ground, a series of isolated populations in suboptimal habitat. If we wanted to save the pandas we had left, we needed to save the bamboo forests they lived in: save what's left and then join them up by a process known as habitat recreation.

China began to look after the remaining forests, instead of trying to catch and cage pandas. The Wolong National Nature Reserve was established in 1963; in 1975 it was expanded from 77sq. miles (200sq. km) to 770sq. miles (2000sq. km). Research on wild pandas showed that the panda's way of life is in fact very sound and effective. Unlike most of their fellow carnivores, a panda's food doesn't run away. It's rather a good way to live. The food is short on nutrients, so pandas have to eat an awful lot, and pass it through their guts at some speed: it's reckoned that pandas will defecate forty times a day. Pandas, like Primates, have an opposable thumb for grasping: they use it to hold bamboo. It's not a thumb quite like our own: it's a spur from the wrist bone that does the same job. This has been interpreted as an object lesson in the make-do, can-do improvisational techniques of evolution, famously explained in another essay by Stephen Jay Gould in *The Panda's Thumb*.

George Schaller, the same man who worked with lions and gorillas (see Chapters 1 and 3), also worked with pandas in the wild. His studies revealed an animal that lives in an energy-saving way on a low-nutrition diet, a little in the manner of sloths. The need for constant feeding has created a largely solitary lifestyle, but the system works with great efficiency. The female is sexually receptive for just a couple of days in the year, but these days tend to be strenuous. Schaller recorded a couple mating forty-eight times in the course of three hours: more or less once every three minutes. What some humans manage in a year, pandas do in the course of a long lunch break. Females produce a cub every two years and, of these, 60 per cent reach their first birthday. In other words, pandas are – inevitably – admirably adapted for the way they live; in fact, stating that truth is more or less a tautology. Pandas don't need human help to survive; all they require from humans is the will to stop destroying the places they are adapted for.

Great efforts have been made to safeguard the remaining wild panda population. The Chinese government is aware that to lose the pandas would be a seriously bad move in terms of international relations. The Chinese issue gold 'panda' coins; one of the mascots at the Beijing Olympic Games of 2008 was inevitably a panda.

WWF, formerly known as the World Wildlife Fund, adopted a panda as its logo in 1961: a perfect and perfectly lovable symbol of what the world would lose with mass megafauna extinctions. Panda conservation has been criticized, most vocally by the British television presenter and environmentalist Chris Packham. He said that expenditure on pandas was 'one of the grossest wastes of conservation money in the last half-century', adding that he would 'happily eat the last panda' if it meant that the money spent on them would go to more appropriate causes.

There is a point beneath the rhetoric: prioritizing the cute and cuddly over the conservation of the most intensely biodiverse and bio-abundant habitats on Earth is surely wrong-headed. But the panda has an immense symbolic value, and its conservation is an example of what can be done where there is the will. The panda population has risen, or at least appears to have risen. In 2016 the panda was downgraded from Endangered to Vulnerable. The definition? They now face a 'high risk of extinction in the wild' rather than a 'very high risk'. There are now forty reserves for pandas in China, up from thirteen in 1998. There are new laws on gun control, and human populations have been moved to accommodate pandas.

Pandas are actually quite good at living in the wild. They have evolved a lifestyle that works, and it has kept them going for about 20 million years. It's the last century or so that's been problematic.

SEVENTEEN
COD

'The piece of cod that passeth all understanding.'

Old schoolboy joke

We began to phase the hunter–gatherer lifestyle from human lives around 12,000 years ago, when we established agriculture in the Fertile Crescent and elsewhere. But we have never completed the job. Not yet, anyway. In one area of life we maintain that ancient lifestyle. It has been estimated that fish provide 15 per cent of humanity's protein intake, and the great majority of this is wild-caught.

We have traditionally exploited the wild creatures of the sea with the naiveté of our ancestors hunting the auroch: in the belief that there'll always be plenty more where that came from, that nature is a bottomless well into which we can send the bucket anytime we feel peckish. The auroch, as we saw in Chapter 7, is now extinct.

Humans started hunting cod in about the ninth century. It's a target that requires a certain intrepidness: cod mostly live some distance from human habitation, preferring cold and deep water. But there are considerable incentives: cod is a sizeable fish and humans find its flesh very palatable. It is a traditional part

Toughest job: Cod Fishing, *painting by Ambroise Louis Garneray (1783–1857).*

of the English dish of fish and chips (in Scotland they prefer haddock). Cod is a major predator; specimens up to 220lb (100kg) have been caught. That's a lot of fish suppers.

But we should define our terms, for 'cod' is used somewhat imprecisely. Technically there are two species, the Pacific and the Atlantic cod; it is the second that most concerns us here. They are related to haddock, whiting and pollock.

Cod doesn't really have a plural. It mostly exists in the English language in the singular: a reference to a food – a substance – rather than living individuals. It – or they – is or are – much eaten in Spain, Portugal, Italy and Brazil. Cod became an important trading item in the seventeenth and eighteenth centuries; the fish could be preserved by salting. The early history of New England in the United States is closely related to the profitability of fishing, and those successful in the trade were referred to as 'the codfish aristocracy'. The House of Representatives in Massachusetts displays a large carved wooden cod. William Pitt the Elder, the prime minister of Great Britain, referred to the Newfoundland cod-fishing grounds as 'British gold'.

This was, then, a large, valuable and apparently infinite resource: something that humans could rely on for nutrition and trade. It was an unglamorous but essential part of our lives. The fishermen who brought such things to the rest of the population were, in a quiet way, heroes: people to admire, people who could walk with a certain swagger. Fishing created a tough, self-reliant culture and gave a sense of identity to coastal communities. They were something: they were fishermen.

Nothing in the history of the world since the Cambrian Explosion of life has been more remarkable than the rapidity with which human intelligence operates. But even as human ideas gallop on apace, improving on the last almost as soon as it is brought into existence, our minds retain the shapes and patterns that our ancestors evolved over uncountable millennia out on the savannah. We still think – perhaps I mean that in the depths of our minds we still *feel* – that nature is vast. We feel that it is both vast and profoundly dangerous and will come and get us in our homes if we give it half a chance, as cockroaches prove (see Chapter 15). We maintain this belief alongside a contradiction: that nature is also vast and beneficent: that it will feed us for ever, so long as we know how to hunt stuff and gather stuff.

The clash between the brilliant modern mind and the ancient ancestral mind can be found in the story of fishing: and the cod fisheries are chosen here as the example in chief.

In the years after the Second World War the technology for fishing improved dramatically. At the same time, the demand for food increased. Fishermen were

Food at a price: Fish Market *by Joachim Beuckelaer (1568).*

able to fish deeper, for longer periods and over larger areas, thanks to radar, sonar and electronic navigation. Vast catches were now possible. Not only that, but the advancing techniques resulted in a lot of by-catch: that is to say, stuff that's useless commercially, caught by accident and discarded. Useless commercially: not useless ecologically. It included the fish that Atlantic cod feed on.

Now here is a much-overlooked fact about cod, about all sea fish: they live in the sea. That means they are very hard to observe. The challenges of studying the ecology of the oceans are ten times greater than any land habitat: the area to be examined is invariably vast and hard to access. Even for those who study the ocean, the situation is difficult. For the rest of us, the business is so opaque as to be meaningless: it is beyond our experience, beyond our easy intuitive grasp. It is extremely hard to get even an approximate idea of population. In short, the real life of the cod is something that passeth all understanding.

The fishing industry is full of terminology like 'stocks' and 'harvest', as if it was something that humans had instigated themselves and had always controlled. Oceanic ecology is infinitely complex and elusive and, above all else, the fishing industry needs to be sustainable: that is to say, it must be capable of reproducing itself. That means sustaining its population at the same level across years and years of fishing.

That has ceased to be the case. In 1992, the Canadian government declared a moratorium on cod fishing in the Atlantic. The European Union and other interested non-EU nations have sought to limit the catch of North Sea cod since the 1970s; the waters have been overfished since the 1960s and still are today.

It was an issue that has raised the most bitter disputes. Fishermen declared that scientists were wrong; there were plenty of fish in the sea. Scientists countered that fishermen had to think like this – because they were professionally obliged to visit only the places where fish could be found. Unlike scientists. Fishermen never tried to fish the empty spaces from which the fish had vanished.

All this is summed up in a single grandiose phrase: the tragedy of the commons. You do what benefits you as an individual, or as a family: not what benefits the larger population. Or to put that another way, if I don't do it, somebody else will. The same phenomenon is to be found in the issues of air pollution, water management, forest destruction, use of antibiotics and overpopulation: in other words, just about all the greatest challenges facing the planet and its dominant species.

The politicking that went on around the Canadian moratorium was intense and deeply troubling. Famously, a scientific report was edited for public consumption. One sentence read: 'there are few indications of improvement'. The word 'few' was removed. It seemed there was a desperate need to believe that things were all right: that fishing could carry on, that no harm had been done.

It's been estimated that cod numbers had been reduced by 70 per cent in thirty years; that all fisheries would be unviable in fifteen years. Others have argued passionately that this was far from the case: the seas were still thriving with cod, if only the intrepid fishermen could be freed from government interference and allowed to get out and catch the damn things. There was money involved, there was also the sense of cultural identity. People felt that their livelihoods were being taken away for no good reason: their livelihoods were exactly what the quotas and the moratorium were designed to protect.

The long-term sustainability of the fishing industry is clearly what both sides in the argument want. It seems strange – but perhaps inevitable, even inevitably human – that this debate should polarize and the two sides be irreconcilable.

The IUCN assesses the cod as Vulnerable. Reports in 2015 suggested that cod were now increasing in both numbers and in health. North Sea cod was declared sustainable in 2017, with loud hurrahs all round – and this decision was reversed two years later. Meanwhile, boats from the EU have been fishing for cod in West African waters. It's been estimated that 83 per cent of world fisheries are overexploited: that is to say, fished unsustainably, fished faster than the fish can reproduce. Species that have been fished into scarcity include swordfish, blue fin tuna, wild salmon and herring, which are the principal prey of cod.

So we can consider the anomaly that modern humans still maintain the hunter–gatherer lifestyle: and also wonder whether or not we will be the last generation to do so. What is required, more than anything else, is an acknowledgement that the world's natural resources are finite and the world's human population is continuing to rise. Are our minds capable of making such an advance? The future depends on our ability to make this imaginative leap.

EIGHTEEN
EGRET

'For there is an upstart crow, beautified with our feathers...'

Robert Greene, *Groats-worth of Witte, Bought*
with a Million of Repentance

Who were the first people to realize that the Earth's resources were not after all infinite? Who were the first people to do something about it? An answer to both these questions might be four women of the late nineteenth century. Between them, they founded two of the most influential conservation organizations in the world. And they were all most powerfully motivated by egrets.

Egret is not a formal and biologically distinct group. It's a name given to a dozen or more species in the heron family: long-necked, long-legged, fish-catching wading birds. A good few of these species are pure white; these are the birds we tend to think of when we hear the word egret. In most circumstances, white is a

Cutting a dash: Paris by Night in the Opera Quarter *by Felix Fournery (1907).*

great colour for getting noticed: a flashing white tail is a warning signal in many unrelated groups of creatures, like deer and rabbits. It follows that the white egrets stand out dramatically from a landscape.

But not to a fish. A fish looking up from dark waters towards the bright sky sees only a paler patch. And since, like all herons, egrets have a great talent for stillness, the fish is lured into proximity – at which point the long bill strikes and grabs the fish sideways; it works as a grab, not a spear, and the length of the bill is the bird's margin for error. A little egret has yellow feet: you can watch one gently waggle a single foot as a lure. The fish, unaware of the great white presence above, sees only the movement of those yellow toes and, viewing them as potential prey, is tempted forwards – to its death.

Humans have always found these slim white birds attractive; they occur many times in the classic haiku of Japanese poets. The eighteenth-century female poet Chiyo wrote:

> But for their voice
> The egrets will disappear –
> The morning's snow

Egrets become even more attractive at the times of courtship and breeding, when both sexes grow long, white, decorative feathers, usually referred to as nuptial plumes. These look gorgeous. And in the nineteenth century they became fashionable. Women wore them in their hats.

A very large hat with many feathers: not only beautiful but also a conspicuous display of wealth and status. You can wear an awful lot of feathers at once, because feathers are incredibly light; that, after all, is their job. It was an unbeatable combination, and the demand created a huge global trade in feathers. And of all the many feathers that were fashionable, egret plumes were the tops.

The nuptial plumes are gauzy and floaty and fabulous: the kind of gratuitous beauty that the wild world throws up so often. They were called 'aigrettes', French for egret, to distance them from the death of birds, just as we refer to the meat of cattle as 'beef'. These feathers were popular – essential – on both sides of the Atlantic, for women were required by convention to wear a hat whenever they appeared in public. It's reckoned that 5 million birds a year were killed in the United States to satisfy the trade; many of them snowy egrets from the Florida Everglades. London became a centre of the European trade: one dealer placed a single order for 6000 birds-of-paradise feathers, 40,000 hummingbird feathers and 360,000 feathers from species from the East Indies. Feathers were quite literally worth their weight in gold.

It was, perhaps, the frivolity of the business that inspired these four remarkable women into action: the fact that so many beautiful creatures had been killed for vanity, for showing off. It is unclear how much these women were inspired by the possibilities of extinction – the numbers of snowy egrets were down to critical levels – or whether it was the cruelty of the business, the simple wastefulness.

Either way, Emily Williamson from Didsbury in Manchester decided to do something about it. In 1889 she founded the Plumage League. In 1891 this amalgamated with the Fur, Fin and Feather Folk, which had been founded by Eliza Phillips in Croydon, in south London, also in 1889. They called the organization the Society for the Protection of Birds. It had two rules:

> That members shall discourage the wanton destruction of birds, and interest themselves generally in their protection.
> That lady members shall refrain from wearing the feathers of any bird not killed for the purpose of food, the ostrich only excepted.

That latter was because feathers could be taken from living and domesticated ostriches.

Even better on a bird: snowy egrets in breeding plumage (colour lithograph by Walter Alois Weber, 1906–79).

The campaign attracted the attention of women of high social standing and therefore considerable influence. The first president was Winifred Cavendish-Bentinck, Duchess of Portland. In 1904 the society was given its royal charter and became the Royal Society for the Protection of Birds. This organization now has more than 1.2 million members, employs 2200 people and manages 218 nature reserves.

A startlingly similar thing happened in the United States. In 1896 Harriet Hemenway and her cousin Minna B. Hall founded the Massachusetts Audubon Society. Their shared aim was to discourage the trade in feathers. Two years later there were equivalent societies in sixteen more states, and in 1901 the Florida Audubon Society established the first National Wildlife Refuge at Pelican Island in Florida. By 1905 the National Audubon Society was established. It has 600,000 members and more than 500 local chapters.

It should be remembered that, at the time of the foundation of these societies, women in neither country were allowed to vote or to own property. But between them these four women did a great deal to change the way that humans think about nature. It was clear, with the plundering of egrets for their nuptial plumes, that nature's resources were finite.

There is an obvious question in this new understanding: what are we going to do about it, then? If anything? It was the beginning of the realization that the planet's dominant species had taken control in an apparently irreversible way. What happened to the planet next would not be the result of the inevitable working of natural history or of the ineffable operation of the will of God. It was now clear that what happened next would depend on the decisions made by humans.

We now talk about 'anthropogenic' changes as a matter of routine: the idea that climate change is driven by the activities of humanity is more or less uniformly accepted in science. The term Anthropocene is now widely used, and no longer as a picturesque image: the idea that the planet has entered a new geological age as a direct result of the activities of humankind is now close to becoming an orthodoxy.

It was the egrets, and these remarkable women who turned against the trade in their gauzy feathers, who first came to terms with this, and it is the most colossal event in human history. They realized that human decisions could allow a species of non-human animal to continue to survive, or to hurry towards extinction. It was these women who paved the way for the Environment Movement when it began after the Second World War.

Decisions made by humans will decide the immediate future of the planet: that is now a generally accepted truth. What those decisions will be remains to be seen. The four women of the egrets gave us an example of what can be done. They changed business, they changed culture, they changed philosophy, they changed the way we look at the world.

NINETEEN

DODO

' "In that case," said the Dodo solemnly, rising to its feet,
"I move that the meeting adjourn, for the adoption
of more energetic remedies —"'

Lewis Carroll, *Alice's Adventures in Wonderland*

The story of the dodo is the story of extinction – yes, of course it is, we all know that the bird has gone the way of the dodo and is as dead as the dodo: as dead as anything ever could be. But at the heart of the dodo's story lies our attitude to extinction. Across its 350 years of extinction the moral of the story of the dodo's passing has changed again and again.

The bird was discovered by the Western world in 1598, when Mauritius became a Dutch colony. The last recorded sighting was in 1662, so it didn't take us long. All the same, nobody noticed. Extinction wasn't a valid concept at the time: everybody knew that it couldn't happen. God wouldn't allow it.

The dodo was a singular creature: huge and flightless. An adult stood around 3ft (1m) high, and could weigh up to 40lb (17.5kg). It had a spectacular beak. It was, in its way, a typical island species: island populations tend to throw up unique species, as evolution works on fast-forward. We've already seen that with the Galápagos mockingbirds and Charles Darwin's understanding of them (see Chapter 4). Flightlessness has evolved many times in often-unrelated, island-dwelling species; it's generally accepted that, in the absence of predators, it was more profitable for a species to invest in legs than wings. Flight had become luxury; it is, after all, the most energy-expensive form of locomotion that exists.

Recent studies of the dodo indicate that it was a forest-dwelling species that fed mostly on fruit, seeds, nuts and roots. There were no mammalian predators, so it was safe to nest on the ground. The first humans who first perceived them thought they were ridiculous: absurd: one of life's traditionally ludicrous mistakes. Dodos cropped up in a travel book of 1634 written by Sir Thomas Herbert: 'Her visage darts forth melancholy, as sensible of nature's injurie in framing so great a body, to be guarded with complementall wings, so small and impotent, they serve only to prove her bird.'

The Dutch sailors called them *wahlvogel*, which means tasteless bird. They complained that the longer the dodo was cooked, the less soft it got, reminding

Too stupid to live? The Dodo by Hans Savery (1597–1654).

me of a recipe I was given for cooking guinea fowl – you put the bird in a pan containing water and a stone and you boil them both. When the stone is soft, the guinea fowl is ready to eat.

The name of dodo comes from the Dutch but the exact etymology is disputed. One suggestion is that it comes from *dodoor* (sluggard) – but the anatomy of the dodo suggests that they had big, strong legs and were therefore fast and elusive runners. Another suggestion is that the name was onomatopoeic and sounded like their two-syllable call – though that is pure speculation; no one will ever hear a dodo again. My own preference is for *dodaars* (fat arse): a reference to the bustle of feather the bird carried behind it.

British scientists got interested in the conundrum of the dodo after Mauritius became a British colony in 1814. A monograph entitled *The Dodo and Its Kindred* was put together by Hugh Edwin Strickland and Alexander Gordon Melville in 1848 and was noticed far beyond the scientific community. The bird's oddity caught the public imagination.

So much so that a man who taught mathematics at Christ Church at the University of Oxford gave himself the self-mocking nickname of Dodo. He had a stutter and used to introduce himself haltingly as 'Charles Do-do-Dodgson'. So when he improvised a story for a family of little girls that he took for a picnic on the river, he wove the dodo into the narrative – and in 1865 the story was published as *Alice's Adventures in Wonderland*, with its author hiding behind the pseudonym of Lewis Carroll. The extraordinary success of the book made the dodo a popular figure across the world, not least because of the brilliant illustration by John Tenniel. Everyone knew what a dodo looked like. It was very odd, and it was very, very dead.

The book appeared six years after Darwin's publication of *The Origin*, and caught the first wave of misunderstanding of Darwin's work. The dodo became an emblem of obsolescence: a creature so useless that it had no option but to go extinct, to be replaced by new improved species – like our own. Darwin's revelations were already becoming a new kind of mythology: the notion that evolution was all aimed to bring about the birth of our own species.

So when something was as dead as the dodo, it was because it was no good, couldn't stand the pace of modern life, a flawed concept, discarded by the go-getting forces of life. Images of the dodo tended to show a fat ungainly bird, plainly not fit for anything very much at all. (Tenniel's bird carries a walking stick.) Such images made the idea of extinction acceptable. They also carried a quiet note of species pride: humans were obviously and gloriously not obsolete at all. Now was our time.

But that view shifted in the second half of the twentieth century. I remember my mother in tears as she researched the story of the dodo for the BBC children's programme *Blue Peter*. Her piece reflected the new orthodoxy: that dodos had been killed off by Dutch sailors. Dodos were unafraid of people. They were so naive they walked more or less straight into the cooking pot – and were called stupid for their pains. They were undone by humanity's carelessness and cruelty: bumbling innocents, unable to fly, unable to escape, wiped out by callousness and ignorance.

But this view has also been superseded. Dodos were a rainforest species, so by definition they were not easy to catch: rainforest is a deeply challenging environment for humans. Their wings were well-muscled and probably used for balance as they ran: you really couldn't walk up to a dodo and help yourself. They were also, as we have already seen, capable runners: shooting them with early firearms would not have been easy in thick forest. What's more, the prize was not great: as we have already seen, their meat was not favoured at all.

No doubt a good few dodos did die for the pot, but that wasn't what drove them to extinction. Quite apart from anything else, the permanent human population of Mauritius was seldom above fifty at the time, and they were not out there hunting dodos. Rather, they were opening the island up for farming: and

Alice and the Dodo: illustration from Alice's Adventures in Wonderland *(1889) by Sir John Tenniel, watercolour by Gertrude Thomson.*

systematically destroying the forests for that reason. The real threat to the dodo was not the bellies of sailors but the needs of farmers. Here was habitat destruction going full bore.

Humans also brought a secondary problem to Mauritius: other species of mammal. Rats came as fellow travellers, pigs were brought in, so were crab-eating macaques, a kind of monkey. And they all helped themselves to the ground nests of the dodos.

But it was the transformation of Mauritius for agriculture, most especially for sugar cane, that did most to finish off the dodo. It did the same for a number of other Mauritian species less famous than the dodo: flightless red rail, broad-billed parrot, Mascarene grey parakeet, Mauritius blue pigeon, Mauritius owl, Mascarene coot, Mauritius shelduck, Mauritian duck, Mauritius night heron, several species of reptile and one of bat.

All gone the way of the dodo.

The dodo survives on coins and banknotes of Mauritius, in *Alice*, in the logo of the Durrell Wildlife Conservation Trust, founded by the great writer and zoologist Gerald Durrell (*My Family and Other Animals*, etc., see also Chapter 97). The Center for Biological Diversity in the United States gives an annual Rubber Dodo Award for 'those who have done most to destroy wild places and biological diversity'. Previous winners include the pesticide company Monsanto and the United States Department of Agriculture and Wildlife Services.

TWENTY

DONKEY

*'The old grey donkey, Eeyore, stood by himself in a thistly corner of the
Forest, his front feet well apart, his head on one side, and thought about
things. Sometimes he thought sadly to himself, "Why?" And sometimes he
thought "Wherefore?" and sometimes he thought "Inasmuch as which?"
and sometimes he didn't quite know what he was thinking about.'*

A. A. Milne, *Winnie-the-Pooh*

Donkeys were humanity's first major labour-saving device. They have been
helping us out with the heavy lifting for more than 5000 years. We have
thanked them by using 'donkey' and 'ass' as terms of abuse ever since – but at the
same time we have always had a deep affection for donkeys, for their
perceived loyalty and uncomplaining nature. Jesus rode into Jerusalem on a

What angel wakes me from my flowery bed? Titania and Bottom *by Henry Fuseli (c.1790).*

donkey rather than a king's charger; humbling himself to be exalted. He preferred a push bike to a Rolls-Royce. He was one of us: that is the message of that Palm Sunday journey. The donkey is the ultimately humble animal: one we both despise and adore. A donkey is essentially comic: but behind the comedy we find qualities that we admire profoundly.

The donkey was domesticated from the African wild ass, inevitably in the Fertile Crescent, most likely in Egypt, which is the nearest part of the crescent to the remaining population of wild asses. There are reckoned to be 41 million domestic donkeys in the world; the African wild ass is now Critically Endangered, and with a population estimated by the IUCN at 23–200.

The donkey is eternally defined by the horse. The two species share a common ancestor that lived 4 million years ago. Horses are traditionally seen as noble and beautiful, enhancing the status of those who possess them. They were used for warfare and conquest, and when we make statues of heroes we tend to put them on a horse. A donkey's axis is essentially domestic: a hired help rather than a companion on great adventures. We tend to stratify our societies: our domesticated equids fit into that very pleasingly. Horses are heroes; donkeys are workers: comic, despicable, useful and, above all, lowly.

But it's the fact that they are not horses that makes them so useful. They are adapted for semi-desert conditions, and they can make small quantities of poor food go a long way. They need, pound for pound, much less feed than a horse. They are smaller and more manageable than most horses. They are mostly solitary in the wild. Horses need to be part of a herd and they don't thrive when kept alone; solitude is no big deal for a donkey. Their way of life in the wild may explain their large ears: they keep in touch with each other by sound. A donkey's bray can be heard by humans for about 2 miles (3km); a good pair of donkey ears can double that distance.

Donkeys are long-lived: even in poor conditions, a donkey can live for a dozen years; those kept in the lap of luxury have been known to live into their fifties. They were kept mainly for transport and draught work; in most cultures they took the place of the ox. Working cattle need to take a break to chew the cud; donkeys can keep going if they are fed little and often.

When humans domesticate a species, they take control of the mate choice and organize the breeding to please themselves. In that way you can design the kind of animal you want, appropriate for your needs. Donkeys have been bred in different sizes, from 7.3 hands to 15.3 hands (a hand is 4in/10cm, the traditional measurement of an equid to the shoulder); that is to say, from 31in (79cm) to 63in (160cm). There are nearly 200 breeds of donkey recognized today.

They spread from the crescent into Southwest Asia and reached Europe about 4000 years ago; the Romans later took them all across their empire. There are

Humbled for a season: Christ on a donkey, by Giotto, from the Scrovegni Chapel in Padua (painted 1304–06).

now around 11 million donkeys in China. In the modern world, donkeys are associated with poor communities and developing nations. If you have a donkey to help you with your work, it's because you can't afford anything better. Donkeys fulfil extremely basic needs. They are dirty-job specialists.

Mules have a little more scope. A mule is sired by a donkey on a horse, and is infertile. Mules are stronger and more athletic than donkeys, but less needy than horses. Like the donkey on their father's side, they have a great reputation for stubbornness. By stubborn we mean of course that they are sometimes unwilling to do things that a human wants them to. The wild ancestors of all equids were prey animals: they have evolved to see the world as a dangerous place. Sudden events, unexpected changes and threatening situations – entering a dark and unfamiliar place, for example – can therefore seem an extremely bad idea. You can attempt to persuade the animal in question by gentle means, but traditionally you just wallop 'em. A donkey or a mule – or for that matter a horse – doesn't always take that in good part. Some turn and run, some rear, some kick out; donkeys tend to take root. The way to get round this is to establish a relationship of trust with your equid. That saves time in the long run, but the tactic is not always found in traditional donkey-using communities.

Donkeys and humans have lived together across the millennia; naturally we have established many myths and traditions about donkeys. The Hindu goddess Kalarati, a destroyer of demons, has a donkey as her vehicle; a donkey was a symbol of the Egyptian god Ra. The Greek god Silenus, tutor to Bachus, the god of wine, rode a donkey. The Ten Commandments state that you must not covet your neighbour's ass: for an ass was a kind of wealth.

The word ass was preferred in the English language until the late eighteenth century. It was then gradually replaced by donkey. The reason is probably to do with conversational delicacy: a reluctance to use in polite company a word that sounds like your bottom. The same sort of delicacy has meant that we generally prefer 'rooster' to 'cock' (see Chapter 29 on chickens). The word donkey is perhaps a reference to dun, which is a traditional horse colour, though there is speculation that it comes from the name Duncan. Best not to be too dogmatic about either explanation; you don't want to look like a smart ass.

Donkeys have always split audiences, being both stupid and stubborn as well as humble and lovable. In the biblical tale of Balaam's ass, the donkey can see the angel but Balaam can't: Balaam beats the donkey but the donkey, entranced by the angel, refuses to move. In A *Midsummer Night's Dream*, Bottom is given an ass's head, to make ridiculous the fairy queen Titania's infatuation with him:

> Come sit thee down upon this flowery bed
> While I thy amiable cheek do coy,
> And stuck musk roses in this sleek smooth head,
> And kiss thy large fair ears, my gentle joy.

It should be added that Bottom behaves with great gentlemanliness in this unexpected position, making no attempt to take advantage of the lovestruck queen. He may be a bit of an ass but he has the gentle manners of his kind.

Perhaps the most famous donkey in literature is Eeyore, who appears first in 1926 in *Winnie-the-Pooh*, A. A. Milne's fantasy of childhood. Eeyore is a depressive always eager to share his misery: ' "I might have known," said Eeyore. "After all, one can't complain. I have my friends. Someone spoke to me only yesterday. And was it last week or the week before that Rabbit bumped into me and said 'Bother!' The Social Round. Always something going on." '

In Britain, where donkeys have not been needed for serious work for many years, donkeys traditionally give children rides on the beaches during seaside holidays. The practice began at Weston-super-Mare in 1886; I rode a donkey there myself in the 1950s, though the practice is in decline these days.

But the great affection for donkeys has continued. The idea of a suffering donkey is distressing: and people are willing to pay quite serious money to protect them. The British charity The Donkey Sanctuary had an income of £37.6 million in 2017. Such figures cause dismay among people who support the conservation of wildlife. Conservation is a different cause to welfare: it is concerned with maintenance of habitat and the protection of non-human animals at the level of species; welfare is concerned with individual suffering. Conservation organizations find it frustrating: it can be hard to raise money to protect an area of rainforest, but comparatively easy to find funds to make a donkey comfortable. But something deep in human nature is distressed by the idea of a donkey's suffering: and these feelings dominate the field of charities for non-humans.

TWENTY-ONE
WOLF

*'Akela from his rock would cry: "Ye know the Law –
ye know the Law. Look well, O wolves!" '*

Rudyard Kipling, *The Jungle Book*

Once there was legislation to persecute wolves: you killed a wolf and the government paid you a bounty. Now there is legislation to protect them: you kill a wolf, you have to pay money to the government. No animal better sums up our changing attitudes to nature and to the creatures of the wild world. Where once wolves were deliberately wiped out, as a matter of official policy, they are now being reintroduced. Also a matter of policy.

To clarify: we are discussing a single species of wolf, the grey or timber wolf (*Canis lupus*), which historically covered most of the northern hemisphere, Old World and New, north of around 12 degrees. In other words, wolves were

Three brothers: from The Jungle Book *by Rudyard Kipling.*

colossally successful from Tropics to Arctic. They are boundlessly adaptable: an apex predator with a vivid social life that allows them to work and hunt cooperatively and so bring down very large prey, including moose and elk. We need a further clarification here: the animal called a moose in North America is usually called an elk in Europe; the animal called an elk in North America is closely related to what Europeans call a red deer. Wolves, hunting in a pack, routinely tackle all these species.

Their highly social nature is what gives them their edge. They have intense forms of communication, including facial expressions. They keep in touch over long distances by their famous howling. They are not great sprinters but run down large prey over a distance, separating an individual from the herd and gradually bringing it down with repeated bites. A pack will normally be up to a dozen wolves, with a single monogamous pair at its heart. In season they will produce a litter of pups, and the younger mature offspring assist in their care. They are highly territorial and disputes with neighbouring packs often lead to fights to the death. They have spread across a considerable range of habitats including forest, tundra, grassland and desert; their numbers depend on the abundance of prey.

And wolves will take livestock from human communities. That's why they have been traditional enemies of humanity since humans first began domesticating animals for their meat around 12,000 years ago (though see Chapter 33 for more on that). They are also occasional predators on humans.

These days, perhaps ever since the invention of firearms, wolves are very wary of humans and keep away from them as much as possible. Rabid wolves have been known to attack humans or anything else that gets in their path on an indiscriminate basis; otherwise, wolf attacks on humans have been mostly restricted to strayed children and occasionally women. But there are plenty of records to show that wolves provided a genuine threat to humans; between 1362 and 1918, it's been claimed that 7600 people in France were killed by wolves.

But the idea of predatory wolves surrounding human habitations goes very deep indeed: the idea is still part of the language. In hard times, we do our best to earn enough to keep the wolf from the door: the first necessity of life is to keep these baying carnivores at arm's length. When we are hungry and food is at last provided, we tend to wolf it down: as if wolves had bottomless appetites.

One of the most famous stories of the Western world is 'Little Red Riding Hood', who goes to visit her grandmother and discovers a wolf in her grandmother's bed. The story was first written – perhaps that should be first written down – by Charles Perrault, who formalized the idea of the fairy tale. His version first appeared in 1697 as *Le Petit Chaperon Rouge*, but like all such tales it was perhaps centuries in the making. The idea that there is something out there trying to get us has always been part of the human condition: and wolves play an enormous part in that fear.

Who's afraid of the big bad wolf? A question still asked to this day. It was fear of lions that dominated life for the first humans, but for their descendants in the northern half of the planet wolves took their place. For centuries children were warned not to stray – not to behave in any inconvenient fashion – because, if they did, wolf would get them. Generations have been raised on such threats.

But the real threat that wolves posed to humans came through our domesticated livestock. Wolves were naturally attracted to human settlements because the rubbish that always accumulates around them often provided a snack. Domesticated animals raised the stakes on both sides. Large mammals conveniently kept in confined spaces were a free lunch, nothing less. It follows that the lives of humans depended on their ability to keep nature at bay: disaster followed for those who failed. Death of livestock could mean death of the family; there was nothing frivolous in the contest between wolf and human.

Humans have won, won conclusively – but there is still part of us that's caught up in the same battle. We still believe we must do all we can to keep nature at bay, under control, subject to our wishes – or everything will go wrong. People mow their lawns ferociously, to show they are in control. If we allow one single cell of unwanted life onto our own land, we have failed.

The war on wolves has been waged across the millennia. Aesop told us about the dreadful fate of the boy who cried wolf – nobody came when his call for help was genuine, and so the wolf ate all his sheep and, in later reworkings of the tale, ate the boy as well. The idea of crying wolf – raising a false alarm – is a lively concept in twenty-first-century life. Wolves also stand for deception: a wolf in sheep's clothing is something we must all look out for. Wolves, like the rest of the creatures in this book, played a part in making us who we are.

Fighting off wolves was a duty: one of the essential tasks of humankind. It wasn't just that they killed and ate domestic animals; they also caused them distress, weight loss and miscarriage; they also reduced the quality of the meat from the stress they caused. Solon of Athens, a reformist statesman of the fifth century BC, established a bounty of five silver drachmas for every wolf killed. In England and across Europe, local rulers established bounties: killing wolves was important work and their extirpation was considered an unambiguously good thing. They were gone from England by the sixteenth century, gone from Scotland in 1684, from Ireland in 1847.

On continental Europe wolves couldn't be wiped out: there was too much wild habitat for them to hold out in. But, in the main, areas well populated by humans were free of wolves by the eighteenth century; their persecution continued into the nineteenth century and beyond. Changes in rural life, with the gradual abandonment of peripatetic pastoralism in favour of fixed farming units, actually helped wolves to recover a little: the freer and the farther from humans, the more they could prosper.

Wolves were not universally seen as villains. One of the great foundation myths of Western civilization involves a wolf: Romulus, who went on to found Rome, and his brother Remus were, in the legend, suckled by a wolf. In *The Jungle Book*, wolves are seen as a civilized society – 'the Free People'. They were led by the wise grey wolf, Akela, and they kept the law. In this subtle and complex book, wolf society collapses into anarchy: but Robert Baden-Powell was not interested in nuances. He used *The Jungle Book* as a sacred text when he founded, as a junior version of the Boy Scouts, the Wolf Cubs, now known as Cub Scouts. I was a member of a Wolf Cub pack myself – in fact, I was sixer of the Grey Six, the only position of authority I have ever held in my life. As was the convention, the pack leader, a fine man named Jim Chapman, held the title of Akela.

There are many stories of feral children, and some of them are of questionable authenticity, like the tale of Romulus. But some are true, and wolves have on occasions been involved in raising these children. Marcos Rodriguez Pantaja, who was born in 1946, lived twelve years with wolves in Spain; he was bought by a goatherd when he was seven and when the man died the child went feral. He was discovered and reclaimed for humanity when he was nineteen, but was never properly integrated.

After 1970, with the conservation movement now moving forward, there was a shift of opinion about wolves. In many places, legislation was brought in to protect them; in some places they have been reintroduced. There have always been objections to these steps: some based on practical stock management, others on atavistic horror, but wolves are beginning to find out what it is to be cherished.

There have been some unexpected aspects to their return. Wolves were brought back to Yellowstone National Park, which is mostly in Wyoming in the United States, in 1995 – and they changed the ecology of the park. One result is that there are now nine beaver colonies when before there was only one. This happened because wolves are serious predators of elk (the red deer-related American elk). The elks could no longer hang around in large numbers browsing on willow, aspen and cottonwood: the presence of wolves forced them to be far more active. Beavers were therefore able to take the food the elk had missed, and so they thrived. Beavers, as we will see in Chapter 91, are a keystone species: they change the nature of the ecosystem they inhabit. Wolves recalibrated what is called the trophic cascade: the way one species relates to another in an ecosystem. They made it richer and more diverse.

There has been some heavy research into the behaviour of the Yellowstone wolves. The most intriguing aspect of these observations is discussed by the American academic and author Carl Safina in his book *Beyond Words*. He quotes Rick McIntyre, a ranger in Yellowstone: 'If ever there was a perfect wolf, it was Twenty-One,' McIntyre said. Note that the researchers, wary of anthropomorphism, gave the wolves numbers instead of names.

Civilization starts with wolves: Romulus Suckled by the She-wolf *by Giuseppe Cesari (Cavalier d'Arpino) (1568–1640).*

Twenty-One never lost a fight – and never killed a vanquished opponent. I spoke to Carl about this in the garden – or yard – of his house on Long Island:

> When a human releases a vanquished opponent rather than killing them, in the eyes of the onlooker the vanquished still loses status but the victor seems all the more impressive. You can't be magnanimous unless you've won... and if you show mercy, your lack of fear shows tremendous confidence. Onlookers might feel that it would be desirable to follow such a person, so strong yet inclined to forbearance.

So there's one to conjure with: humans being taught about mercy by a wolf.

TWENTY-TWO
PIGEON/DOVE

'O my dove, in the clefts of the rock, in the secret places
of the cliff, let me see your face, let me hear your voice;
for sweet is your voice; and your face is lovely.'

Song of Solomon 2:14

There are records of domesticated pigeons found in the cuneiform script of ancient Mesopotamia, and in the hieroglyphics of ancient Egypt, both more than 5000 years old. Humans began domesticating them a great deal earlier: 10,000 years ago, it is speculated. I am inclined to think that the process started even earlier, and that an informal domestication of pigeons was part of human life before we invented agriculture.

All domestic pigeons are descended from the wild species of rock dove. Racing pigeons, city pigeons, ornamental white doves: they are all the same species – like Chihuahuas and Great Danes. Wild rock doves prefer to live on and around cliffs and are enthusiastic cave-nesters. Humans also used caves for shelter. Now it's a fact that young pigeons must grow to a considerable size before they are capable of leaving the nest – so all you have to do if you want a very acceptable handful of meat is to get to the nest and help yourself. If you attract pigeons to your place by offering food and if you then harvest your pigeons sustainably, you have a constant supply of meat (at least enough to keep the wolf from the door). Humans lived with pigeons before civilization began, and I suspect that this was by choice and by invitation. Pigeons were the first bird to be domesticated; they were perhaps the first animal of any kind. Perhaps it was pigeons that gave us the idea of domesticating animals in the first place: in which case you could say that civilization was founded on pigeons.

There are more than 300 species of pigeons and doves in existence today, in the family of Columbidae. There is no formal difference between pigeons and doves. Species that bear the word dove in their common name – turtle dove, Namaqua dove – tend to be smaller than those with the name of pigeon – wood pigeon, green pigeon. But when it comes to the rock dove's descendants in dovecotes and city squares, we call the ones we like doves and the ones we don't like pigeons.

Infinite varieties: fancy pigeons in a coloured engraving (German school, nineteenth century).

So at the baptism of Jesus the Holy Spirit descends in the form of a dove, rather than a pigeon. In most depictions of the scene, the bird is an ornamental white dove, the sort bred to look sweet and live in dovecotes in protected circumstances – pure white is a bad colour for a prey species when it's not snowing; it's the easiest colour of all for a predator to pick out and home in on. White doves would not survive well in the wild; but they have been selectively bred for whiteness, to please human eyes. In other words, the bird that represents the Holy Spirit is the same species as the one we routinely dispose of as pests in modern cities: 'All the world seems in tune on a spring afternoon when we're poisoning pigeons in the park…' as the great American satirist Tom Lehrer sang.

But hatred of pigeons is a fairly recent development in human culture, and it's somewhat compromised by our corresponding love of doves. Doves were the symbol of the Mesopotamian goddess Inanna, or Ishtar, who was goddess of love, sex, war and justice. Aphrodite, Greek goddess of love, was associated with doves; you can find them carved in relief in her temple on the Acropolis in Athens.

Doves are often associated with love, and the expression lovey-dovey isn't just there for the rhyme; many pigeon/dove species have strong pair bonds and are most often seen as half of a pair. The expression 'billing and cooing' is an accurate representation of the behaviour of courting doves.

A dove helped Muhammad the Prophet by building a nest in the cave in which he was hiding; his enemies, seeing the nest, assumed the cave was uninhabited. In the Book of Genesis, Noah released a dove from the ark to try to find land. The dove returned with nothing to show for the jaunt, but seven days later Noah and the dove tried again – and this time the dove returned with a sprig of olive in his beak. By an odd transmutation – the idea of a fresh start from the human race always seems a good idea – the dove, bearing what the King James Bible translates as 'an olive-branch', has become a global image of peace. We have chosen the pigeon as one of the most potent and important symbols of them all. Picasso's lithograph of a dove, executed in his most elegant lines, was used as the emblem of the International Peace Conference in Paris in 1949.

But pigeons have also played their part in war, and they have the medals to prove it. Their startling ability to find their way back home has been exploited for the sending of messages, and in wartime such messages were often quite literally matters of life and death. A pigeon named Cher Ami was awarded the Croix de Guerre after the Battle of Argonne in 1918, during the First World War. The pigeon lost a leg and was shot in the chest, but still managed to deliver the message – which saved the lives of 194 men. A total of thirty-two pigeons have been awarded the Dickin Medal, which is given by the British organization People's Dispensary for Sick Animals to non-human animals who have helped to save human lives; you can find three such pigeons, stuffed, in the Imperial War Museum.

The domestication of pigeons continued long after they were no longer required as handy sources of meat. Pigeon racing is a recognized sport, and a potent gambling medium: the birds are released, often surprising distances from their home, and they find their own way back; 1000 miles (1600km) is not beyond their range. This ability has been much studied by scientists and is all the more surprising because rock doves are not migratory. There is still no clear single answer to the birds' navigatory skills. It's possible that they use the stars and the position of the sun, they possibly use sense of smell, and they can sense the magnetic fields of the Earth: so the skill may be a combination of many things.

Pigeons have also been bred for their decorative value, to produce extraordinary varieties that could never exist in the wild. Tumbler pigeons can loop the loop, pouters have a large inflatable crop, and fantails are self-descriptive, while white doves fly most prettily in the sun. When you see these very different birds together it is impossible to believe that they are all members of the same species – which shows how much can be achieved by selective breeding.

Charles Darwin noticed that. He was fascinated by domestication and the way that selective breeding can produce creatures radically unlike their ancestors. And being a man who never did anything by halves, he became a pigeon fancier. Though a gentleman of means with a position in society to maintain, he took to haunting grog shops to listen to the pigeon fanciers. He was enthralled by the way that skilled breeders could see what to him were imperceptible differences between birds, and which bred startlingly different creatures.

God as dove: The Baptism of Christ *by Andrea del Verrocchio, completed by Leonardo da Vinci (1472–75).*

'I love them to the extent that I cannot bear to kill them and skeletonise them,' he wrote of his pigeons. But he performed that grisly task all the same, examining their bones spellbound, trying to find why hatchlings that all look the same could grow into radically different adults. Darwin's work with pigeons makes it to the first chapter of *The Origin*. He shows how profound changes can be made over remarkably few generations by selective breeding – what he called 'artificial selection'. He goes on to reason that the same process operates in the wild world – and that he called natural selection. So the pigeon, along with the Galápagos mockingbird (see Chapter 4), is one of the world's great eureka birds.

Pigeons are pretty easy and cheap to keep and that has made them one of the great laboratory animals. To quote Tom Lehrer again: 'We'll slaughter them all amid laughter and merriment except for a few we take home to experiment.'

There has been a good deal of work on their cognitive abilities, how and what they can learn. Pigeons can be trained to store a library of 1000 images in their minds, and they have responded well to challenges involving numeracy and literacy, showing that pigeons, and by extension other species of birds and mammals, have more complex thought processes than was previously believed. The behavioural psychologist B. F. Skinner demonstrated what he considered to be superstitious behaviour in pigeons: they apparently believed that if they turned round and round, food would arrive.

When humans domesticate – or even just confine – animals of just about any species, given half a chance some of them will get out and form feral populations. And that's just what happened to pigeons: though perhaps nobody could have predicted how good at it they would become. They are a resourceful species; and not at all fussy about food, feeding on plant matter and invertebrates when they can find them, and perfectly able to exploit just about any other food source. Their natural habitat is rocks and cliffs and caves: as the modern city developed, the pigeons found the stone and concrete cliffs as homely as anything their ancestors knew.

At first they were rather liked and encouraged: until comparatively recently you could buy food for the birds from vendors in Trafalgar Square in London or St Mark's Square in Venice. But the pigeons grew more numerous and humans have grown more fastidious: less and less willing to share their living spaces with non-humans. Pigeons can damage buildings and rooftop machinery like air-conditioning units, though they are probably more of a nuisance than a health risk. As a result public buildings are increasingly decorated with a baroque extravagance of spikes to make them unsuitable for alighting pigeons. The aim of the modern city is to be pigeon-free: but for as long as there are McDonald's fries and dustbins it's likely there will still be city pigeons.

TWENTY-THREE
MOSQUITO

*'If you think you're too small to have an impact,
try going to bed with a mosquito.'*

Anita Roddick

A mosquito is an exquisite thing: especially the mouthparts of the female. Pause to admire her next time you get bitten: look down at the insect drinking life from your body and say: 'Creature of wonder, I am honoured to be sharing the planet with you. The forces that made you also made me and I can feel only awe as I see you gathering the blood that will allow you to lay your eggs, fulfil your life and ensure the next generation of mosquitoes.'

They can sense the carbon dioxide you breathe out from 100ft (30m) away; also the heat of your body. Clothing that contrasts with the background attracts them, so does movement. Sweat also brings them in. Once they have located you they insert a proboscis so cunningly put together that it is able to inject an anti-coagulant to flummox your body's defence system and, at the same time, locate a convenient blood vessel and drink. Who could fail to be lost in admiration of such a creature?

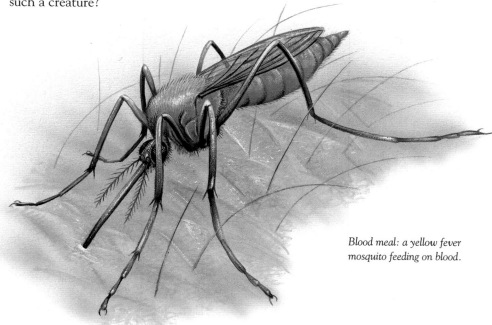

*Blood meal: a yellow fever
mosquito feeding on blood.*

There are around 3500 species in the family of mosquitoes; of these around 200 bite humans. Mosquitoes also bite other species of mammal, birds, reptiles, amphibians, occasionally even fish, and they also bite some crustaceans. The name is from the Spanish, from *mosca* (fly), with a diminutive suffix: little fly. Which is accurate enough: they are in the same class as houseflies, bluebottles, hoverflies and all the other species we loosely refer to as flies. Like most insects, mosquitoes go through four phases: the eggs are laid into stagnant water, from which they hatch as larvae, which are sometimes called wigglers. These live in the water, taking tiny particles of food. They then become pupae, which float, and are sometimes called tumblers. From these they emerge as adults. The males are gentle things that require only a few sips of nectar to keep them going as they attempt to find a female and mate. But the females require a blood meal to make a clutch of eggs or, in some cases, a blood meal to enlarge the size of the clutch. So the females have evolved as highly effective blood-hunters.

It's reckoned that malaria has been a significant part of human life since the dawning of civilization 12,000 years ago in the Fertile Crescent and elsewhere: we will often go back to that time in the course of this book. It seems that permanent human settlements gave certain species of mosquito opportunities they hadn't had before, and so they multiplied – and the disease proliferated.

Malaria is mentioned by Hippocrates, the great physician of ancient Greece. It was prevalent in ancient Rome, particularly in the central city of that empire, and was known as Roman fever. It's speculated that the irrigated gardens and the free-ranging River Tiber created ideal breeding conditions for mosquitoes: stagnant water in which the eggs and the wigglers thrive. Malaria was, for millennia, prevalent over most of the world including Europe and North America. Humans associated it with noxious vapours from swamps: it was this notion – *mala aria* (bad air) that gave the name to the disease.

It has been suggested that half the humans who ever lived died from malaria. When we reckon up the world's deadliest animals, though lions and sharks may be more spectacular to look at, mosquitoes invariably have the numbers – though perhaps it should be stressed here that mosquitoes are not killers but are exploited by the killing pathogens. The WHO estimates that getting on for half a million people are killed annually by mosquitoes, through there are higher figures from other organizations.

Humans finally made the connection between disease and insect at the beginning of the twentieth century, through a collaboration between Sir Patrick Manson and Sir Ronald Ross, who got the Nobel Prize for his work in 1902. After that revelation it was obvious to all that if you can do something about the insects you can do something about the disease. Mosquitoes carry pathogens for other diseases including yellow fever, dengue fever and Zika. They are capable of carrying infected blood from one bite victim to another, but human immunodeficiency virus (HIV) can't survive in a mosquito, so they can't pass that one on.

Malaria is seen as a disease of poverty: if you can't afford mosquito nets and/or closable windows you're going to get bitten every night. But it's also suggested that malaria *causes* poverty, and remains a serious hindrance to development. It is almost a matter of routine. Malaria is treatable, by no means invariably fatal – but it costs many working days and many days of education. The annual healthcare bill for malaria victims in Africa has been calculated at US$12 billion annually; 30–50 per cent of hospital admissions in Africa are for malaria and half of all outpatient visits. The slower development of the southern states of the United States can be explained by the prevalence of malaria in the eighteenth and nineteenth centuries.

Ever since mosquitoes were promoted from a bloody nuisance to potentially lethal adversaries, we have waged war on them as best we can. The draining of wetlands for agriculture helped to keep numbers down, and the eradication of other places where mosquitoes breed has been crucial. In more recent years, the accumulation of plastic rubbish around impoverished human settlements has created an endless series of shallow pools, ideal for raising mosquitoes, so the cycle continues: poverty creates malaria, malaria creates poverty.

The most effective weapon ever used against mosquitoes was DDT, the insecticide developed specifically for mosquitoes. It did an excellent job; it also killed practically every other species of insect. It wasn't even remotely selective and it killed insects in unprecedented numbers, on a scale never seen before in human history. This chemical's greater possibilities were soon realized, and they started to use it for pest control in agriculture. In the 1950s the developed world was saturated in DDT.

This didn't work as the final solution for mosquitoes. Immune individuals failed to die, bred and passed on their immunity, and so they became the triumphant ancestors of DDT-resistant mosquitoes. But it was close to becoming the final solution for the global ecosystem. The poisons were killing more than insects: threatening the health of everything that lived on the planet:

> O what can ail thee, knight at arms
> Alone and palely loitering?
> The sedge has wither'd from the lake
> And no birds sing.

Rachel Carson was a marine biologist, but she expanded her area of concern with the spread of pesticides. The bleak vision from John Keats's poem 'La Belle Dame sans Merci' (quoted above) had stuck in her mind, and she decided to use its basic idea for a chapter title in a book she was writing about pesticides. But she was persuaded to use it as the title for the book itself – and in 1962 she published *Silent Spring*.

Great pioneer: Rachel Carson, biologist and author of Silent Spring *(published in 1962), in woods near her home.*

This is conventionally seen as the start of the Environment Movement (but see Chapter 69). Perhaps it will go down one day as one of the big dates in history: 1914, 1789, 1776. Certainly it was the first time ordinary people (and, in particular, ordinary people in North America) were aware that humans had effectively taken control of nature, and that we now needed to act with some idea of self-control, with a long-term view in mind. It was a book that changed the world: DDT is now banned worldwide and environmental concerns are part of every government's policy. You're no kind of wildlife writer if a publisher hasn't taken you out to lunch and asked you to write the next *Silent Spring*... and even as you sip your wine – second cheapest on the menu – you know that there can't be one. The world can only be enlightened once. We are now perfectly well aware that humans have taken control of the planet: that it's up to us to make the God decisions. No one needs to tell us again. It's getting those decisions right that's tricky.

And that again brings in mosquitoes. With the advance of genetic technology there is now serious talk of deliberately rendering dangerous species of mosquito extinct. The principle is simple; its execution is not. The idea is to produce male mosquitoes only capable of producing sterile offspring, and then releasing them into the wild. This would be a colossal undertaking and would require an awful lot of male mosquitoes.

It also raises complex ethical questions: whether or not humans have a right to 'edit nature'. This can be seen as merely a question of degree: we have been modifying – or editing – nature for 12,000 years. It is the genetic method that changes the game. Our instinctive and emotional response finds this a good deal more sinister than dumping millions of gallons (litres) of poison on the planet.

There are legitimate questions about the effect extinction would have on the ecosystems that mosquitoes currently inhabit: for example, mosquito larvae are an important food for many species of fish. But the counter-argument runs, if the genetic modification affected only one discrete species, then other species of mosquito would fill the vacant niche and do so without biting humans. A further point is that we don't actually know what would happen if we released millions of genetically modified creatures into the environment. It's never been done. As the great scientist Edward O. Wilson said: 'one planet, one experiment'.

TWENTY-FOUR
TIGER

*'Tyger! Tyger! burning bright
In the forests of the night;
What immortal hand or eye
Could frame thy fearful symmetry?'*

William Blake, 'The Tyger'

There are a few line-in-the-sand species. The ones we daren't lose. How could we live with ourselves if we let them go? How would subsequent generations judge the people who lost the last, the very last of a species that really matters to humans? It's not something that bears thinking about.

What are they, then, these non-human animals that seem to count so much more than the rest? The blue whale, two species of elephant, five species of rhinoceros, the lion, golden eagle and bald eagle... and on every single possible list of such species we find the tiger.

Fantasy of ferocity: The Tiger Hunt by Peter Paul Rubens (c.1616).

A map of the historical range of the tiger, superimposed over a map of its range in the twenty-first century, tells us as much about the advancing human population as it does about the retreating population of tigers. As human populations and human demands have grown, so the range of the tiger has shrunk and shrunk again – and, with it, the number of tigers in the world. At the beginning of the twentieth century there were around 100,000 tigers in the world. In 2015 a census put the number of wild tigers at 3890, though it's probably a little higher now; the Indian 2018 census recorded 2967 individuals in India alone, 400 more than in 2014. Tigers once lived from the Black Sea across Central Asia as far as Mongolia, right across South and Southeast Asia, and up north into Siberia. These days they are found in pockets across the southern and eastern parts of their former range. Siberia holds the largest population outside India.

Tigers were once hunted all over India by wealthy British colonialists and by rajahs. You could hunt them on foot, from a blind (or hide) with bait, live or not according to choice, and from elephant back. And, of course, tigers have been known to hunt humans. They have probably killed more people than any other vertebrate species, apart from humans. Tigers are wary of human settlements; perhaps that's been the case since the invention of fire, and certainly since the invention of firearms. But individuals driven to desperation by injury or the destruction of habitat and the loss of prey species have accepted the risks that humans pose. The so-called Champawat Tiger of Nepal and India, a female, is thought to have killed 436 humans before being shot in 1907 by Jim Corbett, the hunter-turned-conservationist. She was found to have two missing canine teeth, and is perhaps responsible for the notion that man-eaters of all species are depraved, defective and aberrant individuals.

In the Sundarbans, the vast swampy area between India and Bangladesh, tigers have long been heavy predators of humans, though humans have never been a major part of their diet. More recently, with tighter regulations about who can get into the place, the annual rate has dropped from fifty-plus to around three. The Sunderbans is an expanse of mangrove covering 3800sq. miles (10,000sq. km), and the place holds around a hundred adult tigers. Why is this population so ferocious? All kinds of reasons have been suggested: the salty water makes them aggressive; high tides wash away the tigers' scent markings and force them to defend territory by means of confrontation; regular cyclones kill people and wash their bodies into the swamp, and so give the tigers a taste of human meat; honey-hunters and fishermen in small boats are just easy prey; the area was never suitable for old-school tiger hunting so the tigers never learnt to fear humans.

The fact is that tigers have always killed people and are still hard at it. I visited a village in northern India where tigers had recently killed livestock and, a few years earlier, killed one of the villagers. Like mosquitoes, tigers kill people. Unlike mosquitoes, it would be – at least comparatively – easy to wipe them out altogether. The extinction of the tiger has already come very close through accident: on purpose we could do the job in no time.

But that is not what we want, it seems. In the main, humans want tigers to carry on living. They have been admired throughout history. The Hindu warrior goddess Durga rides a tiger; in Buddhism the tiger represents anger; there are traditions of were-tigers – men who turn into tigers – in India, Malaysia and Indonesia. In Western culture, Blake's famous poem 'The Tyger' (see epigraph on page 113) is full of admiration for the beast and its ferocity, even though he asks the question: 'Did he who made the lamb make thee?' In Blake's *Proverbs of Hell*, he declaims: 'The tigers of wrath are wiser than the horses of instruction.'

Tigers are ambush predators. They don't hunt, they wait. They sneak up: they need to get close because, although they have great acceleration, they don't have much stamina. Their edge, in their eternal competition with very large prey, is explosive power. They are not social, like lions, and they must hunt alone, but they still take prey that would tax a full pride of lions.

Adult tigers live in individual territories: each female, or female and cubs, has a territory, while a male will attempt to hold a super-territory that takes in two, three or more female-held territories. Young males without territories wander where they can. Tigers hunt large herbivores: deer, wild cattle, wild swine. People have interpreted their ambush technique as a revelation of their essential cowardice. In Rudyard Kipling's *The Jungle Book* Shere Khan, the lame man-eating tiger, is both sinister and despicable.

But we love tigers more than we have ever despised them. The Romans kept them for public entertainments and for personal prestige; in the seventeenth century European nobles kept them for showing-off purposes; by the eighteenth century they were popular animals in zoos. They even became children's toys: Tigger appears in the second of the Winnie-the-Pooh books, *The House at Pooh Corner*, and he is charming and lovable if prone to overenthusiastic bouncing. (He accidentally bounces Eeyore the donkey [see Chapter 20] into the river, but is eventually forgiven.)

So when the Environment Movement began, the possibility of tiger extinction was a powerful thing to place before humankind: an issue that laid bare the extent of the devastation of the wild world. If tigers went extinct, it was surely irrefutable evidence that we were doing something wrong. Nine subspecies of tiger have been recognized; three are already extinct: Bali, Caspian and Javan.

In these early days of the Environment Movement, the stress was all on individual species: save this species, save that species. It's still the most sexy idea in conservation: if we don't do this, this species of charismatic megafauna will go extinct. Fund-raising organizations tend to focus on a single species, knowing its potential loss sways hearts and minds. We will look at the way ideas about conservation developed later in this book, but in the late 1960s the horrific idea that tigers might go extinct was a global rallying point.

In 1969 the annual assembly of the IUCN was held in Delhi; it involved many Indian conservationists, including Kailash Sankhala, director of Delhi Zoo. The plight of the tiger dominated proceedings. They decided to ask all the tiger range states to ban the hunting of tigers and the export of their skins. They also voted to present politicians with significant tiger facts, and a tiger census of India was organized. Guy Mountfort, a British conservationist and a founder trustee of WWF (then the World Wildlife Fund), campaigned for money to save tigers and went on to meet the then Indian prime minister Indira Gandhi. Her response was straightforward: 'I shall form a special committee. A Tiger Task Force. And it will report to me personally.'

This was the beginning of Project Tiger. The census found the population of tigers in India was 2400: the problem was how to keep them. The backbone of the project was the protection of habitat: to create wildlife reserves and make them effective. The world is full of what conservationists call 'paper parks': vast areas designated National Park that begin and end with marks on a map.

But Project Tiger was genuinely committed to saving tigers and it was a global example of what can happen when there is political will for wildlife conservation. There were problems, but the degree of government commitment to wildlife conservation was inspiring and influential. It showed what governments could – and arguably should – be doing.

It has never been easy. 'When I first started to work in wildlife conservation thirty-two years ago, they told me that tigers would go extinct within ten years,' said my old friend Vivek Menon, chief executive officer (CEO) of the Wildlife Trust of India. The Born Free Foundation were making precisely that claim in 2019 – but the fact is that in India, China, Nepal, Bhutan and Russia the population is stable, perhaps even increasing.

Tiger parts have always been popular in Chinese medicine; tiger bone wine is supposed to be good for all kinds of complaints including rheumatism and arthritis (so it makes you young again). Tiger's penis soup is renowned as an aphrodisiac. These beliefs have led to a great deal of poaching of tigers, and created a considerable and lucrative illegal trade. (Wildlife is the biggest illegal trade after drugs and guns.) China initially rejected the Environment Movement but in the 1980s had a change of heart and signed up to the Convention on International Trade in Endangered Species (CITES). In 1993 they banned the trade in tiger parts.

Tiger numbers in India were falling in the early years of the twenty-first century, largely because of poaching, but after a disappointing census of 2007 the government responded with strong measures, establishing eight new reserves,

Burning bright: William Blake's 'The Tyger' from Songs of Innocence and of Experience (c.1794).

relocating 200,000 people to places with fewer people–wildlife clashes. They also invested heavily in anti-poaching measures. No one will ever be complacent about the fate of the tiger or, for that matter, any other species on the planet, but the short-term future of the tiger is reasonably good.

That is to say, wild tigers. Captive tigers are positively booming. There are many tigers kept in captivity; it is reckoned that in Texas alone there are 2000–6000 – it's been said that it's easier for a Texan to own a tiger than a dog that has been certified as dangerous. This is no help at all to the conservation of tigers in the wild: for that, you need healthy wild ecosystems. Protect the ecosystems in which tigers thrive and you are a good way down the road to protecting tigers – and, as a bonus, you protect everything else in the same ecosystems.

TWENTY-FIVE
RAT

*'In some streets a woman dare not leave her baby alone in the
house, even for five minutes. The rats are certain to attack it.
Within quite a small time they will strip it to the bones.'*

George Orwell, *1984*

If you call someone a rat, you mean that person is nasty – and nasty in a
particularly despicable kind of way. It is a low, mean, cunning kind of nastiness,
the sort of thing you wouldn't stoop to yourself. If you rat on someone you have
cheated in a particularly loathsome way. A rat is just about the last thing a decent
human being would wish to be.

How many of us owe our lives to the fact that rats are an awful lot like humans?
I intend no cheap cynicism: we share 90 per cent of our DNA with rats, and that's
why rats are indispensable as laboratory animals. Their similarity to us means that
medication and vaccines can be tested in laboratory conditions – and if they
work for rats they're likely to work for us.

Rats have lived with humans – alongside us, above us, around us, below us – for
the last 2000 years in Europe, and have probably exploited humans in some places
for as long as we have had any form of settlement, however temporary: we store
food, we leave eatable rubbish wherever we go, and rats are skilled improvisers
and generalists: the sort of animal that finds it relatively easy to strike up a
commensal – it means sharing the same table – relationship with humans.

Pest control: the Pied Piper in action on the Hamelin town clock.

We tend to fear rats for their boldness, but it is their wariness – their shyness – that gives them their edge in their relationship with us. They like to keep out of sight, loving corners, dark places and the times when humans are least likely to see them.

But it's time for clarification. We call quite a lot of creatures 'rats'; it's not a scientific term. We talk about bandicoot rats, kangaroo rats, cane rats and pack rats, but they're not under discussion here. Ratty is one of the most beloved characters in children's fiction, from Kenneth Grahame's *The Wind in the Willows*, but he is a water vole, only distantly related to the rats in question here.

There are more than sixty species in the genus *Rattus* and most of them have very little to do with humans, living their wild lives in their old-fashioned way. But two species, the black rat and the brown rat, found that living in and around human settlements was the way forward: and they have prospered from the rise of humanity as few other species on Earth have done. We have already met them in these pages (see Chapter 6 on rat fleas): the fleas that associate with them are vectors for the plague virus.

Ancestral black rats were tree dwellers and adroit climbers; brown rats are burrowers with a preference for low places. This difference allowed the brown rats to outcompete the black in colder climates, being better at finding warm, out-of-the-way places: most notoriously, sewers.

They are rodents: the rodent's unique selling point (USP) is their incisors. These teeth grow all their life: hard enamel at the front, softer dentine at the back. This means that every time they close their mouths they sharpen the chisel points of their teeth. Rodents gnaw: and the rat species in question have used their gnawing abilities, along with their quick intelligence and their adaptability, to change their ancient way of life and become fellow travellers with the planet's dominant species. Our success is their success: we have advanced hand-in-paw with rats.

They have become animals that make human flesh creep. We have been taught to believe that in a city you are never more than 6ft (1.8m) from a rat; an exaggeration that stresses the rat's skill at living with us without being seen. We think of them as particularly dirty animals, though any species of mammal living in close proximity to us is equally capable of contaminating food, leaving droppings, urinating and carrying fleas. It's their proximity that makes them problematic, not their depraved nature. We associate Weil's disease particularly with rats; in fact, you can get it from contact with the urine, blood or tissue of cattle, pigs and dogs as well as rats, and also from contaminated soil. It is primarily a work-related disease, associated with farmers, butchers, vets, rodent controllers and people involved with water sports.

Rats have been seen in a more benign way in other cultures. You can find a fine rat statue if you go along the Philosopher's Walk, from one Zen temple to the next, in Kyoto. Natives of the Year of the Rat in the Chinese horoscope are creative, intelligent, honest, quick-tempered and wasteful. The beloved elephant-headed Hindu god Ganesh has a rat as his vehicle: though perhaps this particular rat is an example of the sort of problem that the god of problem-solving solves.

In Western culture rats are more often seen as evil. The thirteenth-century tale of the Pied Piper of Hamelin – who rids the town of its rats but when they people welch on the payment he steals the children away as well – shows rats only as something you long to get rid of: though perhaps there's a useful (and greatly ignored) moral here: once you start getting rid of things, you can lose more than you bargained for. In George Orwell's novel *1984*, the ultimate torture for the hero, Winston Smith, is his exposure to rats in Room 101. But there's a more benign response to rats in Hugh Lofting's *Doctor Dolittle* books: among his many other good works, the doctor establishes the Rat and Mouse Club, whose members are portrayed with admiration and good humour.

Humans have eaten rats through the ages, when needs must, and in some cultures as a food of choice. They're even something of a delicacy in parts of China and Vietnam. There is a French tradition of grilling and eating rats found in wine cellars.

The sport of rat-baiting – setting a terrier onto confined rats – led to the practice of breeding rats for this sport, and so the process of domestication began. There were genuinely domesticated rats in the nineteenth century, and they have been bred into fancy colours: white, or white with brown hoods. They are said to be tractable, intelligent, amusing and affectionate. The horror and disgust they arouse in others is perhaps part of their charm. Domestic rats are calmer than wild rats, less likely to bite, more tolerant of crowded living conditions and easier to breed. So here, then, was a handily sized, easily maintained mammal, and it was adopted for laboratory work in the late nineteenth century. Rats have been a staple there ever since, adding greatly to the sum of human knowledge. An enormous amount of what we know about ourselves we owe to rats.

They share almost all their disease-linked genes with humans. They have been used for research beneficial to humans in obesity, cancer, cardiovascular diseases, multiple sclerosis and neurological diseases, notably Parkinson's.

They have also been used to help us understand what we mean and understand by the concept of intelligence, most often by working in mazes. It's been found that, if you breed rats who are skilful in mazes with other rats of the same kind, you breed generations of skilled maze-users, implying that intelligence – at least that kind of intelligence – is heritable. There have been thirty Nobel Prizes awarded for research using rats. We have a good reason to be grateful to them.

Rats have caused a great deal of damage to humans and the human world. They have also caused a great deal of destruction to the wild world. This has

Remover of obstacles: the Hindu god Ganesh with his accompanying rat (c.1855).

happened in places where humans introduced them. This has been an accidental process: but wherever humans go rats tend to travel with them.

Rats have been encountered earlier in these pages as playing their part in the extinction of the dodo in Mauritius (see Chapter 19). They have played similar roles in many other islands. Ground-nesting and flightless birds have been pushed towards extinction, and in many cases beyond, by the rats that humans brought to their islands. There have been elaborate and expensive attempts to remove rats from islands famous for large seabird colonies. Between 2011 and 2015 a British team from the Falkland Islands in the South Atlantic – using a ship with three helicopters – dropped more than 300 tonnes of rat poison on South Georgia island. This has been declared a success: the place is now rat-free, which is good news for 65 million birds that annually nest there, including four species of albatross and six species of penguin.

Rats in the laboratory have also been shown to possess metacognition: that is to say, the ability to understand and make accurate assessments of their own thought processes and their own levels of understanding. This was initially thought to be something that separated humans from all other species, but as usual this was a false assumption. Apes were shown to possess it too: and, in more recent experiments, so have rats.

After a series of schooling sessions, rats were offered the chance to take a test. If they passed it, they got a huge reward; if they failed, they got nothing. But if they refused even to attempt the test, they got a small reward. So the rats had to assess whether they had what it takes to pass the test – and that's exactly what they were able to do.

We have always underestimated rats.

TWENTY-SIX
WASP

'If I be waspish; best beware my sting.'

Shakespeare, *The Taming of the Shrew*

I f you're reading this book in a hard copy rather than on a screen, pause for a moment, consider the 150,000-odd words that make up its text, and marvel at the fact that wasps made it all possible. There was an extraordinary explosion in human knowledge with the Renaissance: and for this we can thank the wasps. Wasps changed the nature of human culture.

These days, wasps are mostly looked on as the spoilers of picnics and al-fresco drinks, the capricious dealers-out of extraordinarily – and lingeringly – painful stings. If you find a wasp's nest in or around your home the first thing you do is call in the pest controllers... seldom pausing to consider the revolutionary effect that wasps have had on the spread of human knowledge.

There are more than 100,000 species of wasps and many of them are solitary. It is a large and complex group, with fascinatingly evolved behaviour. Some of the solitary wasps go in for maternal care, laying an egg in a concealed place and provisioning it with small insects and spiders, which they subdue by stinging. These stung creatures are not dead, merely paralysed, and they form a grisly living larder for the hatching grub.

The wasps that concern us most here are from the family Vespidae, which includes all the social species of wasp. To be technical, that should be eusocial: they live in communities based on division of labour, in which breeding, care of the young and other crucial jobs are delegated to particular individuals and/or castes (see also Chapter 12 on honeybees and Chapter 99 on ants).

Wasps have been around since the Jurassic, so the line is 200 million years old. They range in size from a species called fairy wasp no bigger than 0.006in (0.14mm), the world's smallest flying insect, to the Asian giant hornet, which is 2in (5cm) long. They tend to be black and yellow, which is widely accepted as a danger sign: don't attack me, you'd regret it. Many non-related species have the same colours, and some of these are harmless bluffers.

If you are a eusocial species, you need a base. And you have to make it yourself: your needs are too specialized for an ad-hoc natural dwelling to work for you. Naked mole rats, a eusocial species of mammal, do so by burrowing; honeybees,

Many-celled perfection: engraving of a wasp's nest.

as we have already seen, produce wax to make combs; termites create vast mansions by processing and reshaping the soil. And wasps use… paper.

They make it themselves. They chew wood pulp, mix it thoroughly with saliva, then spit it out again – and they use this rather nasty paste to create some of the most exquisite objects you can find in nature. The commonest and most irritating wasp that ever tried to get between you and your drink is capable of constructing a masterpiece: a thing of rhythm, symmetry and fastidious neatness. A wasp's nest is a series of cells in which an egg can be laid and in which the resulting grub will have its being, fed lovingly by the attendant workers, who will forage far and wide, killing small insects and spiders, macerating them and bearing the result within themselves as they return to the nest to feed the goo to the next generation.

A wasp's nest is one of those inanimate objects that strikes awe in the human observer: unmistakably paper, so light in the hands and so beautifully made. You could never do it yourself no matter how many tools you possessed, and yet wasps create them out of the endless millennia of their existence as a species and a family.

蔡倫（公元前?‧一一二一）汉

中国人民邮政

4分

纪92‧8-1   (297)1962

The inventor of paper: Chinese postage stamp commemorating Cai Lun.

And someone – someone in China – saw a wasp's nest and thought that the paper it was constructed from would be handy stuff in human life. Fragments of paper have been found by archaeologists from around the first century BC, but the credit for the job is traditionally given to the eunuch Cai Lun of the first century AD; as humans castrated livestock to ensure docility and loyalty, so some cultures did the same with humans. Some eunuchs, like Cai Lun, became honoured and wealthy men. The Emperor Hedi of the Han dynasty had good reason to be pleased with the work of his servant Cai Lun.

It's not certain that Cai Lun was the first person to make paper for writing on. Paper was also used for padding and wrapping, particularly of fine bronze mirrors. But once paper was established as a medium for human self-expression, the stuff took off. The great Chinese arts of poetry, painting and calligraphy were made possible: not only was paper substantially cheaper than silk, which was used previously, but it was also a good deal better for the purpose. So with greater possibilities of expression and transmission, human thought was able to expand.

Paper slowly spread westwards. Its advantages over parchment – which was made from the skin of calves, sheep and goats – were considerable: so much lighter, so that a lengthy work could be carried by hand and read in comparative comfort. It was a way of storing and passing on the most important information that humans possessed: and that was, of course, contained in the central books of religion. As the use of paper continued westward, the teachings of the Quran could now be passed on by means of paper copies. Paper came into Europe in the reverse direction, from west to east; it was brought by the Moors into Spain, and the first European paper mill was established in Spain in the thirteenth century. The scroll was replaced by the codex – individual sheets linked together, yes, in the manner of a book.

The woodblock printing press had been invented in China; the printing press with moveable type was invented – at least so far as Europeans were concerned, the Chinese having a similar system much earlier – by Johannes Gutenberg in 1440; in 1455, 135 paper copies of what is now called the Gutenberg Bible were printed.

This was effectively lighting the blue touch paper. Written information was no longer restricted to handwritten parchments kept in monasteries: it was available to many more people. It has been estimated that, in the next half century, 20 million books were published.

And today, even in the paperless office, the printer still hums about its work. As I research this book, I make inky notes on the paper pages of a notebook, in a hut full of books, and I normally start my day's work after reading the sports section of my newspaper, yes, in a hard copy. The impact of electronic communication has been incalculably vast: but we still need paper, if only for use in the lavatory – and that was a notion the Chinese came up with in the sixth century. The internet merely speeded everything up; the invention of paper changed us for ever.

So next time you are taking a pleasant drink outside, raise your glass rather than your newspaper to the wasp that comes a-calling. We owe them a good deal: they made us who we are. Without them we wouldn't have access to the Quran, the Bible, Shakespeare's plays, *Ulysses*, *On the Origin of Species* – and the book you are reading now.

TWENTY-SEVEN
EARTHWORM

*'Their sexual passion is strong enough to overcome
for a time their dread of light.'*

Charles Darwin, *The Formation of Vegetable Mould Through the
Action of Worms, With Observations on Their Habits*

Charles Darwin devoted the last years of his life to the study of earthworms, and in 1881 published *The Formation of Vegetable Mould Through the Action of Worms, With Observations on Their Habits*. It was a surprise best-seller, with copies vanishing from the shelves faster than those of *The Origin* twenty-two years earlier. He said: 'It may be doubted whether there are many other animals which have played so important a part in the history of the world, as have these lowly organised creatures.'

The soil is an immensely complicated ecosystem, involving incalculable numbers of bacteria, along with great complexities of fungi. The soil – the stuff that plants grow in – is itself a living thing: and presiding over the process are earthworms.

This is an imprecise term. It covers up to 6000 species of annelid worms that live in the surface litter, in the topsoil and deeper down into the subsoil. Earthworms may be regarded as the curators and managers of the ecosystem they inhabit: the keystone species that makes the whole thing work. We look at a mighty oak tree with fully justifiable awe: seldom pausing to think that it owes its sky-reaching magnificence to the bacteria, to fungi and to the actions of very politic worms.

When hunter–gathering humans prowled the savannahs, the fruit and plant matter they ate came to them with the help of worms; when they killed and ate a herbivore, they drew the benefit of the years it had spent feeding on grass and leaves – all stuff that was made possible by the action of worms. When humans began to farm in the Fertile Crescent and elsewhere, they raised crops in soil nourished by worms: it's not too much of an exaggeration to say that these crops were grown in the faeces of worms. All terrestrial life on this planet – and not just human life – would be radically different without worms. It's been estimated that we get 90 per cent of our food with the help of worms; the exceptions being seafood and hydroponic crops.

Animals all: Man Is but a Worm, *cartoon from Punch showing the fanciful evolution from worm to man (1881, artist unknown).*

Earthworms are ecosystem engineers. They affect the soils physically, chemically and biologically. The biological action converts relatively large pieces of organic matter into rich humus: a wormcast contains 40 per cent more humus than anything else in the top 9in (23cm) of soil. Chemically, they make mineral and plant material accessible to plants by effectively bringing it down towards the roots of plants. A wormcast is five times richer in nitrogen, seven times richer in phosphates and eleven times richer in potassium than surrounding soil. The worms do this, not out of the kindness of their hearts but because it is a by-product of the way they live, taking in soil at one end and passing it out at the other; anatomically, each worm is a long digestive tube ringed with circular muscle.

They move through the soil by lengthening and shortening themselves in a process called peristalsis. Their movement through the soil, lubricated by the mucus they secrete, opens it up, creating pores, which aerates the soil and allows water to pass through more readily, bringing dissolved nutrients as it goes – and that is the physical action.

Earthworms lack eyes, but they have photosensitive cells. Daylight is their enemy, they are vulnerable on the surface and so escaping from light is an important strategy. Their sense of hearing isn't up to much: among Darwin's many experiments was to play music to worms, to see if they responded: his wife Emma played piano, his son Frank played bassoon, another son Bernard blew a whistle and a daughter, Bessy, shouted – but the worms paid none of them any mind. They don't, alas, turn into two equally viable worms when cut in two with a spade, though as William Blake, a little bafflingly, wrote: 'The cut worm forgives the plough.' They have thrilling sex lives for so unexotic a beast, being hermaphrodites: after a sexual encounter, both partners go off and lay eggs.

They have been carrying on this quiet and humble task – in one end and out the other – across the millennia, slowly and subtly creating and altering the landscape in which we live. Darwin was fascinated by this aspect of their behaviour and left various objects out in the garden at Down House, to measure the speed at which the worms buried them. They are very leisurely, and very, very thorough. Stephen Jay Gould (see Chapter 16 on pandas) speculated that part of Darwin's fascination with worms was the way they acted over what to humans are very considerable periods of time. They take centuries to bury things; and that's only one example of worm action on an environment. Essentially Darwin was carrying out an investigation into the shallow end of Deep Time: exploring the timescale necessary for evolution, in the Darwinian sense of natural selection, to take place. The worms give us an idea of extended time that we are all just about capable of grasping.

Darwin was mad about them: 'Without the work of this humble creature, who knows nothing of the benefits he confers upon mankind, agriculture, as we know it, would be very difficult if not impossible.' The traditional method of fertilizing fields depends on the action of worms. Muck-spreading – covering the field in the droppings of animals mixed with their straw bedding – depends for its efficacy on worms, who bring the stuff down from the surface and, as said, make it available to the plants. A traditionally managed pasture will hold staggering numbers of worms: in 2½ acres (1 hectare) there can be as many as a million worms, and their biomass – combined weight – is greater than that of the animals grazing the field above their heads.

But agricultural methods have changed. They involve much greater and deeper disturbance of the soil, and the addition of chemical fertilizers, pesticides and fungicides. The tillage destroys the worms' tunnel systems; the switch to ploughing in late summer interrupts worms breeding; fungicides destroy their food; and some species are unable to tolerate the acidification of the soil that comes with use of nitrates.

In a famous example, modern agricultural methods came close to wiping out the Giant Gippsland earthworm in Australia, average length 3¼ft (1m), maximum length three times that. This is now a protected animal. Many arable fields are short of worms: in England, the sight of a plough being followed by squadrons of gulls feeding on the invertebrates exposed by the cultivation is much rarer than it used to be.

This seems to leave agriculture in a bit of a bind. If you are planning to farm in a relatively worm-free environment, you need either to be ever more invasive with your tillage and ever more eager to use chemicals, not knowing if there will come a point when this escalation becomes unviable – or to return to more traditional methods, with which you are not fighting the environment but tending towards the notion of cooperation. Which brings in the question about yield and cost and market and profit.

Matt Shardlow, CEO of the British non-government organization (NGO) Buglife, said:

> The implications of a lack of earthworms are poor incorporation of organic matter, weak soil structure, greater risks of soil erosion and flooding and less food for other wildlife. Earthworm populations require surface organic matter and low pollution from pesticides. The problem with the plough is not the cutting edge, but the hiding of the organic matter; the process removes it from the surface where the worms forage and create the basic framework for a healthy soil.

Worms have served humans well across the last dozen or so millennia; and for 3 million years before that. They have long been seen as a classic example of divine providence: a planet designed with the special interests of one species in mind. Do we now see our future without earthworms? And if so, how well will this work? Time and again, starting from apparently trivial matters, we realize that the decisions made by the current generation of humans will have extraordinary and far-reaching effects on the history of our own species and that of the planet we inhabit. I wonder how our descendants will thank us.

TWENTY-EIGHT
SNAKE

'And the serpent said unto the woman, "Ye shall not surely die;
For God doth know that in the day ye eat thereof, your eyes shall
be opened, and ye shall be as gods, knowing good and evil." '

Genesis 3:4–5

Snakes probably evolved from burrowing lizards, and there are more than 3700 species of them. The human response to snakes is so strong that in Western tradition we don't look on them as evil creatures. We see them as evil itself. Snakes represent a quality external to humanity, the eternal enemy whose one function in the cosmic order is to seek our destruction. In the last book of the Bible, the Book of Revelation, the snake meets its end at last: 'And the great dragon was cast out, that old serpent, called the Devil, and Satan, which deceived the whole world: he was cast out into the earth, and his angels were cast out with him.' Get rid of the snake and – at last – eternal happiness will follow.

On the mantelpiece in my childhood home there was a statue of the Virgin Mary, calm and beautiful, holding the Christ-child in her arms; casually, almost absent-mindedly, she is crushing a snake beneath her heel. The snake was Evil, not as adjective but as noun, evil incarnate, and here was Mary, portrayed explicitly as the new Eve. Instead of listening naked to the flattery of the snake she was decorously clad and smiling faintly as she crushed the life out of the thing: the right kind of female dealing with evil in the right kind of way. Here was an image of ultimate virtue. It showed us both the generosity of God and the goodness of humankind – and Mary proved the point by killing an elongate obligate carnivorous reptile of the suborder Serpentes.

Snakes are built for concealment. Low to the ground, fitting into improbable places, sneaking up on prey; all snakes are carnivorous. Everything about them is long: paired organs like the kidneys are stowed in a line rather than side by side; they have only one functioning lung. Most snakes have no eyelids and no external ears. You can find them on almost every large landmass apart from Antarctica; exceptions include Ireland, Iceland, Greenland, the Hawaiian archipelago and New Zealand.

They cover a huge range in size; the Barbados thread snake is just over 4in (10cm) long; the reticulated python can reach almost 23ft (7m). An extinct species, known only from fossils, was 42ft (13m) long, and was called *Titanoboa*

Knowledge of good and evil: The Garden of Eden with the Fall of Man *(c.1615), by Jan Brueghel the Elder and Peter Paul Rubens (Rubens painted the serpent).*

cerrejonensis. The green anaconda has been weighed at 215lb (97.5kg). The snake's skill at concealment gives most species their edge, along with an acute sense of smell, which operates through their flickering tongues; they taste the air as they go, touching the fork of the tongue onto receptors in the mouth. They are acutely sensitive to vibration, and some species – pit vipers, pythons and some boas – can sense infrared.

Some humans find the movements of a snake disturbing. The flight of birds, a movement utterly unlike anything we can do, fills us with envy, but the eight-legged motions of spiders (see Chapter 57) and the complex locomotion of snakes get people distressed, arousing in some fear and disgust. Absurdly, to our human-centred brains, it doesn't look natural. Snakes have several gaits, though not all of them can do all of them. Lateral undulation is their default mechanism, and requires snakes to get a grip on something with their scales so they can push against it. Side-winding, in which very little of the snake is in contact with the ground at one time, operates when there is nothing to push against; and snakes can operate a concertina movement when there's no room to side-wind. Perhaps the spookiest of all is rectilinear locomotion, in which the snake seems to move just by willing itself along. Even close up it's hard to see how they do it: lifting up the scales of their belly in subtle fashion to pull themselves along.

Some species kill their prey by constriction: *Boa constrictor* is one of the more widely known scientific names. The image of the desperate human wrestling in the coils of a giant snake is archetypal. Such snakes have indeed been known to attack and kill humans, and occasionally to consume them. Babies and small children are within the scope of large pythons. There are accounts of adult humans predated by large snakes; in 2017 a dead adult human was found inside a 23ft (7m) python. Such feats are made possible by an ability all snakes possess: they can swallow prey larger than their own heads by stretching the tendons of their jaws; this is not technically a dislocation, though it certainly looks like one.

But the thing about snakes that really matters to humans is that some of them are poisonous. Or, to be accurate, some snakes are venomous: you must ingest poison, while venom needs to be injected. You can drink snake venom and survive, so long as you don't have a cut or open sore in your mouth. Our natural inclination is to divide snakes into two groups: venomous and non-venomous. Science does it differently: about 700 species of snakes are venomous: the Elapids, which include cobras, kraits, mambas and sea snakes; Viperids, which include vipers and rattlesnakes, and some species of Colubrids, notably the boomslang.

This isn't one of the chapters in which I will be mocking human fear of a wild creature. The WHO reckons that 100,000 people die from snakebites every year. But killing people is not what venom is for. It's there to disable prey. Most snake venoms are complex; some contain toxins that attack the nervous system and others attack the blood. It is invariably designed for very rapid dispersal – after all, it's not much good if the animal you have bitten can stagger off to die somewhere else. But the venom has a secondary role of self-defence: and there are around 250 species that can kill a human with a single bite. This animal is dangerous: it defends itself when threatened.

As humans established settlements and permanent communities, so the threat of death by large mammals decreased: lions and wolves were increasingly kept away by noise, smell, fire and the powers of retribution open to a large human population. It follows that the most significant vertebrate killer of newly civilized humans was snakes, which can easily get into human habitations without being seen. Lions were the great terrors of early hunter–gathering humans: once we started farming the new fear was snakes.

There's something in humans that regards snakes as unfair. Unsporting. Not the right way to kill people. Not something we really bargained for. A lion that kills people isn't ideal, but at least we can relate to that: lions seem to act on the same level as we do. But slithering, low, legless and poisonous, or rather venomous: how could God allow such a creature to exist? Obviously he did no such thing: so the snake must be the devil's work: evil itself.

The snake is of course a sexual symbol as well as a religious one: a snake introduces naked Eve to knowledge, thereby tempting Adam from his duty: well, the story is capable of many interpretations, and Freudian mythology is just one aspect of it.

Benign cobras can be found in Cambodian sculptures, shading the meditating Lord Buddha with a spread hood. They have been associated with guardianship of tombs and treasures and temples. There is a temple full of living pit vipers in Penang, Malaysia, packed with eager tourists and devenomed snakes. In Hindu and Buddhist symbolism a snake represents death and rebirth: it sheds its skin and emerges from the shell of its former self, renewed, reborn.

Asclepius, son of Apollo in Greek mythology, saw a snake bring another back to life by bringing it healing herbs; Zeus killed him with a thunderbolt before he could impart this wisdom to humans, because humans must not, of course, seek to rival the gods. The staff of Asclepius – a stick with an entwining snake – is a medical symbol to this day.

Dragons and the giant sea serpents of mythology are closely related to snakes; evil in Western tradition. But Chinese dragons are more benign and are the emperor's symbol of power and strength. In Western literature snakes tend to be evil or the instruments of evil: Sherlock Holmes came up against one in 'The Adventure of the Speckled Band'; Nag and Nagaina are enemies of the mongoose Rikki-Tikki-Tavi – and by extension the human family he guards – in Rudyard Kipling's *The Jungle Book*. Snakiness and evil were conflated in the *Harry Potter* saga, and Lord Voldemort, Harry's mortal enemy, is profoundly snaky, with the snake Nagini as his familiar. Elsewhere in *The Jungle Book* and *The Second Jungle Book*, Kaa, the great python, is a loving if enigmatic friend of the boy-hero Mowgli; while the bicoloured python rock snake saves the Elephant's Child from the crocodile in Kipling's *Just So Stories*, children's stories strongly influenced by Darwin.

But the best snake in literature is in John Milton's epic poem *Paradise Lost*. Satan takes the form of a phallic serpent – 'pleasing was his shape/And lovely' – to seduce Eve into eating from the Tree of Knowledge. He praises her 'celestial beauty' before telling her that, after eating from the tree, she and Adam will be 'as gods'. This exuberant portrayal of serpentine evil caused William Blake to remark that Milton 'was of the devil's party without knowing it'. The snake's malice takes humans from the Garden of Eden. They are driven from paradise – terrible remorse on their faces as they flee across the ceiling of the Sistine Chapel – towards a life of pain and suffering. The garden represents idealized memories of the hunter–gatherer life; the pain and suffering are the endless labour that agriculture and civilization require of all save the elite. In the end, the snake represents not only evil but also despair at what we have become.

Meanwhile, real snakes continue to live their lives in the living world and they are still, just as it says in the Bible: 'more subtle than any beast of the field'. That's because they are so good at concealment they can mostly carry on without coming across humans. And when they do so, even the most harmless of species seldom fails to elicit a hint of fear from the humans they encounter.

All is one: the meditating Buddha shaded by a many-headed cobra at Wat Khaek, Thailand.

TWENTY-NINE
CHICKEN

'What, all my pretty chickens, and their dam,
At one fell swoop?'

Shakespeare, *Macbeth*

I was in the jungle in India when I saw a red jungle fowl. So what, you may be tempted to ask, but bear with me. He – clearly a male, from the extravagance of his plumage – strode into a small clearing, and did so rather cockily; note that word. And there is scarcely a single human being on the planet unable to identify the bird. Chicken – rooster – cockerel – chook – cock – and everyone is more or less correct because the red jungle fowl is the ancestor of every domestic chicken that ever ate or was eaten.

We were thinking about getting some chickens at home, so we read up about them and were rather put off. It was all so complicated. There seemed to be so many diseases to worry about. Keeping them alive from one week's end to the next sounded like a full-time job. 'Don't you worry about that,' a neighbour kindly told us. 'They only put that stuff in books to fill 'em up. Just throw a bit of corn at them every now and then and they'll thrive.'

Thrive they did. It really is that easy. And that is exactly what has made the chicken the most numerous bird on Earth. They're big, full of meat, they lay eggs and, above all, they're dead easy.

The ancestral red jungle fowl has a range from India across the bottom part of Asia: Malaysia, Indochina, South China, Philippines, Indonesia. Chicken bones have been found in graves in Heibei province in northern China from 7400 years ago: this is the first undisputed domestic chicken, because Heibei is miles away from the range of wild red jungle fowl. But the chances are that humans and chickens got together long before that; estimates put that a couple of thousand and more years before the Heibei chook.

No doubt the process of domestication was unplanned and ad hoc: jungle fowl come to the village to feed on scraps, so you catch one or two and eat them. Then you feed the birds on scraps on purpose and they stay. Catch and clip the wings and they'll stay much closer – the wild red jungle fowl is not much of a flier, a ground bird that flutters gamely up onto a low branch, away from ground predators, where it roosts for the night. So you breed the biggest and tastiest cocks with the

Fine birds: A Cock and Turkey Fighting, in a Park Setting, with Other Fowl, *eighteenth-century painting by Melchior de Hondecoeter.*

biggest and tastiest hens, and in remarkably few generations you have a domestic rather than a jungle fowl.

Chickens took their time about conquering the Earth. There are no chickens in the Old Testament, the *Iliad* and the *Odyssey*. But this wonderfully convenient source of domestic protein spread both west and east, travelling with humanity as part of civilization and as part of the provisions for the journey. No doubt they were domesticated several times over.

It was some time before chickens played anything like a central role in society. They were peripheral, a bonus. It's been suggested that they were more appreciated for the sport of cockfighting than for their food value. Cockfighting continues in parts of Asia and elsewhere. I once attended a series of fierce and bloody contests in Manila; I was taken there by two of the sweetest people you could ever meet. They yelled themselves hoarse.

These days you meet chickens at every village you stop at in the developing world: often long-legged and bony little survival machines, literally scratching out a living around the legs of protective humans. Feed them and the females will contribute an almost daily portion of protein in the form of an egg.

Because of that daily staple you'd think twice about killing a hen. Chicken was traditionally a treat. I remember, from my own childhood, the Sundays when, instead of having an uninspiring cut of meat like shoulder of lamb, we would have a whole roast chicken on the table: an entire animal, dead, golden, sizzling and all for us.

But even as we revelled in such days of celebration, the world was changing for chickens and for their relationship with humans. It had always been necessary to keep chickens outside, because sunlight – vitamin D – is crucial to their health. There was no other way to do it. But in the years after the Second World War a new kind of feed was developed: packed with vitamins and antibiotics. That meant you could keep chickens indoors. You no longer needed space.

Here was revolution: farming removed from all immediate contact with the natural environment. These days about three-quarters of all the chickens consumed come from factory farms and batteries, and about two-thirds of the eggs. A chicken can live five years easily; up to ten is not beyond its range. Battery chickens kept for meat go from egg to slaughter within six weeks. They live in very small spaces indeed, and are fed the liquids and solids required to bring them up to slaughter weight. In some operations you can get twenty birds to a square yard (sq. m). Often the ends of the beak are snipped off to discourage fighting. These birds are bred for extreme docility. They sit in their own ammoniac waste until they are slaughtered; there's a pre-slaughter wastage of 3–5 per cent.

All this represents a radical shift in thinking. We are no longer treating chickens like animals, like fellow, warm-blooded vertebrates; we are treating them like plants, kept in a single place and treated copiously with chemicals before being harvested at the appropriate moment.

If you travel along Highway 25 in Kentucky you can stop at Sanders Café. This was opened in 1932 by Harland Sanders, an entrepreneurial fellow who two years earlier bought the filling station there. He sold the first café franchise in Utah in 1952, by which time cheap, battery-produced chickens were available. By 1963 there were 600 outlets of Kentucky Fried Chicken (KFC) and Sanders had been given the honorary title of Colonel. He sold out in 1964.

These days KFC is owned by Yumi Brands, which also owns Pizza Hut, Taco Bell and WingStreet food outlets. There are 22,000 KFC outlets in the world in 123 countries. The recipe of eleven herbs and spices remains a secret; the Colonel's own phrase 'finger-lickin' good' has been trademarked. The grave of Colonel Sanders is in Louisville; it is much visited, and people will often lay on it a bucket of fried chicken.

One-third of Americans eat fast food every day; two-thirds of Americans are overweight; two-fifths are obese. It's now widely accepted that it's perfectly okay to take on more food than the body can usefully process. The fast-food industry worldwide spends US$20 billion on advertising. Chicken was a treat; now it's a

Cock, cockerel or rooster? Japanese hanging scroll by Sō Shiseki (1715–86).

staple. Once it was a luxury; now it is seen as something close to a basic human right. Chickens were once a healthy option; now they are part of a global health crisis.

We are so used to chickens that we don't even have a proper name for them, any more than we do for cattle or for ourselves. A chicken originally meant a young bird, just hatched; chickens were little things covered in golden fluff. A male bird (other than ducks and geese) is usually a cock, but that word has collected all kinds of phallic connotations, so it has mostly been dropped. In North America they prefer rooster, even though both sexes roost every night, if they have the chance. Cockerel is the choice in England; that used to mean a young cockbird.

These days chickens are archetypes of neuroses, from the cackling flight behaviour of free-ranging flocks: they believe that the sky is falling, a panicked or overstressed human runs around like a chicken with its head cut off. There are more admiring phrases concerned with the males: cocksure, cock-a-hoop, crowing with delight.

Jesus is shown as both a lamb and as the good shepherd, but in Matthew 23:37 he says: 'O Jerusalem, Jerusalem, thou that killest the prophets, and stonest them which are sent, how often would I have gathered thy children together, even as a hen gathereth her chickens under her wings, and ye would not!' I know of no pictorial representation of Christ as the Good Chicken, but the cock plays an important role in the Gospels. Jesus foretells that Peter will deny his Lord three times before the cock crows: and so it came to pass. These days many church steeples have a cock as a weather vane: dominating the countryside, the cocks stand as an awful warning to us all: deny Christ and you too must weep bitterly.

This book is concerned with the changing attitudes of humans to our fellow animals. I can think of no animals whose role in human society and human thinking has changed so much as the chicken.

THIRTY
MONKEY

*'Thou hast been with the Monkey-People – the grey apes – the people
without a Law – the eaters of everything. That is a great shame.'*

Rudyard Kipling, *The Jungle Book*

Monkeys have disturbed human beings ever since we decided we were unique. They ask us overwhelming questions about who we think we are, where we think we came from and where we think we're going. Their similarities to us are impossible to ignore, nagging us with constant suggestions that the idea of human uniqueness requires qualification.

Monkey is not a precise zoological term. It has traditionally been used to cover any species of primate with an obvious relationship with humans. It has frequently been applied to apes, including gibbons, gorillas, orang-utans and chimpanzees. In the film *The Return of the Pink Panther*, Peter Sellers, as Inspector Clouseau, encounters a street musician with a chimpanzee and tells him in the course of their argument: 'It is your minky, therefore it is your money!'

The term 'monkey' has been used haphazardly for centuries. With slightly more precision, it is these days most often applied to two taxonomic groups that aren't, in fact, all that closely related: the Old World monkeys and the New World monkeys: that is to say, all primates excluding lemurs, galagos (bushbabies) and similar creatures, apes (including gibbons) and – should you wish to categorize them differently – humans.

Old World monkeys come in a single family of about 130 species; New World monkeys are also around 130 species, but in five families. Most of them are tree dwellers: the forward-facing eyes and opposable – grasping – thumbs of primates are adaptations for life in the trees. You need very precise stereoscopic vision to judge distances between branch and branch, and it helps if you can get a good hold of the branch you're on and the branch you're aiming at. (Gibbons, technically apes rather than monkeys, have much reduced thumbs and travel through the branches by hooking rather than grasping; the swinging means of travel, the one that makes such a good zoo exhibit, is called brachiation.)

Many species of New World monkeys have a fifth limb: a grasping prehensile tail capable of supporting their entire weight. These include the aptly named spider monkeys, which move with an unearthly many-limbed certainty that can remind you of spiders.

Devastating shrewdness: Two Chained Monkeys *by Pieter Bruegel the Elder (1562).*

Some species of monkey have come down from the trees to become ground dwellers, recapitulating the transition made by our own ancestors when the forests dried up and the savannahs were created. Baboons return to the trees to roost at night, and will also forage for fruit and edible flowers in the branches, but most of their day is spent on the ground, looking for food and continuing their intense and complex social lives, with constant quarrels and reconciliations. When predators appear they will flee into the trees – but when they are in a tree and the danger is human they will race down as if they were doing a fire drill, having learnt that humans sometimes carry guns and staying in a tree makes you quite literally a sitting target.

Monkeys are a diverse bunch, with the smallest, the pygmy marmoset, no more than 4½in (115mm) without the tail, and weighing no more than 3½oz (100g). The largest is the mandrill, a species of baboon famous for its rainbow bottom; a big male can reach 3ft (1m) in length and weigh 79lb (36kg). The large males of all the baboons, well fanged, are imposing creatures; a friend of mine, with whom I have shared a number of alarming African adventures, swears that the most frightening moment of his life came when a male baboon got into his car.

The close similarities between humans and monkeys have meant that monkeys have frequently been used for research, though in smaller numbers than rats, because they are more expensive and more difficult to acquire and to look after. But they are cheaper and easier than apes (like chimpanzees) and they reproduce more quickly, which can be an asset in certain kinds of research (like genetics). Monkeys have been used for research on drugs and disease, including hepatitis and HIV. Their closeness to humans – physiologically and psychologically – makes them extremely useful laboratory animals, but the same traits raise a number of ethical issues. The broad range of Western thinking on the issue is probably best summed up as something we'd rather not know anything about. We are deeply disturbed by it, should we hear any details, but perhaps most of us would be happy to take a proven drug if it would make us better. It's been estimated that, across the world, there are 200,000 non-human primates used in scientific research, most of them Old World monkey species like the rhesus macaque. Most are bred for the task, though some are wild-caught.

Monkeys have travelled into space. Albert, another rhesus macaque, was fired up in an American-launched V2 rocket in 1948. The rocket reached 39 miles (63km) in height; Albert died of suffocation. The following year, Albert II, another rhesus macaque, became the first primate in space – that is to say, he went beyond the 62-mile (100km) line – reaching 83 miles (134km); he died on the way down from a parachute failure.

Western culture tends to use monkeys to represent qualities we deplore in ourselves: indiscipline, folly and self-indulgence. But Hanuman, the Hindu monkey god (who also appears in Buddhism and Jainism), is a model of virtue, notable for his strength of purpose, his affection and his loyalty. He plays a major role in the Hindu epic *Ramayana*, and is a great support of Lord Rama. In one famous incident, Rama asks him to find a certain herb on a mountain; on arriving at the mountain, Hanuman can't work out which herb he is supposed to collect – so he brings back the whole damn mountain. Nothing by halves, that's Hanuman. He is a very different monkey to the ones we find in Western literature, like the Bandar-Log, the foolish leaderless rabble of the monkey people in Rudyard Kipling's *The Jungle Book*.

The similarity between monkeys and humans was at first seen as one of life's oddities, perhaps a whim of the Creator or an example of his house style. But as the Enlightenment spread across the eighteenth century, humans required some way of organizing monkeys into a framework we could accept – one that didn't require religion to explain it. Linnaeus (see Chapter 8 on blue whales and Chapter 11 on platypuses) published the first edition of his great work *Systema Naturae* in 1735; his garden at Uppsala in Sweden is still there and it still contains two huts for monkeys. It's clear that the conundrum of the monkey was one that exercised him very much.

He had no hesitation in placing humans and monkeys into the order of Primates: close together as closely related species must be. Perhaps strangely, his findings never created vast controversy: rather, the issue was left hanging over science and the world. Linnaeus wrote to a friend: 'It is not pleasing that I place humans among the primates, but man knows himself. Let us get the words out of the way. It will be equal to me by whatever name they are tested. But I ask you and the whole world for a generic difference between man and simians in accordance with the principles of Natural History. I certainly know none. If only someone would tell me one! If I called a man an ape or vice versa I would bring all the theologians against me.'

This problem finally exploded with the publication of *The Origin of Species* in 1859. Darwin scrupulously avoided talking about human descent in this book, but the unasked question it provoked was unavoidable. The opposition was summed up by the future prime minister Benjamin Disraeli: 'Is man an ape or an angel? Now I am on the side of the angels.'

There is a still more famous monkey: or, a least, a notional monkey that came into being in Tennessee in 1925. In what became known as the Scopes Monkey Trial, a teacher, John T. Scopes, was prosecuted for teaching evolution to his pupils. It was a set-piece: a show trial set up by the American Civil Liberties Union. Scopes was found guilty and fined $100; this judgment was overturned on appeal. This wasn't so much a trial between atheists and Christians as one between modern and fundamentalist Christianity. Despite this, the Butler Law, which Scopes had broken, stayed in place until 1965.

Since the trial, the fundamentalists of the United States have sought, with decreasing success, to outlaw or modify the teaching of evolution by means of natural selection. In 1987 an attempt to force schools to 'teach the controversy' failed; in 2005 their attempt to force the teaching of 'intelligent design' – the idea that the processes of life require a hands-on supreme being – also failed. These days you can teach creationism in American schools – the idea that God made the world and so forth – in civics, current affairs, philosophy and comparative religion. But you can't teach it in science. The Pew Research Center found that one-third of Americans believe that humans have always existed; 60 per cent of

Americans accept evolution, but 24 per cent insist on intelligent design. The figure for Europeans who accept undiluted evolution is closer to 80 per cent.

The most famous image of a monkey stands, or rather crawls, at the far left-hand edge of the popular image of evolution: following in turn a knuckle-walking ape, a stooped early hominid and finally the upright and glorious modern human male.

This has its origin in a series of images usually called 'The March of Progress', though in fact it was titled 'The Road to *Homo sapiens*'. This was a series of illustrations for a Time-Life Book *Early Man*, published in 1965. Rudolph Zallinger's vivid and lively images show fifteen human forebears marching inexorably towards the modern human male. The text begins: 'What were the stages of man's long march from ape-like ancestors to sapiens? Beginning right and progressing across four pages are milestones of primate evolution…'

This unforgettable and much-parodied sequence of images – I am a horseman; I have a tee-shirt bearing the usual set of four images followed by a fifth, a human mounted on a horse – is responsible for many of the misunderstandings about evolution. The fact is that evolution is not progressive, nor is it directed towards a goal. Evolution is not about the creation of humans from fundamentally flawed material, like turning base metal into gold. It's about having what it takes to survive long enough to become an ancestor.

Which leaves us with one more classic monkey-misapprehension. It was recently expressed by the American actor Stephen Baldwin: 'Evolution isn't true, because if we evolved from monkeys, how can they still be here?' But we didn't evolve from monkeys. Monkeys and humans share a common ancestor. About 8 million years ago, our lines split, and we took different evolutionary paths. Modern monkeys are not flawed humans, humans who didn't make the grade: they are fully viable modern beings, just as we are.

Self-flattery: traditional image of human evolution.

THIRTY-ONE

ARCHAEOPTERYX

*'Had the Solnhofen quarries been commissioned – by
august command – to turn out a strange being à la
Darwin – it could not have executed the behest more
handsomely – than in the Archaeopteryx.'*

Hugh Falconer, letter to Charles Darwin, 1863

Darwin thought the best way to deal with objections to his theory of evolution by means of natural selection was to answer them before they could be raised. Throughout *The Origin* he is at immense pains to point out: 'yes, I know you could make this objection or that objection, but you'll find that if you think about it, it doesn't really stand up because…' The entire sixth chapter is taken up with such matters, as its title, 'Difficulties on Theory', suggests. Darwin returns to the biggest objection of them all in Chapter 9, devoting it entirely to the 'Imperfection of the Fossil Record'. He writes: 'The distinctness of specific forms, and their not being blended together by innumerable transitional links, is a very obvious difficulty.'

Or to put that another way, where were the missing links? Where were the creatures halfway between one kind of thing and another kind of thing? If Darwin was right, such creatures must exist. But, alas, that doesn't mean we must inevitably find them: the fossil record is a chance-driven thing: all creatures die, but only one in several million becomes a fossil, and of these only one in several million gets found.

Behind these questions was, of course, the search for a creature that was, beyond all possibility of denial, halfway between an ape and a human. This idea of finding and then glorying in the greatest fossil of them all was the motivation for the Piltdown Forgery of 1912, which was not discredited until 1953; it combined a relatively old human skull with the jawbone of a modern ape, probably an orang-utan. The Missing Link is found in fiction in many versions, perhaps first in Sir Arthur Conan Doyle's *The Lost World*, in which humans and the human-like apes – the reasonable and the bestial – fight for mastery.

The lack of transitional fossils was deeply vexing to Darwin. It gave his legion of objectors, and all those who disliked the notion of human descent from non-human animals, a cudgel to be used at will. You say that birds have a common ancestor with modern reptiles, do you? Show me a fossil that's half-bird and half-lizard and I might start listening – till then, certainly not.

First wing: A flock of Archaeopteryx in flight. Reconstruction made at Jurassic Earth, Germany.

The Origin was published in 1859. Two years later, a Bavarian doctor came into possession of an interesting fossil from a nearby quarry; it is speculated that it was payment-in-kind for a bill. About sixty-three years earlier, a German actor named Alois Senefelder invented lithography, the art of printing from stone; he used his invention for printing theatrical literature. He lived in Solnhofen, also in Bavaria, where you can find the best lithographic limestone in the world: beautifully sensitive to detail. So the quarry was worked for lithographic slates, and every now and then a fossil would turn up. Because of the nature of the slate, the fossils found there showed uncanny detail and, unusually for fossils, sometimes soft tissue as well as bones.

The Bavarian doctor's fossil had feathers. They were there in astonishing, in dizzyingly fine detail. He sold it to the Natural History Museum in London and got £700 for it; a small fortune.

The creature preserved in this marvellous stone had jaws with sharp teeth, and claws on the forelimb. It also had a bony tail. So it was a lizard, or rather a small dinosaur. But it had feathers, and those feathers were attached to what was unmistakably a wing. So – was it a dinosaur? Or a bird?

Yes.

This then was Archaeopteryx: which means in Greek 'first wing'. The Germans called it the *urvogel* (original bird). There was no room for doubt; under the impossible detail of this impression you could see not vague outlines of feathers but the barbs, barbicels and barbules – finer and finer details of the feather – alongside the bony head, the teeth and the long tailbone, all features that no modern bird possesses. Here was a half-and-halfer that could not be denied.

The timing of the thing was uncanny. If your theory is true, then where are the transitional forms? Er – here, actually.

Darwin was a man of immense self-restraint. That was what made him a revolutionary: he never went off half-cocked, never made a public statement unless he had worked years proving it in advance. So he greeted the clinching fossil with measured calm: never gloating, never saying 'I told you so', certain that everybody already knew he had done so. It was the truth that mattered to Darwin, not his prestige: that was his nature. So he mentions Archaeopteryx, when the context is right, mostly in the many subsequent editions of *The Origin*, and he does so calmly and gently, not to confront his critics, merely to put them right in a kind – and ultimately ruthless – manner.

It was as if God had switched sides and given his full support to the atheists – or, at least, as if he had taken the ground from the feet of biblical literalists. Here, with a near-miraculous sense of timing, was exactly but exactly what Darwin and his supporters had been longing for.

Since then another ten Archaeopteryx fossils have been discovered. In 1985 there was an attempt, led by the British astronomer Fred Hoyle, to prove that the first fossil and what is known as the Berlin specimen were forgeries; but this failed to stand up.

Since the finding of Archaeopteryx a number of feathered dinosaurs have been discovered, and it is now clear that feathers did not evolve for flight. A feather is a modified reptilian scale, and it's usually suggested that this was an adaptation for bodily warmth; as we will see later (see Chapter 46 on ducks), feathers are better than anything else in the world as lightweight insulation. Flight was a secondary use. This change-of-use thing is a frequent occurrence; our lungs are adapted from the swim bladder, the device that allows bony fish to hold

Unmistakable bird: Archaeopteryx fossil.

themselves stationary in the water (see also Chapter 80 on goldfish and Chapter 98 on vaquitas). So you can choose your own Just So Story about the first feathered creatures to fly. Did they do so by escaping from predators, running on two legs with feathered forelimbs spread out – and find themselves taking to the air and continuing in giant moon-walking bounds? Or did they climb trees and parachute down, eventually gliding to distant trees, and then adding powered flight to their repertoire?

Some have wondered about the extent to which Archaeopteryx was airworthy, but investigations in 2018 appeared to show that the – bird? reptile? – animal, anyway – was capable of powered flight, if no great master of the art.

But, above all, Archaeopteryx was the creature that showed the world that the most significant thinker in human history was right. Darwin's great book gave his idea legs; Archaeopteryx gave it wings.

THIRTY-TWO

HOUSEFLY

Am I not
A fly like thee?
Or art not thou
A man like me?

William Blake, 'The Fly'

On the best nights we dream of flight, roaming the world weightless and free, gravity forgotten. Freud said such dreams were actually about sex, not something that has ever diminished their pleasure. We seek to fly in sport – gymnastics, ski-jumping, all the horsey sports – and we have a special admiration for our favourite flying creatures: albatrosses, eagles (as we've already seen, in Chapter 10), robins, monarch butterflies, emperor dragonflies…

But perhaps the finest flier of them all is the one that has just landed on your sandwich, that watches you from the ceiling as you prepare a meal, that is gathering in numbers around the place where you dispose of unwanted rubbish. You will never find a better aerialist when it comes to precision and the ability to make the transition from flight to stillness.

This is the housefly. More than one species of fly come into human habitations, and we call most of them houseflies, or sometimes bluebottles – bigger, noisier and, of course, bluer. But a single species dominates, and does so across the world. That is *Musca domestica*, or housefly.

And they are all superb. Take a close look next time you see one: those large compound eyes that look like an aviator's goggles. If they almost meet in the middle, the fly is male; if more separated, female. These eyes add up to an excellent bit of kit, operating with three separate ocelli, which are simple eyes with a single lens. (Simple doesn't mean poorly functional; we humans are equipped with a pair of simple eyes.) Flies can process visual information seven times faster than we can, which is a considerable advantage when moving in three dimensions in confined spaces.

Their flying and their landing gear are equally remarkable. At the end of each leg there is a pair of claws and, below them, two adhesive pads. When they land on a ceiling – a glass ceiling is no problem to them – they will fly direct to the chosen landing site, perform a half-roll, absorb the impact with their two front legs and then stick on with the back four. From there they can move about with freedom, always three legs in contact with the surface and three in motion. For further adhesion they exploit a phenomenon in molecular physics called Van der Waals force, which involves weak, short-range electrostatic forces.

Houseflies have two wings; most insects have four. This is the USP of the order of Diptera, the two-winged flies, and there are an awful lot of them: more than 120,000 species already described and no doubt there are many more awaiting discovery. These include mosquitoes (see Chapter 23), craneflies and hoverflies; the last are rather favourites of mine, lovely things in their unshowy way, with mesmeric flying skills, including total mastery of stationary flight.

The Diptera are marked by a great adaptation. The rear wing that most insects possess has in them evolved into a pair of organs called halteres, and these are what give this group such devastating control in flight. The halteres give them precise knowledge of their own attitude in the air, and every pilot will tell you

Always ready: Fruit Piece with Peaches Covered by a Handkerchief *by American painter Raphaelle Peale (1774–1825).*

Looking ready to buzz: prehistoric fly in Baltic amber.

how much that matters. Houseflies can beat their wings at 1000 times a minute, but they are built for precision rather than speed: 4½mph (7km/h) is about standard. They are not great travellers, but 7 miles (12km) is within their compass, and there is a record of 20 miles (37km).

It's really not clear what gave the housefly the edge among all those other members of their order. They subsist on a liquid or semi-liquid diet, but they can soften solids with their saliva before ingesting it. They can live two or three days without a meal, but they survive a good deal longer with access to good food, especially sugar. Males can exist and fulfil their biological destiny on very little, but females need a protein meal if they are to make eggs. Flies play an important ecological role in breaking down organic matter and making it available – in the form of their edible selves – to other creatures, many of them more sympathetic to humans. You don't have swallows unless there are plenty of flying insects in the ecosystem. Houseflies have organs of taste on their legs, so they can walk over potential food and assess its usefulness – that's why you often see flies rubbing their legs together, to clean these taste organs for the next time.

Their life cycle operates on the usual four-part insect schedule, but does so remarkably quickly: females can lay around a hundred eggs in a batch, up to 500 in a lifetime. Maggots hatch out, often within a day, and, shunning the light, they pupate in a few days, and a day after that they become adult flies. The whole process can take as little as six days, though twelve would be more usual. The female is sexually receptive within twenty-four hours: they really do breed like flies. Adults can live for up to four weeks, but can shut down and go into diapause – like a hibernation – to outlast bad times.

Diptera fossils have been found as far back as the Permian era 250 million years ago; the first fossils of the *Musca* genus date back 70 million years, so they survived the great extinction event of 65 million years ago that finished the dinosaurs. They evolved in the Central Asian steppes and, because of their curious affinity for humanity, they have colonized every continent apart from Antarctica.

But their manner of life has threatening implications for humans. Houseflies frequently share food with humans: their mode of consumption requires them to place saliva on the food before eating – or rather drinking – it; also, a fly will defecate almost every time it lands. They often move to human food from faeces and from rotting food, places rich in pathogens. Flies have been associated with sixty-five diseases including cholera, dysentery, typhoid, leprosy, conjunctivitis and salmonella. The best answer to this is what is sometimes called 'cultural control': high standards of hygiene pre-empt contamination via houseflies. Keep food covered, keep kitchens clean and dispose of food waste wisely so you don't create easy breeding conditions.

Humans frequently associate flies with death, partly because their own death seems such a small thing to us, and partly because a fly walking on a human face implies that the human is dead. Emily Dickinson wrote: 'I heard a fly buzz when I died.'

The ubiquity of houseflies has helped to create a general intolerance for all insects in humans. As a result of that, we have waged war on insects and done so with too great success, one that is beginning to change the world, to our considerable cost.

The phrase 'he wouldn't hurt a fly' is used to imply an extreme measure of gentleness: as if even the most kindly people would be perfectly prepared to injure a fly. But as we continue the war against insects we should be prepared to face the consequences of living in a world short of insects and other invertebrates (see Chapter 27 on earthworms and Chapter 65 on bumblebees, in particular). They are part of the planet's structure and we destroy that at our peril.

So let us leave the last word on flies to the Japanese poet Issa:

Don't kill that fly!
See how it wrings its hands!
Its feet!

THIRTY-THREE
DOG

*'Deliver my soul from the sword, my darling
from the power of the dog.'*

Psalm 22:20

We do all we can to keep houseflies out of homes, and with them, as we have already seen in these pages, fleas, rats, cockroaches and mosquitoes. And what is the first non-human animal we welcomed inside? A wolf. We have filled our homes and our lives with wolves: as a species they are perhaps the best beloved non-humans on the planet (see Chapter 21).

All domestic dogs are descended from wolves. A wolf's scientific name is *Canis lupus*; the domestic dog is usually classified as *Canis lupus familiaris*: a slightly different kind of wolf. We cuddle these wolves, offer ourselves to be licked by these wolves, encourage our children to play with these wolves. As a result, the domestic dog is the most widely distributed terrestrial carnivore on the planet (meaning from the order Carnivora, rather than just eaters of meat). They are reckoned by some to be the first animals to be domesticated: there's a wolf, let's have him in the house!

No doubt the process was gradual and incremental. Human settlements create feeding opportunities and wolves were attracted to them in pre-firearm days: it's easier to scavenge than hunt, and no one objects too strongly to a free meal. This had two instant advantages for humans: the wolves cleared up rubbish – and, more importantly, they operated as an early-warning device, telling the community when there were large mammals or unfamiliar groups of humans approaching.

Wolves also have puppies: and puppies have an instant appeal for humans. Wolves are animals with profound social grasp, which involves an instinctive understanding of the way a dominance hierarchy works. They submit to a dominant personality: lick, roll on their backs and generally make a big deal of their willingness to accept a subordinate place. A wolf would make a very good pet – at least until it started to acquire independence and ambition to climb higher up in the pack. I met a man who, when he was a boy, had an African wild dog puppy – a quite distinct species, but with a very similar social set-up to wolves. It was the best pet he ever had.

God and dog: Vision of St Augustine *by Vittore Carpaccio (c.1502).*

You acquire a wolf puppy, it grows into a young animal. From there it is a short step to taking control of the breeding: and that is what domestication really means. And as we have already seen, the process of domestication makes marked changes in every species that it touches: cattle, pigeons and chickens come in domestic forms that can be very different from their wild ancestors – in size, in appearance and in behaviour. Those who first worked with the wild cattle, the aurochs, would not have treated them in the relaxed and easy way that my neighbour Bobby handles his herd of dairy Jersey cows.

In 1959 a Russian scientist named Dmitri Belyaev started a long experiment with foxes – silver foxes that he bought from a Canadian fur farm. He bred the nicest males with the nicest females. He picked those that approached a human with cheerful curiosity, rather than those who backed off yelping and snarling. This policy continued through successive generations, and the nice foxes produced ever-nicer foxes – or, to put that another way, foxes that were happier and more submissive around humans. But behaviour wasn't all that changed.

With the constant selection for docility, the appearance of the animals also changed. They acquired white spots on their coats, curly tails, floppy ears and shortened skulls. It was as if Belyaev was breeding foxes that were born puppies and stayed puppies.

The parallel with dogs is staring us in the face: wolves and foxes are both in the family Canidae. If you consistently breed from nice wolves, you will end up with nice doggies. And that is exactly what we have done. Around 10,000 years ago there were only wolves howling all around us: now when my friend Thomas pays a visit, Magic, his black Labrador guide dog, comes too and, when released from duty, navigates his way round the house from stroke to pat via the occasional hoovering mission around the kitchen: as benign a presence as you could hope to have in your house.

There is a great liking for the traditional theory that dogs were domesticated to aid Man the Hunter in his endless quest for protein, but that was surely a later development. The semi-domesticated dogs' first task was to become ad-hoc guardians of the human settlement, encouraged to hang around by scraps left out first accidentally and then on purpose. Throw out food and you have your own guard dogs, there to make the night less perilous: and as we have seen, getting through the night was the great challenge for early humans. Dogs have always been capable of giving comfort to humans: that, I suspect, was their first gift to us and the most enduring one. Comfort includes warmth: take a dog – a nice tame dog – into your dwelling on a cold night and the place becomes warmer. There is a saying among Australian Aboriginal people of the 'three-dog night'; one so cold that you require them all to keep warm.

Not that dogs, even half-tame dogs, wouldn't have been a help in a constructive and intelligent hunt. Dogs outperform hunting humans in many areas, most particularly smell. They can find an animal we would otherwise be unaware of. The drawback is that they also chase them off – but when you have cunningly placed a few of your fellow hunters out ahead, your dog or dogs will drive the frightened beast straight into the rest of the hunting party. All you have to do to learn the technique is, preferably from a safe distance, to watch a pride of lions at work.

Dogs have been part of human culture all over the world, and across the millennia. In ancient Mesopotamia a dog was the symbol of Ninisia, goddess of healing and medicine; Cerberus, the three-headed dog, guarded the gates to Hades in Greek mythology. Dogs have been part of human life for hunting, as guardians and, crucially, because we like to have them around. When Odysseus returns to Ithaca twenty years older than when he left and disguised as a beggar, only his dog Argos recognizes him: an example of fidelity he will also find in his wife Penelope. Members of the Dominican religious order pride themselves on their faith; their name translates as 'dogs of the Lord'.

Beware of the dog: mosaic from the street at the House of the Tragic Poet in Pompeii.

In short, then, dogs consistently get a good press. This is despite the many demotic phrases about a dog's life: I wouldn't treat a dog like that, I wouldn't give that to my dog: consistent implications that a dog is worth a great deal less than a man in a thousand different ways.

Pompeii was destroyed by the eruption of Vesuvius in AD 79: perhaps the most famous of all the remarkable things preserved by the terrible event is a notice that reads *Cave canem* (beware of the dog). It even comes with a nice doggy mosaic. My favourite dog in art is in Carpaccio's *Vision of St Augustine*, in which the saint stares raptly at something we can't see, perhaps God Himself, while his little white doggy stares equally raptly at his master, perhaps seeing his master as his own god, or perhaps wondering when the old boy will snap out of it and come for a walk. Though I shouldn't forget the famous sequence *Dogs Playing Poker*, by the American artist Cassius Marcellus Coolidge, or the thousands of dogs that populate the crowd scenes of L. S. Lowry.

Dogs have made many appearances in fiction and film: Toto in *The Wizard of Oz*, Lassie, Jip in *David Copperfield*, Fang in the *Harry Potter* books, Snowy in the *Tintin* cartoons, Nana in *Peter Pan* and – my favourite doggy book – just about the entire cast of *101 Dalmatians*, in the book by Dodie Smith. Jack London wrote about dogs with a serious literary intention. Uncle Matthew, Nancy Mitford's hunting, shooting and fishing milord in *The Pursuit of Love*, talks of one of London's books: 'I have only ever read one book in my life, and that is *White Fang*. It's so frightfully good I've never bothered to read another.'

Dogs have been bred for different functions: collies and sheepdogs for herding; hounds and pointers for hunting; retrievers and spaniels for shooting (or at least picking up animals that have been shot); terriers for killing rats. More than any other domesticated species, dogs show the extraordinary ways in which artificial selection – Darwin's phrase – can change the appearance and the behaviour across the generations. (Darwin was a great dog-lover. When he was sixteen his father gave him a terrible wigging: 'You care for nothing but shooting, and dogs, and rat-catching, and you will be a disgrace to yourself and your family.')

Dogs have been bred to extremes. A Yorkshire terrier can weigh as little as 3½oz (100g), and an English mastiff can weigh 345lb (155kg). A Great Dane has been measured at 3½ft (106cm) at the shoulder, a Yorkshire terrier at 2½in (6cm). The disparity in height between a Chihuahua and an Irish wolfhound is around 30in (75cm). Some breeds have been specially designed to bring out the cuteness response, already mentioned in these pages (see Chapter 16 on pandas, and Chapter 52 on seals): dogs with flat round faces and big eyes are a bit like human babies and so bring out the urge to nurture. Breeds like boxer, bulldog, Pekingese and pug, with their foreshortened muzzles, suffer from lifelong respiratory problems: snorting, snuffling and snoring are normal for these breeds. There are many other examples of selective breeding for traits damaging to the dog: dachshunds (initially bred to chase badgers) and basset hounds (originally sized-down bloodhounds with immense scenting ability) are now bred to extremes, so long and so low to the ground that they live severely circumscribed lives and tend to suffer from spinal problems. It's important to add here that many people in the dog-breeding world object strongly to such examples of exaggerated breeding.

At the same time, the companionship of dogs has been more and more appreciated by humans. Increasingly, and especially after the Second World War, it became acceptable to have dogs in the house, living with people, being considered as 'one of the family'. Previously it was usual to keep dogs outside in a kennel, hence the expression 'in the dog house', meaning that you are not welcomed back into the family circle. As a result of these changes, humans require dogs that are still more remote from their wild ancestors. Frequently they are neutered: anything that reduces the tendency to scent-mark, roll in dung and fight is good for family life. Dogs that shed hair are becoming less welcome, especially with the growing problems of allergic responses in humans, so some breeds of dog have been crossed with poodles (which have woolly coats that don't shed): cockapoos and Labradoodles are increasingly popular pets.

All this is in marked contrast to dogs in the developing world. Here dogs tend to be feral or 'community-owned': such individuals comprise three-quarters of an estimated global dog population of 900 million. In the developed world, more uses for dogs have been found: helping the blind, the deaf, the disabled; offering psychological help; sniffing for drugs and explosives; paying visits to care homes for old and for young people.

Dogs have been bred for many purposes, a good few of them now redundant. But we still find new reasons for keeping dogs. They have been trained to give medical alerts, predicting and giving warning of epileptic fits and diabetic attacks; dogs have been successfully used in the early diagnosis – by smell – of cancer. There are new leisure activities and sports involving dogs, including agility, flyball, cannicross and freestyle – which is heel-work to music, basically dancing with your dog.

Dogs represent, more clearly than any other species, the human need for non-human species: the human need to reach out beyond our own to get the best out of life. Let us risk using a word abhorrent in scientific literature: in humanity it seems that there is a deep need to love an animal of another species and, perhaps more importantly still, to be loved back. The position of the twenty-first-century dog in the developed world comes down to that. Dog is love.

THIRTY-FOUR

BEAR

'If you go down in the woods today
You're sure of a big surprise.'

Anne Murray, 'The Teddy Bears' Picnic'

If dogs are the best-loved non-humans in the real world, bears are the best-loved non-humans in the world of the imagination. We have established a great shared fantasy of the loving and beloved bear. It is quite distinct from real bears, who are real carnivores and effortlessly capable of killing humans. It's best not to get the bears of the imagination confused with the bears of reality: a guide in British Columbia told me of a client who told him: 'I want a bear to lick honey off my nose.'

There are bear species living in North and South America, Europe and Asia: large of body with stocky legs. They walk with a certain flat-footed doggedness that perhaps reminds us a little of ourselves: they go in for plantigrade locomotion, just as we do: walking on the soles of the feet, heel to the ground, rather than on their toes or their nails/hooves like most mammal species.

There are eight species of bear alive in the world today. We have already dealt with the panda (see Chapter 16), and we will save the polar bear for the very last (see Chapter 100). The other six species are American black bear, Asian black bear, sun bear, sloth bear, spectacled bear and the brown or grizzly bear: that's the same species and it's found in North America and in Europe. It is this last species we usually think of when we hear the word 'bear': that massive head with curious round ears stuck on as afterthoughts. You see one full face and it looks almost absurdly like a child's drawing; but when you see them in profile the long muzzle comes as considerable shock: clearly full of excellent carnassial teeth. A living bear can go from a bear of fantasy to a bear of uncompromising reality in the turn of a head.

Bears do attack humans on occasions. They excite a particular form of horror in the hopelessness they inspire in us: they can run faster than we can, and can climb trees better, so, if one makes a set at you, you're a goner. Research in the

Close enough: brown bear and cubs, Katmai National Park and Preserve, Alaska.

Too cute to shoot: cartoon showing Theodore Roosevelt refusing to shoot a bear cub, by Clifford Berryman (1902).

1990s showed that an average of two people a year are killed by bears in the United States and Canada, compared to fifteen people who are killed by dogs. The best way to keep safe from bears is to let them know you are around, rather than risk surprising them; in thick bush my guide repeatedly shouted 'Hey bears!'

They are seriously imposing animals. To meet one face to face is a powerful experience, and they have imposed themselves on the human imagination as a result. There is evidence that they were worshipped by our Cro-Magnon ancestors 35,000 years ago: there is something both human and superhuman in their appearance and movements, and also in their omnivorous diets; the only European bear I have seen was gorging on bilberries on a mountainside in Slovakia. In some cultures, bears represent the spirits of the forefathers: people in Finland, Siberia and Korea have shared similar beliefs.

We like to call ourselves after bears: bear names include Bjorn, Arthur, Bernard, Auberon and Ursula. The self-identification with bears is a trait that runs through human history; there's always been a part of us that wants to be a bear: both in their size and ferocity and in their charm – Ursula means little bear.

That doesn't mean that we have invariably treated bears with kindness and respect. Bear-baiting was a recognized sport in England from the sixteenth century until the nineteenth. The practice involved tormenting the bear and setting other animals against it; we still speak of a particularly rowdy place or event as a 'bear pit'. Bears have also been taught to dance, which sounds a great deal jollier than the reality. A bear cub is acquired, generally by killing the mother. Its claws are trimmed or removed, and so are some of the teeth. A ring is placed through the nose. The bear is placed on a metal floor with a fire beneath: it is forced to its feet via a rope and staff attached to the nose ring and, as it shifts its position from one hind paw to the other to escape the heat, it learns to dance. It is a public entertainment still found in some parts of the world, though generally illegal.

Bear bile has great value in Chinese medicine; it is used for treating liver and gall bladder problems. The bile produced by bears is high in ursodeoxycholic acid. This digestive fluid can now be synthesized, but there is still a considerable market for the real thing; it has been estimated that 12,000 bears are kept in bear-bile farms in China, Vietnam, South Korea and Myanmar, usually in horrible conditions. The international trade in the substance is illegal.

The fantasy of the quasi-human bear has a long oral history. It was perhaps first written down by Robert Southey in 1837, in his story *The Three Bears*; Goldilocks got into the title some years later. The ground was well prepared for the explosion of bear-loving that began at the beginning of the twentieth century.

This coincided with a widespread human reimagining of childhood: we were beginning to see our early years as a special and even a blessed time, rather than a long-winded preparation for adulthood. In 1902 a story about the United States president Theodore Roosevelt gained currency. He refused to shoot a bear. Roosevelt had shot a good many bears in his time and remained an unappeasable hunter for most of his life, though he was also a great pioneer of the conservation of wild places and wild creatures. But he turned down the chance to bag an American black bear that had been cornered by dogs and beaten with clubs. He wanted only that some other person put the animal out of its pain; for him, it wasn't a sporting shot.

This incident appeared in a cartoon in the *Washington Post* by Clifford Berryman, with Roosevelt portrayed as a hunter of mercy. In subsequent versions the bear became younger, smaller and cuter, emphasizing the president's big-heartedness. Then followed one of those curious coincidences: manufacturers of toys in Europe and America both decided, separately but more or less simultaneously, to make bears as children's toys. Morris Michtom, the American, went so far as to call his own work 'Teddy's bear', with full presidential permission, for all that Teddy Roosevelt never cared for this diminutive. Richard Steiff of Germany, the great pioneer of children's toys, exhibited his own bear at the Leipzig Toy Fair in 1903.

The teddy bear had lift-off: and it became an indispensable part of our new view of childhood. John Walter Bratton composed an instrumental piece called 'The Teddy Bears' Picnic' in 1907; in 1932 Jimmy Kennedy added the words. And as toy bears became popular, so they changed. Their faces grew flatter, their muzzles smaller or non-existent, their eyes bigger. In short, they became cuter. As dog breeders did with certain kinds of dog, as Walt Disney did with Mickey Mouse, so the toy makers did with the teddy.

The cult of the teddy bear has inspired many lovable bears in children's fiction. Perhaps the most famous is Winnie-the-Pooh, already met in these pages as the friend of Tigger (see Chapter 24 on tigers). A. A. Milne published *Winnie-the-Pooh* in 1926 and *The House at Pooh Corner* two years later. Both books are fantasies set in the mind of a child. Pooh, though he admits that he is 'a bear of very little brain', is loyal and devoted and determined to see the best of things. He is a classic *idiot savant*. After the trouble caused by Tigger's arrival in the forest, Pooh, his friend Piglet and the human child (and master of the forest) Christopher Robin have a conversation:

> 'Tigger is all right really,' said Piglet lazily.
> 'Of course he is,' said Christopher Robin.
> 'Everybody is really,' said Pooh. 'That's what I think,' said Pooh. 'But I don't suppose I'm right,' he said.
> 'Of course you are,' said Christopher Robin.

The tradition of the bear as a good, kind, naive creature capable of telling us humans a few important truths about ourselves continued with the Paddington stories by Michael Bond. *A Bear Called Paddington* was published in 1958, at a time when Bond was employed by the BBC as a cameraman (he worked on a number of programmes with my father). Muted themes in the books – about kindness, tolerance, good manners and a willingness to see the best in people – were made explicit in the films *Paddington* and *Paddington 2*. Paddington embodies qualities that make the family who adopted him and then the community all around him happier, more generous and more loving.

The bears of fantasy represent many good things, but above all they are the comforters of childhood. I remember a family jaunt to Amsterdam twenty years back, when we returned to the hotel and looked up from the street to the lighted window of our room, where our older, then only son, aged six, had placed his bear in the window to welcome us back, turning the hotel into home. Meanwhile, the wild species of bear continue to live radically different lives from the one we have imagined for them.

THIRTY-FIVE
CAMEL

'The camel's hump is an ugly lump
Which well you may see at the Zoo;
But uglier yet is the hump we get
From having too little to do.'

Rudyard Kipling, *Just So Stories*

One of the prevailing characteristics of the human condition is our ability to maintain two contradictory ideas simultaneously and for prolonged periods – even for most of our lives. Perhaps life would be intolerable if we did not. We like to nurture contradictory notions about evolution, believing that it is about the pursuit and capture of perfection – and, at the same time, we also believe that some creatures are ugly, malformed, ill-conceived and, all in all, represent a kind of cosmic or divine error. We take pleasure and comfort in both these views, despite the fact that they are both wrong. There are perhaps unconscious echoes of both views in the pages of this book, despite my awareness of their wrong-headedness.

We can support this elemental contradiction in a single animal. The camel is both the ship of the desert, perfectly adapted for travel in arid country, and a horse designed by a committee: a classic example of natural bungling.

The fact is that evolution is not about seeking perfection, as any bipedal plantigrade human with back pain knows extremely well. But evolution is no more prone to perfection than it is to the production of hopeless cases: evolution is, as we have already seen, about producing living forms capable of becoming ancestors. Why haven't humans adapted to become genetically resistant to cancer? That would be a good step towards perfection, would it not? But cancer most often strikes after childbearing age, so immunity to cancer cannot be a factor in natural selection.

The camel is neither walking perfection nor an abortion of God. It is, at least in Western tradition, a trifle absurd, but as a desert-adapted animal it is of course superb. So we'd better start with the hump. Or humps.

There are three species of camel currently in existence. The dromedary, with one hump, mostly evolved and living in West Asia and the Horn of Africa; the two-humped Bactrian camel is associated with Central Asia; Bactria is an area

Instant exoticism: camels bearing kings. From Scenes from the Life of Christ – Epiphany *(1463–64) by Andrea Mantegna.*

north of the Hindu Kush mountain range. These two species are now entirely domesticated, though there are feral populations of dromedaries in Australia, India and Kazakhstan. The only wild species left is the wild Bactrian camel – recognized as a different species to the domesticated one – which is found in the Gobi and the Taklamakan Deserts; the population is a little over 1000 and it is classified as Critically Endangered.

There are related species in South America: llama, alpaca, vicuña and guanaco. The group evolved in North America; camel ancestors crossed the Bering land bridge (when it still existed) into Siberia and worked their way across Asia.

They are stunningly well adapted to desert life and the hump is, as it were, at the heart of it. These large fatty deposits – they're not full of water, as is sometimes thought – can be accessed at need, providing both energy and hydration. Camels can go up to ten days without needing to drink at all, longer if they have access to fresh green vegetation. Their tough leathery mouths allow them to eat thorny desert plants without discomfort; they can exist on suboptimal fodder for extended periods. When they exhale, the water vapour remains trapped in their bodies rather than being wasted on the desert air.

Their wide, two-toed feet spread their weight well on sand. There is a feature of sandy deserts called fech fech: a firm surface crust that conceals very loose sand beneath. It's a thing feared by drivers of desert vehicles, but the stuff doesn't bother camels at all. They just walk over it. They can close their nostrils in a sandstorm, keep sand from their eyes with an additional transparent eyelid, and their eyes are equipped with thick lashes and their ears with thick hair, more adaptations to keep the sand out. They can lose 30 per cent of their body weight without inconvenience, regaining the weight when they next drink – when they can take on 44 gallons (200 litres) in three minutes.

Camels have been domesticated for more than 3000 years. They provide transport, for both people and baggage, they also provide food, in the form of meat and milk, and also, from their hides, clothing, shelter (tents and yurts), and bedding. A camel with a single rider can cover perhaps 80 miles (130km) in a day, a fully laden baggage camel – they can take as much as 600lb (270kg), though it's more usual to take about half that – about half that distance.

As already related, the Prophet Muhammad hid in the Cave of Thawr to escape persecution after leaving Mecca, where he was helped by the cave-nesting doves (see Chapter 22). He was rescued from the cave by a follower who arrived with three camels: the party then travelled to Medina, making what is normally an eleven-day journey in eight days before Muhammad rode into the city to a rapturous welcome.

There are plenty of camels in the Bible, mostly associated with the patriarchs, especially Abraham. It has been suggested that these camels are an anachronism: that camels were domesticated later than the events described.

In the West, camels have a reputation for being awkward, difficult and downright unpleasant. Rudyard Kipling wrote a marching song for baggage camels in the British Army:

> Can't! Don't! Shan't! Won't!
> Pass it along the line!
> Somebody's pack has slid from his back,
> Wish it were only mine!

I can refute this view from first-hand experience: I once had the pleasure of riding a camel trained to perform dressage moves including a flying change, which is a sort of skipping stride in canter. It was a Bactrian camel so I sat between the humps, and found the animal charming and willing. There was a certain time-lag between the aid and the response, a bit like driving a boat, but I adapted to this and was delighted by the experience. The owners, who performed at horse shows in top hats and swallow-tailed coats, told me that camels were a pleasure to work with: 'but when you try too hard to make them do something they don't like, they just lie down'.

They have a reputation for spitting, which is fully justified: they are able to bring up their stomach contents and project them when they are alarmed: it is a defence mechanism rather than an aggressive one. They are highly social, herd-dwelling mammals and are a rare example of large mammals who can copulate sitting down.

In Western culture they are an instant exoticism: one camel establishes the notion of an alien culture: check the poster for the 1946 Marx Brothers film *A Night in Casablanca*. Camels are an essential part of the Christmas story: the three kings, the wise men from the East, came to the birth of Christ bearing the first Christmas presents. Traditionally they are mounted on camels: the implication is that the birth of the saviour was no local matter but one that affected the whole world, exotic foreigners included. You can find Bactrian camels at the birth of Christ in the painting by Mantegna.

The story is told in T. S. Eliot's poem 'The Journey of the Magi':

And the camels galled, sore-footed, refractory,
Lying down in the melting snow.

Galled implies rubbed by harness as well as grumpy; and refractory means stubborn and pig-headed. Eliot reimagined the Christmas story, but not ancient Western prejudices about camels. This notion is continued in Anthony Powell's twelve-novel sequence *A Dance to the Music of Time*, in which a character produces a novel entitled *Camel Ride to the Tomb*. Its author, X Trapnel, explains: 'I grasped at once that was what life was. How could the description be bettered? Juddering through the wilderness, on an uncomfortable conveyance you can't properly control, along a rocky, unpremeditated, but indefeasible track, towards the destination crudely, yet truly, stated.'

Good metaphor, but unfair to camels.

THIRTY-SIX
PENGUIN

'One can't be angry when one looks at a penguin.'

John Ruskin

We find glory in eagles, majesty in lions, cuddliness in bears: and all these conclusions tell us more about our own species than about any other. In penguins we find comedy – and, out of that, a touching sort of dignity, the dignity of the clown who, in spite of all his troubles, must always get back on stage to please his public.

I have never been comfortable with that way of looking at wildlife. I remember secret tears of rage when, at my primary school in the late 1950s, they showed a Walt Disney nature film loud with music designed to make everything funny. The silly laughter of my classmates made it worse: it was the film's success that pained me. They had succeeded in making animals funny and that seemed to me a kind of blasphemy.

But there's no escaping the fact that some animals awaken a widespread comic response in the great majority of humankind and, of them all, penguins are very high on the list. Are they the world's favourite birds? Very possibly. They look like little men in dinner jackets. They waddle like fat little clowns and yet they swim like underwater jet fighters. They are comic and brilliant at the same time: and, what's more, they're just a bit like us.

And that's the heart of the matter: they're not comic because they're funny birds, they're comic because they're funny people, or rather because they parody humanity with an apparent fussy dignity made ridiculous by a silly walk and a constant willingness to break from the walk and toboggan on their bellies – and all the time with the same deadly serious expression on their faces. Their unawareness of their own comedy makes them all the funnier. Apparently.

There are 18–20 species of penguins living on Earth today, depending on which scientists you agree with, but one of them counts above all others. That is the emperor penguin: the impossible stoic of the Antarctic. United they stand: huddled together to protect their eggs from winds of 125mph (200km/h) and temperatures of –40°C (–40°F). They do so for two months without a scrap of food and, when they are finally relieved of incubation duties by their partners, they must march up to 75 miles (120km) to the sea before they can get their beaks round something to eat.

As a result of both the comedy and the heroism – both traits as interpreted by humans – it seems that penguins are birds that humans cannot help but care about. They attract curiosity, wonder and laughter: both admiration and a powerful protective urge. London Zoo acquired its first penguin – a king penguin – in 1865 and it was an instant sensation. Other species of penguin followed.

Penguins became standard zoo attractions: sure-fire crowd-pleasers. A zoo wasn't a zoo without penguins to rejoice in: and always the public thrilled at that transition: that contrast between the comic waddler on land and the agile blade in the water: Clark Kent to Superman in less time than it takes to find a phone box.

Penguins are entirely southern hemisphere birds, though they are not restricted to the Antarctic; a good half of the species prefer temperate waters and one, the Galápagos penguin, is tropical. They are all flightless: and they are all superb swimmers. Their wings have been adapted for underwater use, so that they swim through the water with the action of a bird flying through the air. Most species can cruise in the water at around 7mph (11km/h), but they can accelerate to almost twice that at times of need. They are fish-eaters. Smaller species stick close to the surface and seldom dive for longer than a couple of minutes, but emperor penguins have been known to reach depths of 1,850ft (565m), and stay down for more than twenty minutes.

Penguins are birds and so they evolved for the air, but they have used the equipment for flight and transformed it, so that they can live in the manner of fish. Water is where they are at their best, not comic at all but wonders of speed and agility. Perhaps, to a penguin, a human looks as comic in the water as a penguin does to us when it walks on land.

With the advances of wildlife film-making penguins became still more deeply embedded into human consciousness. We could watch them swim in their native waters: we could see them pop from the water like champagne corks from a bottle when a leopard seal or an orca – killer whale – was about.

And, above all, we learnt about the extraordinary life of the emperor penguins: at more than 3ft (1m) tall, they are the biggest penguins still living (there are fossil penguins that would have stood 5½ft/170cm). We could see the huge huddled herds of males, each incubating an egg through the worst of Antarctic winter: the most southerly breeding birds on the planet.

Heads bowed to the violence of the austral winds, they stand like living snowmen or snow penguins, younger birds in the middle of the huddle, the birds on the outside always shuffling round, working their way to a warmer spot inside while others take a reluctant turn on the rim.

And then that extraordinary journey back to the sea. In 2005 the feature-length documentary March of the Penguins became an unlikely global hit. It is impossible not to identify with the gallant reckless journey-makers walking

Eternal zoo favourites: penguins in London Zoo irrigated with a watering can during a heatwave in 1927.

gamely with legs ill-suited to this purpose and, when they can, slithering the down slopes on their bellies.

In 2018 penguins became a global news story when footage of a penguin rescue was shown on a BBC wildlife documentary, *Dynasties*, with David Attenborough doing the commentary. There was a group of emperor penguins trapped in a gully: fallen down, blown down, unable to escape. Being flightless, the only way out was a near-impossible climb.

You could see chicks dying. You could witness desperation. You could also see members of the film crew in tears. I would have shed them myself, had I been there: as I have wept before when out in the glorious and harrowing wild. 'Tragedy is part of life,' as Attenborough himself said: something we humans know as well as any penguin.

But that awareness doesn't make us heartless. Quite the opposite. True, it is important – essential – for anybody in the wild not to interfere. That rule counts double for wildlife film crews who spend months out there with a pressing need for the killer-shot. But when the camera crew saw the plight of the trapped penguins, they agreed to abandon the doctrine of indifference and get involved. They dug a shallow slope, they created steps: and then they filmed it all. At this point the film was comically speeded up, so that we could see the penguins marching like clockwork soldiers (in their best mess-kit) up the slope and back into the welcoming freezer of their home.

There was a more or less universal chorus of approval for this drastic bit of intervention. It didn't disrupt a hunt; it deprived no creature of a meal; it didn't upset the balance of nature. And besides – and here is the unwritten subtext of the whole drama – they were *penguins*. Would the film crew have reacted so strongly had they been rats or slugs or spiders or city pigeons? Had they been small brown birds or snakes or nematode worms? And if they had intervened in such cases, would they have been met with such widespread and heartfelt approval?

We have always been ready to love penguins. The Swiss-British co-production *Pingu* was made from 1990 to 2000 for the Swiss and from 2003 to 2006 for the British: the adventures of the Plasticine penguin were brilliantly poised so that children and their parents laughed together. In 1932 the first Penguin biscuit was sold and it became a popular snack in Britain; the name and the picture on the wrapper made it an instant favourite. Three years later, a publishing house decided that there was a widespread market for cheap serious books. To make these books friendly and accessible they placed an image of a lovable penguin on the paper cover.

Humans love penguins, and yet many species of penguin are Endangered, for a complex suite of reasons. This disturbing truth prompted Rory Crawford, from Birdlife International's Marine Programme, to send out an impassioned and exasperated blog: 'It's almost beyond belief that we are in a situation where some of the world's most-loved birds are heading for extinction – and that urgent conservation action is in serious need of funds. This is PENGUINS we are talking about – the beloved star of films, cartoons, cuddly toys, biscuits – not some obscure beast!'

Britain is involved in this. The country still has plentiful and important areas across the world called Overseas Territories, many of which lie on the far side of the Equator. They are important to Britain for historic and strategic reasons. They include the Falkland Islands, which are also important for penguins. British people have a special responsibility for penguins: but then so does everybody else. All things being, as they are, connected.

THIRTY-SEVEN

OCTOPUS

'His companion began to be a little disappointed. Doc's enthusiasm for the octopi indicated that he was not as flexible as she. And no girl likes to lose centre stage to an octopus.'

John Steinbeck, *Sweet Thursday*

They were, I now saw, the most unearthly creatures it is possible to conceive. They were huge round bodies – or, rather, heads – about four feet in diameter, each body having in front of it a face. This face had no nostrils – indeed, the Martians do not seem to have had any sense of smell, but it had a pair of very large dark-coloured eyes, and just beneath this a kind of fleshy beak. In the back of this head or body – I scarcely know how to speak of it – was the single tight tympanic surface since known to be anatomically an ear, though it must have been almost useless in our dense air. In a group round the mouth were sixteen slender, almost whip-like tentacles, arranged in two bunches of eight each.

This from *The War of the Worlds*, written by H. G. Wells and published in 1898: the invasion of the Earth by Martians who looked like octopuses. The notion of an inhuman intelligence with tentacles – sinister, slimy and inspiring deep unease – has been with us ever since we thought about the possibilities of beings from other planets – and, indeed, long before. We have always been able to summon up monsters of all kinds from our deep human need for things to frighten us.

The octopus has long been a favourite beast to base such fantasies on, for the combination of the human-like intelligence clearly visible in its eyes and its utterly alien eight-limbed mode of life. The kraken in Norwegian folklore is more octopus than squid – though we'll meet the giant squid later in these pages (see Chapter 90). Medusa, the Gorgon with snakes for hair, is clearly a fantasy springing from octopuses. (Let us establish octopuses as the plural at this early stage; octopi is incorrect, for the word comes from Greek not Latin; the Greek gives an insufferably pedantic octopodes.) A Hawaiian creation myth has the octopus as a creature left over from a previous universe, which has a pleasing poetic truth about it. When Odysseus had to sail between Scylla and Charybdis,

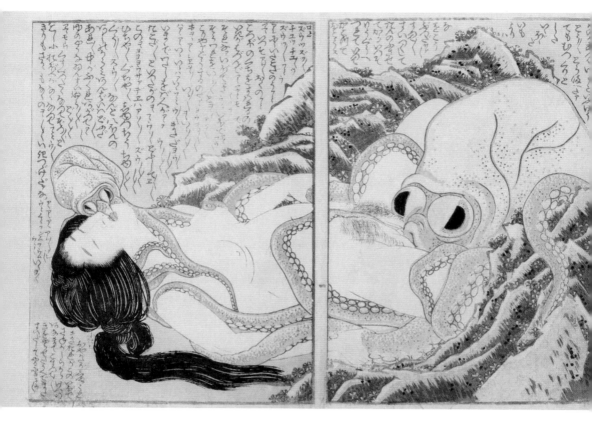

Imagination haunter: The Dream of the Fisherman's Wife *by Katsushika Hokusai (1814).*

Scylla, a kind of super-octopus, had twelve feet and six heads, each on a long snaky neck. She devoured six of his shipmates.

Octopuses have long been familiar to humans, perhaps most notably in the Mediterranean, where octopus remains a sought-after food, but at the same time they are disturbingly alien, utterly unlike fish. It is the combination of their very obvious and colossal distance from humanity, and their equally obvious intelligence, that makes them so fascinating to us. They are reckoned, with the related cuttlefish, to be the most intelligent of all invertebrates, and to be a good deal smarter than many vertebrates.

The animal kingdom is not divided into two: vertebrates on one side, invertebrates on the other; that is like dividing the world into molluscs and non-molluscs. There are around thirty-four different phyla in the animal kingdom, of which Chordata – or vertebrates – is one. Another is Arthropoda, to which insects, spiders and crabs all belong. Yet another is Mollusca: this includes slugs, snails, oysters – and octopuses.

There are probably more than 300 species of octopus: two eyes and a beak, like Wells's Martians, but with the mouth at the centre of eight limbs; limbs is more correct than tentacles or arms or, for that matter, feet. They are soft-bodied, so they can change shape and hide in relatively compact spaces. They have excellent sight and a highly complex nervous system, along with the highest brain-to-body mass ratio of all invertebrates, and greater than many vertebrates. But their intelligence is not just about the brain; their nervous system gives the limbs some kind of limited autonomy. In other words, octopuses are intelligent in a way we can't begin to imagine.

They live in every ocean and in many different habitats, and do so as bottom dwellers, as free swimmers and in coral reefs. They can make a leisurely crawl across the ocean bottom, moving in a rather delightful – as opposed to spiders (see Chapter 57) – sprawl of limbs. They can also swim, and they escape from predators at pace by means of jet propulsion, expelling water through their breathing siphon.

There is huge variation in size: the giant Pacific octopus can reach 33lb (15kg) with a limb span of 14ft (4.3m); other still bigger specimens have been claimed. Meanwhile, the star-sucker pygmy octopus (*Octopus wolfi*) has a limb span of 1in (2.5cm).

The British sci-fi television series *Doctor Who* introduced a race of aliens called The Ood, which is basically an octopus in a suit. The Doctor has two hearts: a real octopus has three. They also have chemoreceptors in their limbs, so (like the housefly in Chapter 32, but with an entirely different system) they taste by means of touch.

And they have a genuine intelligence. It was once thought that use of tools was one of the things that separated humans from all the rest of the animals: octopuses show that tool use is not even restricted to vertebrates. They have been observed using coconut shells to construct a home. In the laboratory they have been tested in mazes and shown to have good short- and long-term memories. They can distinguish shapes and patterns and – though this is contentious – they are capable of observational learning – that is to say, they can watch a task, imitate it and then master it. They can recognize individual humans. An octopus in a tank was exposed to two lab assistants. One fed it, the other routinely irritated it with a wire brush. Soon, at first sight, the octopus hid from the one and flaunted itself before the other.

Anecdotal evidence of octopus intelligence is also remarkable. An octopus in a tank played with a bottle, squirting it from the aerating system, and did so more than a dozen times until he got bored with it. Octopuses regularly climb out of their tanks to seek prey, company and sex. Octopuses have been known to climb aboard ships in search of prey, and they regularly get in – and out – of lobster pots to steal the bait.

Alien as octopus: illustration from The War of the Worlds *by H. G. Wells (1906).*

What is remarkable about octopuses is not just that they are smart, and not just that they are smart in a different way from humans. It is that their smartness evolved quite separately from our own. Intelligence is not something passed up through vertebrates, increasing all the while, up into mammals, especially the apes, and ending up with us. With octopuses, intelligence evolved quite independently. Just as insects, birds and bats can fly (and the extinct pterosaurs could fly), but all evolved flight quite separately, without reference to each other, so certain vertebrates and certain molluscs independently evolved intelligence.

Octopuses occasionally turn up with a curious erotic charge. This is most obvious in Hokusai's woodcut of 1814, *The Dream of the Fisherman's Wife*, in which she is comprehensively ravaged by an enormous octopus. The erotic octopus also turns up in James Bond, first in a collection of short stories by Ian Fleming in a volume initially entitled *Octopussy and The Living Daylights*, and subsequently in a film (Roger Moore period) called just *Octopussy*.

Since Wells's Martians invaded the Earth, many thousands more aliens have attempted the same trick. It's amazing how many of them have tentacles and large, obviously intelligent eyes. Alien intelligence fascinates us: and octopuses represent a form of intelligence that is farther from humanity than anything else we know.

THIRTY-EIGHT
DOLPHIN

'When I see a dolphin I know it's just as smart as I am.'

Captain Beefheart

'On the planet Earth, man had always assumed that he was more intelligent than dolphins because he had achieved so much – the wheel, New York, wars and so on – while all the dolphins had ever done was muck about in the water having a good time. But conversely, the dolphins had always believed that they were far more intelligent than man – for precisely the same reason.' This is from Douglas Adams's *The Hitchhiker's Guide to the Galaxy*, first published in 1979 – and it nails the transformation of dolphins in human culture. Humans had always liked and admired dolphins; it's hard to find evidence of a bad press for dolphins in any mythology. But after the 1960s their reputation soared ever higher: and they became famous for the combination of their lofty intelligence and their joyous, apparently perfected way of life. Dolphins were seen in some circles as examples for humans to follow: so that we might use our own intelligence to reach a comparably joyous and peaceful way of living.

Dolphin is not a precise term in science: once again, we find a clash between folk taxonomies that work well enough in an informal sort of way, and the precise and unambiguous classification that science must aim for. There are getting on for forty species in the family Delphinidae. This includes six species we normally think of as whales, the most famous being the orca or killer whale. There are also half-a-dozen species of river dolphins, which come in four or five different families: we shall consider them later, in the chapter devoted to the baiji or Chinese. The common names of the river dolphins reflect the river system they inhabit(ed): for example, Ganges river dolphin, Yangtze river dolphin (see Chapter 74). Porpoises lack the beaked face and the extravagant leaping behaviour of dolphins, and are classified in a separate family.

When we use the world dolphin informally, we usually mean a member of the thirty-plus species of ocean-dwelling mammals and, in particular, the common bottlenose dolphin, which is found in the Atlantic. This is the species most often used in displays of trained dolphins in 'dolphinaria'. In the eponymous children's television series of the mid-1960s, Flipper was a common bottlenose dolphin, trained to behave in a way that reminded us of ourselves. The programme was called 'an aquatic *Lassie*', and featured many sequences of Flipper, raised half out of the water, communicating avidly with the two young human brothers, all the time with an apparent smile on his face.

The ancient ancestors of mammals left the water behind to become land dwellers but there is no pattern to evolution save opportunity. So the ancestors of dolphins – hoofed, plant-eating mammals – took to the water again nearly 50 million years ago and began to live semiaquatic lives. The nearest land-dwelling relations to whales and dolphins are the two species of hippopotamus, which are still semiaquatic. But dolphins went all the way.

They are fast and agile, and so capable of catching fast agile prey, getting a good grasp of fish and squids with their conical teeth. Their swimming motion is up and down, mirrored in the backbone-bending way that some large mammals run. (Fish move by bending from side to side.) Dolphins find their prey and each other by sonar: they emit loud clicks through a fatty organ in the head called a melon and listen to the echoes. They live intense social lives, and in many species that leads to highly developed skills in communication and to the development of what we categorize as intelligence.

I once spent an hour surrounded by a super-pod of around 1000 spinner dolphins off the coast of Sri Lanka: they constantly leapt from the water, rotating two, three or even more times on a horizontal axis before slapping back down into the water, and I thought then that any human watching such a spectacle must long to leap from the human condition and sport with dolphins for ever. We have a strong tendency to empathize with dolphins, and these days one of the standard bucket-list wishes is to swim with dolphins: to join for a moment their perfect aquatic society… even if we know that this is a sentimental idea that tells us more about ourselves and our longings than it does about dolphin society.

Most of the creatures of the deep live lives remote from us, in ways unimaginable to our land-loving brains. But dolphins routinely come to meet us halfway: for they must emerge from the sea to breathe. They are happy in shallow waters around human settlements, places that no large whale can easily deal with. They appear regularly above the surface as they travel in a series of undulations, and often at speed they leap from the water to catch a breath, in the mode of travel known as porpoising, perhaps because it's energy-saving – it's more energy-efficient to travel through thin air than thick water – or it may be an expression of exuberance. Or, of course, both.

Their intelligence implies an eagerness to learn, and they often come and check out humans who venture to sea in boats. They were considered good omens by the ancient Greeks. Taras, the son of the god Poseidon, was shipwrecked and rescued by a dolphin; thanks to this, he was able to return to land. He did so and founded the Greek colony of Taranto in Italy, a community of seagoing people.

Joy-bringers: Cupid riding a dolphin, detail from mosaic, Utica, Tunisia (fourth century).

Dolphins acquired the trick of bow-riding: using the water displaced by a boat's passage to give themselves more speed for less expenditure of energy. They bow-ride because a free ride is never a bad thing, but it also seems pretty clear they sometimes do it just for sport. Dolphins often seem to go beyond the need to earn the daily crust: it's as if they share with us humans the belief that quality of life also matters. Unquestionably, dolphins are familiar with the idea of enjoyment: and they express this in a way that humans can't help but relate to. As a dolphin leaps from the sea, so the heart of the watching human leaps in response.

Humans and dolphins seem almost naturally on good terms with each other. There are many stories of dolphins helping human swimmers by keeping them afloat and able to breathe; dolphins understand about the need to breathe air while swimming in the water every bit as well as we do. There are also stories of dolphins driving sharks away from swimming humans. Humans and dolphins have often been known to cooperate when catching fish, to mutual profit.

Cultural transmission: mother dolphins instruct their offspring in the use of sponges.

Hal Whitehead and Luke Rendell recount one example in their excellent *The Cultural Lives of Whales and Dolphins,* in which a sailor fell overboard during a sailing race in the Caribbean. The race was stopped and the competing boats searched for him without success. But a pod of dolphins found him, kept him company and repeatedly approached and swam away from the man, who had been treading water for a couple of hours. On one of the searching boats the skipper wondered if the dolphin movements were intended as a signal – so they investigated and found the man in time. The book's authors move on from there to ask if dolphins have a sense of right and wrong: 'When whales and dolphins go out of their way to help other creatures with their needs that looks pretty moral, at least on the surface.'

Big questions arise from our increasing intimacy with dolphins. In a population of Indo-Pacific bottlenose dolphins off the coast of Australia, mothers have been observed teaching their young how to use tools: specifically, pieces of sponges used to protect their sensitive noses when digging on the seabed. That is, in a form impossible to deny, culture: in this case, something *deliberately* passed on from one generation to the next. Dolphins also have a clear sense of personal identity. Bottlenose dolphins adopt a signature whistle, in effect a name. They develop it in their first year and retain it for the rest of their lives. Other members of the pod will mimic this whistle.

We have only begun to scratch the surface of dolphins' capabilities and the complexities of their lives. One of the problems is that it's very difficult and very expensive to establish a working environment with captive dolphins, and the more we learn about them, the more ethical questions are raised by the very idea of keeping them captive. Dolphins teach, learn, cooperate and grieve. Like us they understand about death. The brain-to-body mass ratio in some species is second only to humans; some parts of their complex brains have been found before only in humans.

The angelic dolphin is a favourite notion in many forms of New Agery: the idea that dolphins have achieved a level of perfection beyond human grasp. It is likely that such ideas began with the experiments of the extraordinary John C. Lilly. He invented the isolation tank, in which he sought to explore the nature of the human brain while deprived of almost all sensory stimulation, sometimes with the use of lysergic acid diethylamide (LSD). Lilly also attempted to communicate with dolphins, sometimes by using LSD himself. The idea of the joyous dolphin is helped further by the way we see a smile on the face of the bottlenose dolphin, and happy laughter in the opening of their beaks.

Sometimes the response to such pleasant whimsy is to oppose it with contrasting notions of a violent and dystopic dolphin society. Certainly mature male dolphins often carry the scars of conflict, but imposing human fantasies – whether hostile or benign – onto wild populations will not help us understand dolphins. The more we learn about them, the more complexities we find in their nature, their intelligence and their society. Dolphin intelligence is a comparatively recent discovery – we have known that the higher apes are pretty smart for a great deal longer – so dolphins have the glamour of newness.

Ocean-going dolphins are threatened by a suite of human activities, which includes pollution of the ocean with plastic, chemicals and noise; the noise upsets their delicate sonar. They are also routinely killed as by-catch in fishing operations: dolphins, though brilliant swimmers, are air-breathers and they can drown. The notion of 'dolphin-safe' labelling for tuna came in the 1980s, but dolphins are still caught in fishing nets. They also get entangled in the nets spread to guard fish farms.

The widespread realization of dolphins' intelligence and their consequent similarity to humans has caused a measure of rethinking about the place of humans in the world and our right to do as we please with what we find there. The spectacle of captive dolphins now provokes anger, their accidental slaughter inspires deep regret. Perhaps if we are ever to reassess the way we live and the way we treat non-human animals – and by extension the way we run the planet – we will do so by means of dolphins. Dolphins are the gateway to a greater and deeper understanding of ourselves and our place in the world. That is true whether we are responding emotionally or with full scientific rigour. Both responses are valid in their different ways.

THIRTY-NINE

RHINOCEROS

*'But you'll never become a rhinoceros, really you won't…
you haven't got the vocation.'*

Eugène Ionesco, *Rhinoceros*

You get an eye for habitat and a feel for what species you're going to find there. It comes from knowledge, from experience, from what you have been shown by local experts, from the habit of being around wildlife. You know that in this sort of wood you'll find pied flycatchers, but in that sort of wood you won't. You also know that in that kind of wooded savannah you will find black rhinoceros – and when you don't it's something of a shock to the mind. You feel that something has gone amiss. And that's because it has.

You will find, in many chapters of this book, the explanation that this or that creature has declined, or gone locally or even globally extinct, because of 'habitat destruction' (for example, see Chapter 19 on dodos): the creature no longer has a place where it can make a living. The relentless pressure of an ever-expanding human population forces wildlife to the margins, so that some species are only found, if at all, in special protected areas.

But in many places in Africa there are endless miles of habitat that's perfectly suitable for rhinoceroses. All that's missing are the rhinoceroses. That's what makes the issue of the survival of the rhino so crazy, so full of contradictions. Rhino conservation, at least in Africa, should be the easiest job in conservation: instead, it is bewilderingly complex, and the rhinos continue to decline at a frightening rate.

There are five living species of rhino: two in Africa and three in Asia. Until 10,000 years ago there were two species in Europe, including the woolly rhinoceros; you can see images of them on the walls of the Chauvet Cave in France. The two African species are confusingly called white and black rhinos; in fact, both are grey. It's been suggested that the idea of whiteness comes from the Afrikaans *wyd* (wide), for the broad square mouth that contrasts with the hook-lipped mouth of the black rhino, but that's disputed. The white is primarily a grazer, the black prefers to browse vegetation from trees and bushes. Neither species has front teeth, both pluck vegetation with their lips and the different shapes of their mouths reflect the difference in diet – and therefore habitat.

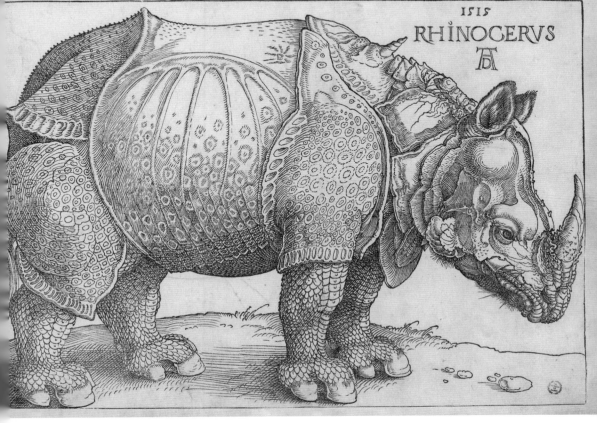

An unending fascination: The Rhinoceros *by Albrecht Dürer (1515), who never saw the real thing.*

The three non-African species have indeed been reduced by habitat destruction. The Indian rhinos' historical range stretches from Pakistan to Myanmar, but they are now found only in a few protected areas, most notably Kaziranga in Assam, where I once saw eleven rhinos feeding together. The Sumatran rhino has declined sharply, to around 200 individuals in 2011, and is now extinct on the Malaysian peninsula and Borneo, while the Javan rhino, the smallest of the five, is down to sixty.

The word rhinoceros is from the Greek and means nose-horn: the abbreviation rhino merely means nose, so is less than glamorous. Rhinoceroses are all herbivores, with a comparatively modest brain-to-body mass ratio. They're often considered 'prehistoric'. The reason for this is, I suspect, their resemblance to one of everybody's favourite dinosaurs from childhood, Triceratops, the one with three horns and huge bony neck-plate, already met in these pages in fancied combat with *Tyrannosaurus rex* (see Chapter 13). That makes us think of all rhinos as throwbacks: as creatures lucky to survive the tough processes of evolution. That is a misreading: rhinos are fully viable, twenty-first-century creatures, more than adequately adapted to the task of eating, socializing, mating, rearing young and becoming ancestors, if only humans would allow them the space to live and the opportunity to do so without being shot.

Rhinos were crucial to the 1960s revolution in human thinking about wildlife, about non-human animals. I can still remember the shock of the revelation that rhinos were in serious danger of extinction: and that it was all the fault of humans. It was at the dawn of the Environment Movement in 1961, and the organization now known simply as WWF released figures that showed the pell-mell depletion of the world's population of rhinos. In the UK, the tabloid newspaper *Daily Mirror* carried the news on the front page: alongside a nice picture of rhinos appeared the word 'DOOMED', in huge letters, followed by 'to disappear from the face of the earth due to Man's FOLLY, GREED, NEGLECT'.

Before this, rhinos had mostly been seen as creatures of utter savagery, highly dangerous to humans and legitimate targets for any hunter intrepid enough to have a pop at one. I remember as a schoolboy reading an account of a man who shot a rhino so that he could eat its tongue for breakfast; he left the rest for the hyenas. It was presented as a pretty dashing and admirable thing to have done.

It's hard these days to explain the devastating nature of the WWF revelations about rhinos: the impossible idea that something as big and as fierce and spectacular as a rhino was vulnerable, that all five species were marching towards extinction, all because of little old us. We were forced to change our ideas about rhinos; we were also forced to change our ideas about ourselves. Humans were now more terrible than rhinos: and that, to revert to the language of tabloid newspapers, is official. It was an early warning of the fact that humans were and are altering the planet in ways that would soon become irreversible. Rhinos changed our worldview.

This shift in understanding hasn't helped rhinos very much, though it has slowed down the rate of decline. Three of the five species are assessed at Critically Endangered. There was a startling incident in 2018 which highlighted this alteration in worldview. The white rhino is divided into two subspecies, the northern and the southern. (A subspecies is a population that is genetically and morphologically different to another subspecies, usually isolated geographically, but capable of interbreeding where ranges overlap.) In March of that year, Sudan, the only surviving male northern white rhino, died at Ol Pejeta Wildlife Conservancy in Kenya, leaving two females. There was no ducking the fact that the first of the long-predicted rhino extinctions had come about. Despite the surviving females, there would be no more rhinos coming from that lot. The northern white rhino has gone.

Rhinos have always fascinated us: perhaps because of their glorious unlikeliness. Albrecht Dürer famously drew an Indian rhino from a description and a rough sketch by someone else: the beautiful result is perhaps an image of what we would like rhinos to look like. We have always admired and perhaps envied rhinos: for their power, the imposing presence and, above all, for the weaponry they bear on their noses. We want that horn for ourselves. That, at base, is the reason why all five species of rhino are in trouble; and why I have walked for miles in perfect black rhino habitat in South Luangwa National Park in Zambia knowing I would never see a rhino. They went extinct in that place in the late 1980s.

A revelation of destruction: front page of the British newspaper Daily Mirror, *1961.*

In Yemen and Oman the horns are used for dagger handles: a potent emblem of status. Rhino horn is part of traditional Chinese medicine: not, as has often been thought in the West, as an aphrodisiac, but as a treatment for disorders of the blood, fevers and convulsions; it is seen as a powerful, life-saving medicine.

These two factors have placed continuous pressure on the world's rhino population. Rhino horn has always been right at the top of the illegal trade in wildlife: it really is worth its weight in gold. It wasn't until 1993 that China became a signatory to CITES, under which trade in rhino horn and many other animals and plants – alive, dead, in parts – is forbidden. In October 2018, China announced that it was going to permit the domestic use of tiger bone and rhino horn for 'scientific, medical and cultural purposes'. The announcement was met with dismay in much of the world and the decision was postponed. It seems that China still can't get a handle on the way that many nations seem to take, at least when convenient for themselves, a serious attitude to issues concerning wildlife and the possibility of extinction.

The situation with rhinos was greatly complicated at the beginning of the current century by the emergence of Vietnam as a prime market for rhino horn. Increasing prosperity (at least for some) has made rhino horn – hitherto an unachievable status item – available to newly rich citizens. They use it to treat both cancer and hangovers, and it is a hugely desirable item, pushing up the price of rhino and overheating the world market.

Rhino horn is made from keratin, the same stuff as your hair and your fingernails. Laboratory trials have found that it has no medical properties whatsoever, but the demand for the stuff is still driving a thriving industry in poaching. In South Africa in 2018, an average of two rhinos were killed every day by poachers. Domestic trade in rhino horn was banned in South Africa in 2009, but the ruling was overturned in 2017. People who keep rhinos in game farms in South Africa claim that poaching will dwindle and vanish as soon as the rhino-horn trade is made legal: but then they would do.

There have been some grotesque attempts to deter poachers. These include de-horning wild rhinos, and also treating the horns of living rhinos with pink dye, and at the same time introducing a lice-killing substance that causes nausea when consumed by humans. Others have questioned the ethics of this, and also its practicality, since rhinos have died during the anaesthetic necessary to carry out this operation.

It's all the most terrible mess. The fact that rhino horn has no medical value is in a way beside the point: the knowledge that a huge and powerful animal has been killed just for you is, perhaps, the most potent and meaningful part of the business. That and the staggering expense: if it costs that much, it really must work.

Meanwhile, I walk the rhino-free trails of the South Luangwa National Park dreaming of rhinos: the time they lived there, and the time they might live there again. I do so knowing that in North Luangwa National Park thirty-four black rhinos were introduced in 2003, and the population is now established and growing.

FORTY

NIGHTINGALE

'*For here I've heard her many a merry year –*
At morn, at eve, nay, all the live-long day,
As though she lived on song.'

John Clare, 'The Nightingale's Nest'

We humans envy many other species. This trans-specific envy has inspired and driven many aspects of human behaviour: we envy the majesty of a male lion, the speed of a horse, the flight of a bird. We envy the rhinoceros his horn; we envy the egret his breeding plumes. And perhaps more than anything else, we envy the birds their song.

I have chosen the nightingale to represent this envy and what became of it. I have done so because the song of the nightingale is the most spectacular and complex birdsong in the world: a repertoire of 250 phrases made up from more than 600 units of sound has been recorded in a single individual. The accepted theory is that birds with a complex repertoire attract mates and successfully defend a territory because the song demonstrates their experience and knowledge. That makes it a sincere and honest indication of the excellence of the bird doing the singing (more on this issue in Chapter 79 on peafowl).

There's something about that reductionist view that intrigues me. We greatly admire human musicians: so here's a guitarist's joke.

Q: How many guitarists does it take to change a light-bulb?
A: Two: one to change the lightbulb, and one more to explain how Jimi Hendrix would have done it better.

The best human musicians command immense respect, and during their lifetime they have access to many of the best things in life, including the best sexual partners. But that's not why they make music, nor why they do it supremely well. They do it because they are lost in the music. Surely male nightingales sing because they must sing: and female nightingales listen and choose the singer that pleases them most.

The song of the nightingale is thrillingly loud. They sing through the day as well as the night, but most of the competition shuts up during the hours of darkness, so nightingales have the entire concert hall to themselves. A nightingale's voice will easily carry for a mile on a still night. The song involves a series of clear passionate whistles, often rising to an almost unbearable peak of intensity. This is mixed in with other, more challenging and less melodic phrases, and also a deep throbbing drumming. The range is astonishing, and the intensity of the performance of a nightingale in full flow is unbeatable. To perform at such a level is remarkable enough: to continue to do so for day after day makes deep demands of the cockbird's physical stamina and musical inventiveness.

Humans have always responded very deeply to birdsong. We envied the birds' melodic abilities so much that we stole them. We stole them as we stole the rhino's horn and the egret's plumes (see Chapters 39 on rhinoceros and Chapter 18 on egrets): but this time we did so without harming the creature involved. We stole the melody and added to it something that all placental mammals – all mammals that are neither pouch-bearing marsupials like kangaroos nor egg-laying monotremes like platypuses – have in common: that is to say, rhythm. We humans all learn rhythm in our mother's womb. We spend nine months listening

Can't paint the song: nightingale in The Birds of Great Britain *(1862–73), painted by John Gould.*

to the pounding, four-four beat of the maternal heart, and that remains part of us when we emerge into the world and start to breathe for ourselves.

Our love of rhythm gives us a taste for rhythmic phrases, and as we developed speech so we began to make poetry. And when we added rhythm to the melodies of the birds, we had what we humans recognize as music. When we added the rhythmic words and repeated phrases of poetry to the rhythm and melody of music, we had song.

We were able to make music with our voices, perhaps even before we developed language – and perhaps such music was part of the development of language. As time passed, we invented other ways of making sounds. We made musical instruments: and the first musical instruments we fabricated were flutes: instruments that sound like birds. You'd have thought it would be easier to invent a plucked instrument from the stretched gut in the food store, or to construct an instrument that produces different notes by means of percussion, in the manner of a xylophone. But the earliest musical instrument ever found is the Divje Babe flute, found in Slovenia and dated at 43,000 years old. And some say even older. It was carved from the femur of a young cave bear, a species now extinct. It even has a couple of holes, though some have suggested that these were made by the teeth of carnivores, and their symmetrical nature is pure coincidence. But even if we reject this as a musical instrument, the next earliest instruments are also flutes. A flute made from a vulture's bone was found in Ulm, in Germany, and is 35,000 years old. There are other flutes of slightly more recent date – in fact, the oldest instruments that are not flutes are less than 5000 years old and are found in Ur in the region of Sumer, in Mesopotamia, in a collection of lyres and harps – and flutes.

There is a neat poetry in making a flute from the bones of a real bird: and since bird bones are hollow it is reasonably straightforward to do so. It is not that hard to imagine how our ancestors worked out how to do it: I remember examining the bones of a warthog on a windy day in Zambia. The bones had been hollowed out by invertebrate bone-gnawers, and when I held one up for a closer look it caught the wind and whistled, as a bottle does. Two bones, two notes. A musical instrument. I had invented music: or at least recapitulated that moment in human history, quite by chance.

Our African ancestors would not have heard a nightingale in full song. They would have been more likely to respond to the melodic whistles of orange-breasted bush-shrike, black-necked oriole and Heuglin's robin (now more usually known as white-browed robin-chat). The nightingale, as a supreme performer, must serve here for all the songbirds that have moved human beings with their song and sparked in humans a desire to emulate or even outdo.

Nightingales have been celebrated across the centuries for the beauty of their song. Though it is intended primarily for female nightingales, it also gives deep pleasure to humans. It's been suggested that the male's song provokes a chemical response in a female's brain, something John Keats seems to understand in his 'Ode to a Nightingale':

My heart aches and drowsy numbness pains
My sense, as though of hemlock I had drunk…

This poem, like many another poem about birdsong, like many another nature poem, is more concerned with the poetic heart and the poet's exquisite emotions than the actual bird. Keats writes of the bird singing in flight, something no real nightingale ever did; they sing only from deep cover.

For years it was thought that it was the female who did the singing. In fact, only male nightingales sing, while the females have the full range of alarm and contact calls. In Greek mythology Philomel was a princess violated and mutilated by her sister's husband; she is transformed into a nightingale and nightly sings a lament. It was thought that the female nightingale sings her heartbreaking song while impaling herself on a thorn. There is a poem about this, by an anonymous writer and set to music on the lute by John Dowland in the sixteenth century:

The dark is my delight;
So 'tis the nightingale's.
My music's in the night;
So is the nightingale's.

My body is but little;
So is the nightingale's.
I like to sleep next prickle;
So doth the nightingale.

The nightingale narrowly outcompetes the skylark as the most famous songbird, and the one that poets most often turn to. Both songs, both species, are unmistakable and spectacular.

We need to use a little imagination to see why birdsong is so important in human development. We have only had readily available music for a century. Before that we had to make it ourselves, or find someone willing to do it for us – or listen to birds. We had much greater access to the song of birds than we do now, with more green spaces and far less ambient noise. Birdsong filled an early human need for easy beauty: but as our lives grew more urban, we needed a way to bring the music to us. We will explore that in Chapter 81, which is centred on the canary.

These days we fill our world with music, sounds that we record and constantly play back. We hear it in pubs and bars and restaurants and shops and public places; we play music to ourselves through headphones in cities and in forests. I frequently see people running or walking in places loud with birds, with their earphones blocking out all natural sounds. We stole music from the birds: perhaps having done so, we no longer have a need for birds. Certainly we seem to be remaking the world on that premise.

FORTY-ONE
PIG

'Neither cast ye your pearls before swine.'

Matthew 7:6

Everyone loves a good songbird; everyone hates a cockroach. But pigs split the crowd right down the middle. Pigs are the favourite food of the world's most populous nation; pigs strike horror and/or disgust in those who follow two of the world's major religions. To be called a pig is a serious insult; pigs are beloved characters in literature, film and television.

The pig was domesticated from the species we call wild boar, one that includes wild sows: both wild and domestic pigs are usually classified as one species *Sus scrofa*. Their natural range takes them across Eurasia into India and Japan; they can live anywhere there is a productive habitat and plenty of water; pigs need to drink at least 3 gallons (13 litres) a day. It's been speculated that the species was domesticated twice over, in Asia and again in Europe. Pigs are characterized by a long snout, which is supported by a prenasal bone: this supports a nose that is, at the same time, hugely sensitive and a highly effective digging implement. Wild swine are intelligent and highly sociable, and those traits can be found in domestic pigs whenever it's possible for them to express such things. People who work with pigs tend to have a great affection for them; these include my father, who spent a long winter as a swineherd.

Our ancestors probably managed wild herds of swine – following them, feeding them, accustoming them to human presence, killing selected individuals – 12,000 years ago. Pigs were domesticated fully – taken into human communities and their breeding controlled by humans – 5000–7000 years ago. It's comparatively easy to keep pigs, though it's a good deal harder to keep them well. Pigs are omnivorous, and the tradition of feeding them on scraps and waste food goes back for centuries, no doubt for millennia. In China, parts of India (notably Goa) and Korea there is a tradition of the pig latrine, though in most places it is dying if not already extinct. This comprises a small refuge in which humans can defecate, with a pigsty below. The pigs eat the succession of warm meals offered by the local populace: an elegant way of solving problems of hygiene and nutrition. Of course you still have the (lesser) problem of the disposal of pig droppings, but they make excellent fertilizer.

Pigging out: Sow with Piglets in the Sty *by John Frederick Herring Jnr (1815–1907).*

Pigs have often been kept as respected adjuncts to family life: a single pig, fed largely on a family's leftovers and requiring minimal care, is slaughtered when the time comes, producing at the same time a family feast and a small cash bonanza. In Ireland the family pig is traditionally referred to as 'the gentleman that pays the rent'. Pigs are prized for the amount of nutrition they bring – another old saying is 'you can eat everything from a pig except its squeak'. Brawn, or head cheese, is made from the heads of pigs.

We traditionally associate pigs with filthy habits and monstrous appetites: dirty pig, greedy pig, let's go to McDonald's and pig out. In the 1960s it was fashionable to refer to the police as pigs; some years later I attended a football game (American football) between the New York Police Department and Miami Police Department. The event was billed as the Pig Bowl; I was told that the first word was an acronym: Pride, Integrity, Guts.

A good number of the pig's reprehensible habits can be put down to bad husbandry. A pig kept in a confined space has no option but live in its own filth; and, when fed on humanity's leftovers, can only eat this unwanted food or starve. Pigswill, or just swill, describes a truly horrible plate of food. An animal that is bred to be largely hairless – all wild pig species are hairy – will roll in mud for protection from heat, from flies and from cold. Pigs don't sweat: the idea of 'sweating like a pig' is a fantasy, an assumption that since pigs have so many undesirable habits they probably have a good few more as well. Pigs have a reputation for eating their own young; in *Ulysses* Stephen Dedalus says: 'Ireland is the old sow that eats her farrow.' Such cannibalistic behaviour doesn't represent the norm, or even the occasional depravity of pigs: it is related to the stresses suffered by pigs – lively, intelligent, sociable as we have already established – kept in confined and solitary conditions.

To observe wild pig species is to see a different side of porcine life. Warthogs, an African species in the same family, are trim, agile and devoted to the concept of family. They have a great turn of speed, though they can't keep it up for long, so they take refuge in holes. They go in backwards; if disturbed, rather than wait and see what happens, they will burst out as if fired from a gun, ferocious tusks leading the way. I have seen free-ranging domestic pigs at Knepp Castle in Sussex, in England: their habit of digging for food creates a dynamic environment, quite different from one in which the surface in undisturbed.

Wild swine went extinct in Britain in the seventeenth century, but the fad for 'wild boar' meat in restaurants created a trade in which wild swine were kept captive. Being powerful and capable beasts, they escaped and have established a series of feral populations. Some applaud this as the return of a once-extinct part of British fauna; others see them as a horrific imposition on a community. Either way, they are changing the ecology of the woods they live in, disturbing the soil and creating new habitats.

These feral pigs have polarized views – but polarizing views is what pigs are good at. In Chinese astrology, a native of the year of the pig is stylish, a perfectionist, hardworking, generous, tolerant and kind. But in both Islam and Judaism, a pig is forbidden. More than that: loathed, despised, anathema. The fact that pigs are forbidden is mentioned no fewer than four times in the Quran; more than anything else, a pig exemplifies what is *haram* (forbidden). The Bible also forbids the eating of pork, and orthodox Jews won't touch the stuff. Fundamentalist Christians, who follow the notion of biblical literalism, generally ignore this prohibition, preferring a pick-and-mix version of biblical instructions.

Model of efficiency: pig latrine,
Han dynasty, late second century AD.

Broadly speaking, prohibitions on homosexuality in Leviticus and elsewhere are honoured but bacon sandwiches are fine.

The strength of feeling about pork is startling. The proximate cause of the Indian Rebellion, formerly known as the Indian Mutiny – when the Indian people rebelled against the British colonialists in 1857 – came when a rumour ran through the sepoys (native Indian troops) that new cartridges issued to them had been greased with pork and cow fat. The soldiers were required to bite the cartridges to prepare them for firing. This was horrific to Muslims, to whom pork was forbidden, and also to Hindus, to whom the cow was sacred. That offered a rare unifying cause, seeming to exemplify the contempt of the colonialists for religions other than their own – and so the ultimately unsuccessful revolt began.

The presumable reason for the prohibition is that pork 'goes off' quickly in conditions that suit the microorganisms that cause the process. Pork can convey the larvae of roundworms to the consumer and cause trichinosis, with symptoms that include diarrhoea, abdominal cramps, muscle pain and fever. You get this by eating raw or undercooked contaminated pork; proper cooking kills the parasite. It is particularly important not to feed pigs on uncooked meat. Smoking, salting and microwaving are not reliably lethal to the parasites.

One of the reasons that pork is more dangerous to humans than other kinds of meat is the extraordinary similarity between humans (and other primates) and pigs. These similarities are in anatomy and physiology, including organ placement, fat distribution and hair cover. This is not a matter of direct relatedness but the kind of coincidence called convergent evolution: the same solution to the same problem via a different route (like the evolution of intelligence, see Chapter 37 on octopuses). All of which makes pigs very useful to scientists as 'translational research models' when working on issues relevant to humans.

But the resemblance is deeper than that. Comparative studies of human and pig genomes show some startling similarities at the genetic level. A certain amount of DNA has been retained by both species since we shared a common ancestor 80 million years ago. In other words, pigs can be harmful to humans not because of their distance from us but because of their proximity.

Despite our negativity about pigs, we also have a huge affection for them: usually portraying them in fiction as parodic humans. Piglet, in *Winnie-the-Pooh*, is constantly and courageously confronting his own timidity; Gub-Gub in the *Doctor Dolittle* books is a cheerful companion; Babe (in the eponymous book and film) is a pig that works as a sheepdog; Wilbur is the hero of *Charlotte's Web*; while Peppa Pig and, most grandly of all, Miss Piggy are television stars. Miss Piggy assumes the role of the diva without the physical equipment to make the performance convincing, and so becomes an endearing character.

The most famous of all fictional pigs appear in George Orwell's *Animal Farm*, a fable about Soviet Russia and, by extension, all utopias. The wise pig Old Major invents the doctrine of Animalism, which is about freedom and equality for all the animals on the farm. This doctrine is taken farther by Snowball and Napoleon; Snowball is murdered and Napoleon establishes a reign of terror, reopens negotiations with the human race and permits the pigs, but only the pigs, to wear clothes. The book has one of the best last sentences ever written: 'The creatures outside looked from pig to man, and from man to pig, and from pig to man again, but already it was impossible to say which was which.'

FORTY-TWO

CHIMPANZEE

'Chimpanzees, gorillas, orang-utans have been living for hundreds of thousands
of years in their forest, living fantastic lives, never overpopulating, never
destroying the forest. I would say that they have been in a way more successful
than us as far as being in harmony with the environment.'

Jane Goodall

I remember being taken to watch the chimpanzees' tea party at London Zoo: chimpanzees dressed in clothes with mugs and teapots and plates and food and fruit juice: parodic humans mimicking human behaviour and making a right mess of it, and of themselves. How everyone laughed. Was I the only one who failed to see the joke?

I can't remember the year, but I'd like to think it was 1960. That was when an Englishwoman called Jane Goodall, who had trained as a secretary, went to Tanzania to observe wild chimpanzees. What she discovered over more than fifty years of study changed our understanding of chimpanzees, changed our understanding of what it takes to be human, and changed our ideas about the discontinuity between humans and other members of the animal kingdom.

In many of the preceding chapters we have looked at the way that humans are *like* other species of animal. But we are not like chimpanzees. We *are* chimpanzees. Near as damn it, anyway: we have 95–98 per cent of our DNA in common, and we are often subject to the same diseases. Creatures with that degree of relatedness are normally classified in the same genus: and it's been suggested, on many occasions, that we should do the same with chimpanzees and recognize them as *Homo troglodytes*, rather than as *Pan troglodytes* as we do today. Chimpanzees split from their common ancestry with gorillas 7 million years ago; they split from humans 4–6 million years ago. In other words, chimpanzees are more closely related to us than they are to gorillas. Jared Diamond's book of 1991, subtitled *The Evolution and Future of the Human Animal*, is titled *The Third Chimpanzee*. To get there, he lists the chimpanzee, the bonobo or pygmy chimpanzee – and us. Those who can draw a hard-and-fast line between what is human and what is chimpanzee can also dance with the angels on the point of a needle.

Chimpanzees are a versatile species, capable of making a living on the savannah, in deep rainforest, in montane forest, in swamp forest and in dry woodland. They are omnivorous, but fruit is always central to their diets. Nuts, leaves and roots make up

Redefine humanity: Jane Goodall, pioneer in animal behaviour, at work in Tanzania.

most of the rest, and around 2 per cent of their diets is meat and insects. Chimpanzees are intensely and vividly social: it's been said that 'one chimpanzee is no chimpanzee': they define themselves by their network of relationships with others.

Any study of any society, human or non-human, tends to bring from us an instant need to make value judgements and comparisons with the way we live ourselves. When looking at the society of chimpanzees, there is a tendency to read from it our own virtues and/or our own shortcomings. Chimpanzee society is strongly hierarchical, and the question of who gives way to whom is always on the agenda. Some read in this a fierce and uncompromising way of life; Frans de Waal, whose researches were mostly based on his study of a community at Rotterdam Zoo, said that it was essential to consider not just the rows but the making-up: reconciliation was crucial to their society. He wrote a book called *Peacemaking Among Primates*: the title clearly invites us to consider ourselves in this structure of falling out and making up. (I wonder how many human children are conceived as part of the process of making up.)

Chimpanzees operate in what is called a fission-fusion society: small temporary groups within a larger group. The mother–infant bond is intense and long, up to seven years of close association; males will have a close relationship with their mothers until one of them is dead. Females emigrate from their family group while males stay behind. People who have studied chimpanzees in depth, knowing each individual within a large group, report coalitions, often temporary and expedient, along with complex political awareness. Females come into oestrus and tend to mate with several males. Communication is central to society: facial expressions, posture and vocalizations are all important. The most important sound is generally described as the pant-hoot: the most imitated chimpanzee sound, and indicative of excitement. It's been suggested that individuals can be identified from their pant-hoots; it is perhaps a kind of signature. The pant-hoot was once much imitated at British football (soccer) matches, used to disparage footballers of colour. It is still found in some footballing cultures today.

Goodall arrived at Gombe Stream National Park, backed by the scientist Louis Leakey, who is best known for his anthropological work; he and his wife Mary established the fact that humans evolved in Africa. (It was their team who discovered the famous footprints at Laetoli Gorge in Tanzania, see Chapter 1.) Since Goodall had no scientific training, she improvised her method from circumstance; for example, giving individual names rather than numbers (unlike the Yellowstone researchers who named the great wolf Twenty-One, see Chapter 21).

Goodall's early observations brought nothing new. But she was an intense, indefatigable and very accurate observer, and she succeeded in establishing herself as a neutral presence, more or less literally part of the landscape. So this immense – and, crucially, one-sided – intimacy was able to develop, and she began to see things that had never been seen before. A chimpanzee she called David Greybeard showed her three shattering things in quick succession. First, she saw him gnawing at a carcass: chimpanzees had never been seen eating meat before. Then she saw him use a length of grass to fish for termites in a hole too narrow to insert a finger; it was thought back then that humans were the only creatures on Earth to use a tool. Then she saw the same chimpanzee stripping a twig of leaves before using it to hunt for termites: not just using but actually making a tool. She reported back, and Leakey famously cabled her: 'Now we must redefine tool stop redefine man stop or accept chimpanzees as human.'

She returned to England to face scientists who rejected her claims as 'anecdote and speculation', and 'no real contribution to science'. But no revolutionary contribution to human understanding is ever accepted straightaway. Goodall went back to Gombe to continue her studies and, bit by bit, what she discovered there became orthodoxy. She was filmed at work by National Geographic and *Miss Goodall and the Wild Chimpanzees* of 1965 was viewed by 25 million Americans and became a global hit.

As her studies continued, it became clear that, on occasions, chimpanzees will actively hunt prey, most particularly red colobus monkeys, though they were also observed hunting baboon, bushbaby, duiker, bushbuck and warthog. Chimpanzees

Craftsman at work: chimpanzee using a tool to extract and feed on insects in Tanzania.

will also sharpen a stick with their teeth and use it to spear bushbabies from their holes.

As you would expect from the nature of chimpanzees and their society, hunting is a cooperative business. Four distinct roles were observed: the drivers, the blockers (who work from the bottom of a tree to block off escape routes), the chasers and finally the ambushers. The footage of a successful hunt of red colobus monkeys is distressing for humans, not just because chimpanzees had for centuries been accepted as pure vegetarians, but also because it looks a lot like cannibalism, for all chimpanzees and red colobus monkeys are quite distantly related.

Humans have also learnt a very great deal from captive chimpanzees. The notion of language as proof of human uniqueness has been challenged by chimpanzees: firstly with an individual named Washoe, who was taught American Sign Language. He acquired a vocabulary of at least 350 signs, and taught some of them to his adopted son Louis. On seeing a swan he signed 'water' and then 'bird'. He put together a near-sentence when a doll was placed in his drinking mug: 'Baby in my cup'. On another occasion he signed: 'You me out go'. He received the signed answer 'OK but put clothes on'. Washoe immediately put on his jacket. One of his regular teachers suffered a miscarriage and was absent for some time. On her return she signed to Washoe: 'My baby died'. Washoe signed back 'cry'. He then traced the track of a tear on her face. This is an astonishing bit of trans-specific empathy: chimpanzees don't weep.

The value and meaning of this is eternally disputed, and every conclusion drawn from it will inevitably be challenged. One certainty: the capacity of a chimpanzee is greater than we had bargained for. This gives rise to a considerable degree of discomfort when it comes to the use of chimpanzees in research. Such work is banned in Austria, New Zealand, Netherlands, Sweden and the UK. Stuart Zola, head of Yerkes National Primate Research Laboratory in the USA, told *National Geographic* magazine: 'I don't think we should make a distinction between our obligation to treat humanely any species, whether it's a rat or a monkey or a chimpanzee. No matter how much we may wish it, chimpanzees are not human.'

But if chimpanzees are not human, they are a great deal less like rats. Most of us would instinctively take a hierarchical rather than a binary view. Instead of saying it's either human, and we treat it one way, or non-human and we treat it another, we tend to see a gradation of concern, with humans requiring a higher level of concern than chimpanzees, who in turn require a higher level of concern than monkeys, and monkeys higher than rats. When it comes to the put-to, we don't instinctively accept that the animal kingdom can be divided into two – humans on the one hand, everything else on the other.

And still more things from chimpanzees have come through to challenge our notions of uniqueness. Chimpanzees in the Taï Forest in Cote d'Ivoire use stones to crack nuts; those in Gombe Stream National Park, although they have both nuts and stones, do not: indicating that this is a business of cultural exchange rather than genetic impulse (or 'instinct').

Chimpanzees laugh. Or, to be more cautious, seem to have a response similar to that of laughter in humans. When play-fighting and chasing or tickling, with human handlers and with each other, chimpanzees have a laughter-like response.

And what about religion? Is that the last division between humans and the rest? Goodall has witnessed some extraordinary behaviour across the decades she has spent with chimpanzees. Here she describes a male approaching a waterfall:

> As he gets closer, and the roar of the falling water gets louder, his pace quickens, his hair becomes fully erect, and upon reaching the stream he may perform a magnificent display close to the foot of the falls. Standing upright, he sways rhythmically from foot to foot, stamping in the shallow, rushing water, picking up and hurling great rocks. Sometimes he climbs up the slender vines that hang down from the trees high above and swings out into the spray of the falling water. This 'waterfall dance' may last ten or fifteen minutes.

I invite readers to draw their own conclusions from this. But no matter what they are, the fact remains that there is more to chimpanzees than we ever suspected, and they are far closer to us than we can easily understand. We have no need to search for a 'missing link' that unites us humans with our fellow animals. They stand before us, communicating, exchanging ideas, creating, caring, cooperating, laughing, perhaps even praying.

FORTY-THREE
ALBATROSS

'At length did cross an Albatross,
Through the fog it came;
As if it had been a Christian soul
We hailed it in God's name.'

Samuel Taylor Coleridge, *The Rime of the Ancient Mariner*

We are not like albatrosses. But we would like to be: and that's one of the reasons why these birds have always had such a hold over us. It's about flight: and more especially about the human envy of flight.

And albatrosses fly. More than any other birds, they are masters of soaring. Soaring is the art of gaining height without power: without the power of a propeller or the power of a pair of beating wings. Albatrosses can cover 600 miles (1000km) in a day and do so with scarcely a beat of their great wings; wandering albatrosses have a wingspan of 12ft (3.5m).

Their wing joints lock, so they can rest on the wind as easily as you do on your sofa; in gliding flight their heart rate is scarcely higher than it is at rest. They mainly use two techniques: dynamic soaring and slope soaring. The first involves turning into the wind to gain height, for it's the air flowing over an aerofoil surface that gives lift. You rise, but as you do so you lose speed and must drop down… picking up speed from your descent, so you can turn back into the wind again and climb right back up where you started, establishing an effortless rhythm of rise and fall that is infinitely more economical than powered straight and level flight. Slope soaring involves the use of wind deflected upwards, off a cliff face or, with albatrosses, off the enormous waves that dominate the southern ocean where most species of albatross spend most of their time. Wandering albatrosses work to a glide ratio of 22:1: that is to say, 72ft (22m) of forward travel for every 3¼ft (1m) lost in still air. That is staggeringly efficient.

There are usually reckoned to be twenty-two species of albatross, ocean birds that come to land only to breed. They are faithful to the site and to each other. On land the limitations of their extravagant and uncompromising build can make them look comic: landing and taking off are both difficult skills to master, as they have a high stalling speed, and they must use a runway to build up or to lose speed.

A hellish thing: the mariner shoots the albatross in an engraving after Gustave Doré of 1870 for Samuel Taylor Coleridge's The Rime of the Ancient Mariner.

They are startlingly unlike other seabirds, with that recklessly long-winged silhouette. Creatures that stand out from the rest almost invariably attract human attention and prompt us to build myths around them. Seagoing people, especially those who reached the southern ocean, couldn't help but respond to albatrosses. So it was fancied that albatrosses were the souls of drowned sailors. In Hawaii, albatrosses were seen as embodiments of the current generation's ancestors; Maori people made flutes (see Chapter 40 on nightingales for more on bird-flutes) from the wing bones of albatrosses; their length makes them perhaps the best bone flutes ever made.

There was a taboo against killing albatrosses, though this was by no means universal. Seaman routinely shot and ate albatrosses without feeling or fearing any major ill effects. But the notions of the special nature of albatrosses, and the idea that proper reverence should be shown towards them, was brought out with dizzying brilliance in Samuel Taylor Coleridge's *The Rime of the Ancient Mariner*, first published in 1798.

In this extraordinary poem, the ancient mariner tells of the disasters that befell him after his terrible action:

> And I had done a hellish thing
> And it would work 'em woe
> For all averred I had killed the bird
> That made the breeze to blow.

The albatross had led the ship out of a maze of icebergs, but afterwards he was slain by the mariner's crossbow and all kinds of dreadful things began to happen. The worst of it is that the mariner is not killed with the rest of the crew, but must horribly live on.

This poem has given albatrosses a position of honour in the English-speaking world. It is not necessary to have read the poem to understand why it's a bad thing to have an albatross round your neck or, for that matter, to accept that albatrosses are in some way special and therefore worthy of generous treatment. Above all, albatrosses are not birds to be wasted.

There is a benign version of the albatross encounter in *The Voyage of the Dawn Treader*, the third of C. S. Lewis's *Chronicles of Narnia*. An albatross leads the crew and the ship out of the darkness and horror of the Island Where Dreams Come True: as it circles the mainmast, the bird whispers to Lucy, 'Courage, dear heart', and she knows that the albatross is Aslan himself, who normally appears (see Chapter 1) as the Great Lion.

So when it became widespread knowledge that we were losing albatrosses at a rate of 100,000 a year and that there was also serious trouble on the remote islands where albatrosses breed, there was a stronger than usual sense that something should be done.

Albatrosses are long-lived birds. Long life is a strategy of its own: these birds operate at a leisurely pace unknown to frenzied sparrows. A pair of albatrosses will raise a single chick roughly every couple of years, while a pair of small birds might get off a dozen or more chicks in a single year. Albatrosses exist at a pace that can't work *without* a long life or, rather, many long lives. If birds are constantly dying before their natural span, the albatross lifestyle and albatross numbers become unsustainable.

And this was and is happening on two fronts. The first problem is long-line fisheries: albatrosses come to steal the bait and get snagged on the hooks and drown. Drowned in immense numbers. At the same time there is a state of crisis on the breeding islands: rats and mice accidentally introduced to these remote islands, mostly by whaling ships in the nineteenth and early part of the twentieth centuries, have reached huge numbers, eating the eggs and the still-living chicks of albatross.

The Agreement on the Conservation of Albatrosses and Petrels was ratified by thirteen nations and came into force in 2004. Participating nations include the UK, Australia, New Zealand, South Africa and Brazil. The problem is enforcement: making sure that people already involved in difficult and dangerous work are prepared to take extra trouble that will bring them no immediate profit. The initiative is reckoned to be a success: though the problem remains, especially with illegal fishing boats.

The Albatross Task Force was set up by Birdlife International and the RSPB in 2005. Their research showed that 90 per cent of seabird deaths from long-line fishing are easily preventable: bait the hooks below water, bait them at night, dye the bait blue, fly bird-scaring devices over the lines.

By 2018 the island of South Georgia in the South Atlantic had been cleared of rodents: rat-free for the first time in 250 years (see also Chapter 25 on rats). The place will in season hold 15,000 pairs of black-browed albatross, 12,000 pairs of grey-headed albatross and 1700 pairs of wandering albatross. The process of clearing took seven years and cost £10 million, and it was mostly done by means of private funding rather than government initiative. It seems that humans really want albatrosses to survive. There are other programmes on other islands designed to work the same trick.

The 2017 television series *Blue Planet II* caught worldwide attention with a sequence of an albatross feeding plastic to its chick. The problem of plastic waste in the oceans, a threat to many forms of sea life, has since become a cause of widespread concern. It takes no great leap to accept that rubbish is a bad thing: the fact that the waste products of humanity are turning up in places thousands of miles from their origin is deeply troubling. That vivid image of the plastic-fed albatross created a huge surge in awareness.

Albatrosses have always been birds that humans have liked, admired and envied. That hasn't helped them when there has been a clash of interest between them and us. But like so many other animals in this book, we have changed our view of them. We still like them, but where we once envied their glorious independence from land and their affinity for the wildest water of the planet, the most fearsome place to humans that exists on Earth, we now see that, for all their strength and power and ability, they, like everything else, are within our power. If we humans want them to survive, then we humans must act accordingly. That is a tale that will be told again in many different forms before we reach animal number 100.

FORTY-FOUR
PASSENGER PIGEON

*'Before sunset I reached Louisville, distance from Hardensburgh fifty-five miles.
The pigeons were still passing in undiminished numbers and continued to do so
for three days in succession.'*

John James Audubon

The vulnerability of many non-human species is an accepted fact of life these days. Concern for threatened animals is part of the way we see the world in the twenty-first century. The idea that humans must take active steps to look after whales, tigers and albatrosses if we wish to keep them is widely accepted. If we don't take these steps we will lose them: that's hardly controversial stuff these days. There are a million clashes and disagreements about the extent to which we should do this looking after, but the basic premise – that humans can cause species extinction – is more or less globally accepted.

We talk about anthropogenic extinction: extinction caused by humans. It's no longer something we can duck. The possibility of extinction is not just scare talk: it's happened before our eyes. For this leap of understanding, we must thank one species above all others, and that is the passenger pigeon. The bird that died.

Humans became an agricultural species 12,000 years ago and an industrial one in the late eighteenth century – but we still live with hunter–gatherer brains. We still believe that nature is both infinite and hostile: that we can help ourselves to what we want and it will be replenished, and that we are forever underdogs, doing our plucky best against the outrageous odds of nature.

But it became necessary to attempt some adjustment to this ancient and ancestral worldview with the passenger pigeon. This was the most numerous species of bird in North America: and perhaps the most numerous species of bird that has ever existed on the planet. And in the course of half a century or so we did for the lot of them.

Message from Martha: Martha, the last passenger pigeon, who died in Cincinnati Zoo in 1914 at the age of twenty-nine.

Blackening the sky: passenger pigeon flocks recalled in Falling Bough *by Walton Ford (2002).*

Ancient reports of passenger pigeons boggle the mind. There was a flock in Southern Ontario in 1866 that was 1 mile (1.5km) wide and 300 miles (480km) long. It took fourteen hours to pass and it contained – so far as any could reckon – 3.5 billion birds. When they stopped to establish breeding colonies, branches would snap under the weight of nests.

No wonder people thought we would never run out of nature: these are figures beyond comprehension. Our brains simply don't work with numbers like that: like astronomical distances and Deep Time, a flock of such magnitude doesn't really make sense to us.

But this was the way the passenger pigeons worked. Vast numbers were no accident or coincidence: it was their fundamental strategy for existence. They led a nomadic lifestyle, moving from one immense food source to the next, taking, in season, acorns and beech-mast, berries, invertebrates and, when they could get it, cultivated grain. Their name is from the French verb *passager* (to move about). They would pop up, in devastating numbers, feed and move on. There is a record of a nesting colony of passenger pigeons covering 850sq. miles (2200sq. km) containing 136 million breeding adults. The males were especially handsome things: slate-blue above and coppery below; the females a more muted version of the same colour scheme.

They lived all over North America east of the Rockies, across the Great Plains to the Atlantic coast; perhaps a quarter of all the birds in North America were passenger pigeons. Vast numbers gave them certain invulnerability: no predator could affect them very much. It's a strategy called predator satiation, in which superabundant numbers can scarcely be dented by the puny attacks of a few predators, and it is found with wildebeest on the Serengeti, in periodic cicadas that erupt only occasionally but always in very large swarms, and in oak trees, which will at irregular and unpredictable intervals produce a coordinated and synchronized superabundance of acorns. And it worked across the millennia in North America for the passenger pigeons.

But things changed with a rising human population with many different ways of killing pigeons. They were killed to protect crops; they were killed for their meat – and the strategy that had allowed them to prosper was the strategy that led to their downfall. Their superabundance made them the easiest possible target for the increasingly abundant humans below. You hardly needed to be much of a marksman: a random shot from a single cartridge was likely to bring down half-a-dozen birds. There is a record of one double-barrelled blast killing sixty-one: spearing fish in a barrel is a testing sport by comparison.

There were so many of them that birds could be knocked out of low-flying flocks by thrashing about with a pole. The hunting of passenger pigeons became an industry: birds were shot, netted, had their nesting tress burnt down, were asphyxiated by the burning of sulphur, were attacked with rakes and pitchforks, were pelted with potatoes and were poisoned by corn soaked in whisky. They could be lured down by tethered pigeons, who sometimes had their eyes stitched shut to stop them from giving out signs of distress; these were known as stool pigeons, and gave rise to the American expression for a gullible person set up to be cheated. It was an important business, for the pigeons were widely used to feed slaves.

What is astonishing to us as we look back is that hardly any effort was made to stop the killing. Even when it was clear that the birds' numbers were shrinking drastically, the great slaughter continued. Humans were unable to accept the idea that humans could cause extinction. The idea that we had to take responsibility for our actions with regard to nature was shockingly new and, to most people, unacceptable to the point of being meaningless.

By the time there were serious concerns about their extinction, it was too late. John F. Lacey, a Republican Party congressman, brought in what is considered to be the US's first wildlife protection laws in 1900, with a ban on interstate trade in illegally killed wildlife.

But by this time the vast flocks of passenger pigeons were gone and the remains of the wild population were vanishing fast. It seems that the last few passenger pigeons were unable to thrive. Everything about their way of life depended on being one of many; to be one of a few just didn't work. And besides, it wasn't just the hunting; for all that this was the main reason for their extinction. There was also much destruction of habitat: the woodland which provided them with food, shelter and nesting places was being destroyed to create more land for agriculture.

It was a battle and it was about numbers. The devastating superabundance of the passenger pigeon is a thing of the past: the devastating superabundance of human beings was, at the time of the extinction, a thing of the future. But, when it comes to impossible numbers and unprecedented biomass, we humans now have all the records to ourselves. It's our planet and we can do what we like with it. The way we should act on that truth remains an area of debate.

The last passenger pigeon died in Cincinnati Zoo in 1914. Her name was Martha.

There are various events in the history of humanity that reveal the power of our species: controlling fire, inventing gunpowder, harnessing nuclear power, creating vaccines, putting a human on the moon – but in any such list we should include the extirpation of the passenger pigeon. If we can wipe out the most numerous bird that ever existed, then nothing is beyond us.

FORTY-FIVE

TSETSE FLY

'As flies to wanton boys, are we to the gods; they kill us for their sport.'

Shakespeare, *King Lear*

An adult tsetse fly must take a blood meal every three days. They live entirely on the blood of vertebrates, and these prey items include humans and their domestic animals. Like mosquitoes (see Chapter 23) they are the vectors for pathogens that cause a number of dangerous diseases to all the creatures they bite, including sleeping sickness, which is invariably fatal in humans if not treated. Tsetse flies are, then, creatures to inspire fear and loathing in our own species.

But tsetses are also a touching example of maternal care, a rare thing in an insect. They are largely responsible for the vast wild areas that still exist in sub-Saharan Africa, which are the places you are most likely to find tsetse flies. These great national parks are beloved across the world and bring so many tourist dollars into developing nations: for them we must thank the tsetse fly.

Tsetse flies are related to houseflies – they are both calyptrate muscoids – but tsetse flies are noticeably different at rest, because their wings overlap completely, with a scissor action. The second point of difference is harder to see but quite easy to experience: a sharp, forward-pointing proboscis, equipped with a bulb. That is the device that allows them to make a living: they insert it into your flesh and suck. It's a sharp, hard, clean bite, a good deal more painful than that of a mosquito, but it's all over in half a second: for most people there is none of the painful itching that comes from the anti-coagulant in the mosquito's saliva.

They are called tsetses from their name in a Bantu language, Tswana. Africans will pronounce the word exactly as it is spelt, but those of European extraction tend to pronounce it 'tetsy'. Unlike most insects, tsetses lay a single egg at a time, and the female retains it in her own body: and also retains the resulting grub after it has hatched. She will carry and feed it through the first three larval stages, nourishing it with a milky substance that she secretes. She then produces – though 'gives birth' seems a more appropriate phrase – a grub which at once buries itself in the ground and pupates, to emerge a short time later as an adult tsetse. Unlike mosquitoes, males and females both drink blood; it's mostly the males that bite humans. The females go for bigger game.

Beautifully adapted: the tsetse fly is a masterpiece in its way.

Both can carry trypanosomes, which can cause sleeping sickness in humans and diseases that affect domestic animals: for example, nagana in cattle and horses causes fever, lethargy and build-up of fluids. The prevalence of tsetse flies and the lethal effects of the diseases they carry mean that farmers in much of Africa have been historically unable to keep and use draught animals. Mixed farming, which involves arable crops and livestock, has been impossible in very many areas. Here the only option is to till the land by hand: and the hoe is a much less effective instrument than the plough. The tsetse fly, even when sleeping sickness is factored out of the equation, is a cause of the continuing poverty in much of rural Africa; most of the thirty-seven countries in which tsetses occur are poor and debt-ridden.

Diagnosis and treatment of sleeping sickness in humans is complex and requires trained staff. The disease has been in decline since effective modern treatment became available in 1970. In 2009 the number of cases reported dropped below 10,000 for the first time in fifty years; in 2016 the total number was 2804, according to figures from the WHO.

The tsetse fly population was given a huge boost in 1887, when Italian colonists introduced cattle infected with rinderpest. This is a disease endemic to Asia: the African cattle had never met it before, had no resistance and died in great numbers. It's estimated that 5.5 million cattle died, and to make things worse the period of disease coincided with a drought. There was a great famine, and the desperate conditions that existed were exacerbated by outbreaks of smallpox, cholera and typhoid. In these terrible times British colonists took hold of Kenya, and Germany seized what is now Tanzania.

And as the cattle died, there was nothing left to graze their extensive pastures. These scrubbed up, became bushy and reverted to wooded savannah: which was perfect breeding ground for tsetses. Once the tsetses were established, they kept the cattle out, but the native and more resistant grazing animals lived on. Areas affected in this manner include the Masai Mara, the Serengeti, the Okavango Delta and the Kruger: perhaps the four most famous national parks in the world.

It's often said – and it certainly feels that way – that the landscape of the great parks of Africa is unchanged since man first walked upright. But they have indeed changed, as landscapes under the hands of humans – and the mouths and hooves of their cattle – tend to do. The current landscape reflects humanity's comparatively recent absence. Some of the world's greatest treasures, then, are things we owe to the tsetse. Tsetses have been called the world's best game rangers; Norman Carr, who started wildlife tourism in Zambia, used to say: 'Thank God for the tsetses – without them we'd all be cattle farmers.'

FORTY-SIX
DUCK

'Why a duck? Why a no chicken?'

Chico Marx, *The Cocoanuts*

Some of the non-human animals we domesticated have played a central part in the development of humanity. Of those already met in these pages, dogs, chickens and cattle stand out, and we shall come to the horse in a few pages. Other species have played a comparatively minor role, but one still full of significance. Ducks have always been something of an acceptable bonus to humanity, at least in Europe and North America. But in Asia they have played a much larger part.

Once again we are not talking about a precise biological term. Ducks are members of the family Anatidae, which includes swans and geese. They're all waterbirds, and we tend to call the smaller ones ducks. Not that we're at all consistent in this: the shelduck and the Egyptian goose are quite closely related and roughly the same size. Still more loosely, we might refer to many unrelated waterbirds – grebes, divers (loons) and rails (moorhen, coots, etc.) as ducks.

Wild ducks can be found all over the world, right into the sub-Antarctic: they nest on the island of South Georgia, already visited on these pages in the context of both rats (see Chapter 25) and albatrosses (see Chapter 43). They include the South Georgia pintail, which is found nowhere else.

Most of the species we call ducks are either dabbling ducks (collectively Anatini) or diving ducks (Aythyini). Both of them, when feeding on the water, duck their heads beneath the surface, and that is why they have been given their name. Dabbling ducks duck their heads beneath the surface but remain visible, heads down, tail in the air:

All along the backwater
Through the rushes tall,
Ducks are a-dabbling,
Up tails all!

This is from a poem written by Ratty in Kenneth Grahame's *The Wind in the Willows* and, as a rough description of the dabbling species in action, is surely unimprovable.

The diving ducks, as you would imagine, duck their entire bodies beneath the surface. It follows that they can exploit deeper water than the dabblers, so, at least when they are feeding, you tend to find dabblers like mallards close to the shore and diving ducks like tufted ducks out in the middle. There are other groups of what we normally call ducks, including the sawbills, who use their serrated beaks to catch fish (these include mergansers, goosander and smew) and the stiff-tails, like ruddy duck and white-faced duck.

Is this quite wise, Jemima? Illustration from Beatrix Potter's The Tale of Jemima Puddle-Duck *(published in 1908).*

And it's a shock, I know, but they don't all quack. In fact, the only duck that makes the sound that we think of as a proper duck's quack is the female mallard; the male makes a quite different rasping sound. The prettiest duck noise is made by wigeons; they converse in sweet whistles and the French call them *siffleurs*.

Wild ducks have been a favourite target for shooters since the invention of the shotgun. One of the most significant conservationists from the dawn of the Environment Movement was Peter Scott, whose first love was not conserving ducks but killing them. (More on Scott in Chapter 69.) After his conversion, he founded what is now the Wildfowl & Wetlands Trust. His television programme *Look!* was broadcast between 1955 and 1969 and did a great deal to recruit me, among many others, to the cause of conservation. 'Every man should have a cause, even if it's only bloody ducks,' he said – or at least so I was told. His widow Philippa was unable to confirm that, but, as a moment of self-deprecation in a man who achieved great things, it sounds singularly appropriate. He named his second daughter Dafila, which is the no-longer-used generic name for pintail, a particularly lovely duck. Dafila became a painter of wildlife subjects including, and perhaps especially, ducks.

We have had domesticated ducks in our lives for at least 4000 years; probably much longer. The process of domestication took place at least three times, perhaps more than that: in Egypt and Asia, probably first in what is now Malaysia, and in South America.

Now here's a fact. There are getting on for 150 species in the family Anatidae, but only two of them have been domesticated. The Muscovy duck is native to Mexico and Central and South America, and remains a comparative exoticism. The ancestor of all other domestic ducks is the mallard. Mallards are familiar to most people in the world: the males with brilliant green heads, the females a modest brown with an iridescent blue flash (the males also have this) on the wing. Their natural range is across North America, Europe and Asia, and they have been introduced to many other countries.

They are common in most places where there is open water. They tolerate a wide range of conditions and diet, which is always a useful trait in a creature selected for domestication. They are dabbling ducks – up tails all – and feed, when in the water, by using their beaks as sieves, pushing out the water and retaining the nutritious stuff, in the manner of the whales met in Chapter 8, but rather than krill they eat plants, worms, insects, small fish and molluscs. They will also take plant matter on land.

The ancient Egyptians caught wild birds with a net and bred from the resulting birds. The ancient Romans kept ducks. Under the Ming dynasty (1368–1644), the Chinese developed Peking duck, force-fed on grain to make them fatter. The process of domestication changed the nature of the ducks, as domestication

Faintly comic: Ducks, hanging scroll by Ren Yi (1840–95).

almost invariably does. Wild ducks are essentially monogamous (if sometimes unfaithful); domestic ducks are uninhibitedly polygamous. Wild mallard drakes defend their territories aggressively; domestic ducks are far less stroppy.

Many breeds of domestic duck have been developed from the mallard ancestors, a lot of them plain white, which, as said before in these pages (see Chapter 22 on pigeons), is a bad colour for a prey species, because it stands out from the background. Domestic ducks will happily breed with wild mallards if they get loose, and around human populations domestic/wild hybrids are common – often the drakes have an exaggerated white collar, and are sometimes known as vicar ducks.

It is comparatively easy and cheap to keep chickens; harder and more expensive to keep ducks. It follows that, in Europe at least, ducks have been traditionally considered rich man's food: haute cuisine rather than peasant cooking. The breast and thighs are the cuts most often eaten, and duck served with orange is considered a classic of Western cooking.

The situation is different in the warmer parts of Asia, where ducks can be kept in rice paddies. The combination of duck-keeping and rice-growing as an integrated piece of farming goes back for more than 600 years. The ducks clean, weed, remove pests and fertilize the paddies, and can of course be harvested in their turn, while the females give an almost daily gift of an egg. This style of farming went out of fashion with the introduction of industrial chemicals in the

1960s, but, as the ecological cost of such experiments has been shown to be considerable, it's coming back into fashion. The various breeds of runner duck are ideal for this: they can be marched briskly to the paddy fields of choice, where they tend to remain, as they have been bred to be flightless. These breeds have become very popular in the developed world because their walk and general mien give humans pleasure; there has been a great rise in the keeping of domestic ducks, not for hard-nosed nutritional reasons but for the pleasure they bring while still alive.

The feathers of ducks are also exploited. Those that lie beneath the impermeable outside or contour feathers are called down, which is the finest insulator in nature. Down has 'loft': the ability to expand from a compressed state and to trap air. Ducks can sit comfortably on freezing water because their contour feathers keep them dry, and beneath them the down feathers – which only work when completely dry – keep them warm. Eider ducks use their own down to line their nests, and this can be collected by humans; we still refer to a bedcover as an eiderdown. But most of the down we use today comes from domestic ducks and is a by-product of the meat trade; 70 per cent of it comes from China.

We take a traditional delight in the comic qualities of ducks: the way their waddle resembles an obese human, their self-preoccupation, their quacking. A British psychologist, Richard Wiseman, has researched a thesis that ducks are the best animals for human humour. Groucho Marx's signature walk was called the duck-walk; the routine, quoted in this chapter's epigraph (see page 210), is triggered when Chico misunderstands the word viaduct. 'If you're going to tell a joke involving an animal, make it a duck,' Wiseman said. And as the duck said to the waitress, put it on my bill.

There is a long tradition of comic ducks, of which Donald is merely the most famous; he began his career for Walt Disney in 1934. In Prokofiev's *Peter and the Wolf*, the duck's part is quacked by the oboe, who continues quacking inside the wolf. Perhaps the most sinister duck story is *The Tale of Jemima Puddle-Duck*, by Beatrix Potter, in which the silly Jemima, dressed in bonnet and shawl, is more or less seduced by the cynical and flattering fox, until she is rescued from her self-created disaster by the sheepdog Kep.

Ducks haunt the game of cricket: a batter who is dismissed without scoring is out for a duck. The usual explanation is that a duck's egg looks like a zero. So does a chicken's egg, but no doubt the comic nature of ducks comes in here. Australian coverage of cricket on Channel 9 used to show a comic cartoon figure called Daddles the Duck whenever a batsman was dismissed for nought. The same notion of the duck's egg occurs more obliquely in tennis, in which the word for nothing is 'love'; it's usually accepted that this comes from the French for egg: '*l'oeuf*.

FORTY-SEVEN

KANGAROO

'Skippy, Skippy,
Skippy the bush kangaroo!
Skippy, Skippy,
Skippy, our friend ever true!'

Eric Jupp

There was a bit of the poet in Charles Darwin. His way of working was to back up the most colossal leaps of the imagination with years of grinding study: first the insight, then the proof. He also had a knack of finding the thought that sums everything up. Here's what he wrote in his notebooks of 1836 about Australia:

A little time before this I had been lying on a sunny bank & was reflecting on the strange character of the animals of this country as comparisons to the rest of the world. An unbeliever in anything beyond his own reason might exclaim: 'Surely, two distinct Creators must have been at work; their object, however, has been the same & certainly the end in each case is complete.'

The continent of Australia – the landmass of Australia plus Tasmania and New Guinea – is remote from all others, the most isolated continent on Earth. That isolation is the reason for the uniqueness of Australian animals. This is the oldest as well as the most isolated continent, and life continued there for many millions of years without outside influence. It was another of those sleeping influences on Darwin's thinking: if you reject the two-creators hypothesis, what are you left with?

Most of the mammals native to Australia – that is to say, not brought there by humans – are marsupials: creatures who give birth to tiny, almost grub-like creatures who then continue their development, nourished by their mother's milk, in pouches. (The rest of us, apart from the egg-laying monotremes like the platypus, see Chapter 11, are placental mammals.) The kangaroo is the classic marsupial. The red kangaroo is the largest species, with large males standing 6ft 5in (1.95m) tall, but the mother gives birth after a gestation period of just thirty-three days. The resulting creature, called a neonate, has functioning forelimbs but only little stumps for hind legs. Once it has reached its mother's pouch it will then live there for the next eight months, spending more and more time outside, and after leaving for the last time will continue to suckle for another four months.

There are native placental mammals in Australia: bats have been living there for 15 million years, and rodents for around 5 million, presumably arriving by rafting, as unwilling and unwitting passengers, on fallen vegetation. But most Australian mammals, and all the older-established ones, are marsupials. (Humans have introduced placental mammals including the feral dog or dingo, red fox, rabbit, horse, water buffalo, dromedary and several species of deer.)

Kangaroos caught the human imagination as soon as Europeans became aware of them. The British like to believe that Captain Cook 'discovered' Australia in 1770, even though the place has had a human population for 65,000 years. Cook wasn't even the first European; Spanish, Portuguese and Dutch vessels got there first, and the place was initially known to Europeans as New Holland. The Dutch navigator François Pelsaert described a wallaby in 1629.

But it was Cook who brought the kangaroo into European consciousness. A specimen was shot when Cook's ship, the *Endeavour*, put up for repair for seven weeks on a beach near modern Cooktown in northern Queensland, after suffering damage from the Great Barrier Reef. Sir Joseph Banks, the trip's naturalist, and Cook both noted the event in their diaries. Banks wrote: 'To compare it to any

Triumphantly recognizable: The Kongouro from New Holland *by George Stubbs (1772).*

European animal would be impossible as it has not the least resemblance of any one I have seen.' He also praised its 'excellent meat'. It was Banks who chose the name, which was adapted from one of the Aboriginal languages. (There is a story, or rather a myth, that kangaroo means 'I don't know', in an Aboriginal language, the supposed reply when Banks asked for its name.) Two skulls and two skins eventually made it back to London on the *Endeavour*, where descriptions of the deer that hopped like a frog met with a certain cynicism.

Banks persuaded George Stubbs, the great horse artist, to paint a kangaroo from the skin and his description: it was the first, and for twenty years the only, image of the kangaroo in Europe. It is pretty accurate, though the ears are too big and the feet wrong. And it was a puzzler right from the start. It didn't fit into Linnaean classification; George Buffon declined to include it in his *Histoire Naturelle*. It was more popular with the public than with scientists. The first live kangaroo made it to London in 1791, and you could go and admire it for one shilling (£4 in 2020 values). James Boswell records Dr Samuel Johnson performing an imitation of a kangaroo: 'Nothing could be more ludicrous than the appearance of a tall, heavy, grave-looking man, like Dr Johnson, standing up to mimic the shape and motions of a kangaroo. He stood erect, put out his hands like feelers, and, gathering up the tails of his huge brown coat so as to resemble the pouch of the animal, made two or three vigorous bounds across the room.'

There are four species of kangaroo, using the term to mean the largest members of the Macropod family: the eastern grey, the western grey, the antilopine and the red. They are all open-plains herbivores, using the tactic of speed rather than concealment to avoid predators.

Since kangaroos were first known in Europe it was conventional to deal with their oddness, and the oddness of all Australian animals, by considering them as primitives: creatures left behind in the evolutionary arms race. It was certainly the idea I grew up with: I did a school project on Australian mammals when I was ten, reading that marsupials were almost a degenerate form of life, an aberration, unlike the line of placental mammals that gave rise to wonderful us.

But I have yet to hear a woman eight and a half months pregnant claiming that the way we placental mammals give birth is plainly and obviously superior to that of the kangaroo. Subsequent studies of the way kangaroos move show that, far from being inferior, kangaroos have the most sophisticated method of terrestrial locomotion that has been developed.

Grazing kangaroos move in what's called a pentapedal form: using tail, hind legs and forelegs to shuffle along. But when they stand up and hop – they are the only large hopping mammals on Earth – they are startlingly efficient. Red kangaroos can cruise at 12–16mph (20–26km/h); they can hit a top speed of 43mph (70km/h); and they can sustain a speed of 25mph (40km/h) for 1.3 miles (2km).

They are able to do this because of the large and highly elastic tendons on the hind legs. These operate on the principle of stored energy; a stretched rubber band has stored energy. In other words, kangaroos get free energy from the very action of bounding: they get power from the twanging of their tendons as well as from their muscles. It follows that movement, and in particular acceleration, is much less effort for a kangaroo than it is for a human or for a horse. Each hop is synchronized with the breathing, for added efficiency.

We will look at marsupial predators in the following chapter, but it's generally agreed that these were less fearsome than the big cats and the canids among placental predators. In other words, though it helps kangaroos to be fast, they don't have to be *that* fast. But Australia is a large and dry continent, and the distances between food resources in the arid centre of the continent can be considerable. Kangaroos have evolved a stunningly efficient method of commuting from one meal to the next.

The singular nature of the kangaroos, along with the isolation of their homeland, has meant that, perhaps more than any other species on Earth, the kangaroo represents a nation. You find a kangaroo bearing the Australian coat of arms (in partnership with an emu); there are kangaroos on the currency, the Kangaroos is the nickname of the Australian rugby league team. People attending cricket matches, especially those against their former colonial masters England, often carry green flags bearing a gold kangaroo and/or inflatable kangaroos. An Australian vernacular expression for someone a little mad is 'a kangaroo loose in the top paddock'.

Kangaroos have mostly prospered from human developments in Australia. Forest clearance for the grazing of sheep and cattle has created kangaroo habitat, and the introduction of artificial water holes for the stock has also helped kangaroos. They have a single stomach, unlike cattle and sheep and other ruminants, but they can regurgitate food for a second chewing. This is relatively strenuous, unlike the easy process of chewing the cud, and so it's less frequent.

There has been a surprising discovery related to the kangaroo's grazing: unlike cattle (see Chapter 7) kangaroos don't give out methane; they retain the methane and convert it into energy. If the same ability could be transferred to cattle, the process of climate change would be slowed down: so it has become an area of research.

There is a traditional chippiness among Australians, though it's far less acute than it used to be. It is based round what is called the Cultural Cringe: the art of making Australians feel bad about their presumed lack of culture, their essentially primitive nature. The kangaroo can be seen as the perfect rebuttal of this: what you thought was a primitive creature is, in fact, the most sophisticated mammal on the planet.

FORTY-EIGHT
THYLACINE

*'Can it be true? Has Thylacinus been seen alive? And in mainland
Australia, not Tasmania? I so want it to be true.'*

Richard Dawkins, 2017

Octopuses developed intelligence quite separately from us vertebrates (see Chapter 37). In the same way, as already mentioned, insects, birds, bats and the extinct pterosaurs all developed flight by completely different evolutionary routes. To recap: the same solution to the same problem was reached by radically different routes: convergent evolution, or just a convergence.

Australian mammals offer an enthralling way to study this phenomenon. The old-established Australian mammals are all marsupials, and they created an evolutionary radiation all of their own, quite separate to that of us placental mammals. We have already seen how kangaroos fill the role of antelopes in grassy plains, though they look very dissimilar (see Chapter 47). But some of the

Marsupial wolf: thylacine by John Gould for The Mammals of Australia *(published 1845–63).*

marsupial solutions are strikingly similar to those of the placentals: mole and marsupial mole; rabbit and rabbit-eared bandicoot; flying squirrel and flying phalanger; groundhog and wombat; mouse and marsupial mouse.

And then the meat-eaters. There was a large cat-like Australian predator that went extinct more than 40,000 years ago, which weighed up to 285lb (130kg). It had powerful jaws and shearing carnassial teeth and is usually referred to as the marsupial lion. There was also a dog-like predator, pretty similar to a wolf, and that was the thylacine or, a trifle confusingly, the Tasmanian tiger. It went extinct in 1936, fifty-nine days after legislation was introduced to protect it.

It is most often seen these days on bottles of Cascade lager, which is brewed in Tasmania. These show a lively drawing, based on that of John Gould in *The Mammals of Australia*, published in three volumes between 1845 and 1863. The thylacines come across as lively, lean-jawed creatures, looking dog-like without being dogs. They have a pattern of stripes on their backs, which explains their misleading nickname.

Richard Dawkins wrote, in his book *The Ancestor's Tale*: 'Zoology students of my generation… had to identify 100 zoological specimens as part of the final exam. Word soon got around that if ever a "dog" skull was given, it was safe to identify it as *Thylacinus* on the grounds that anything as obvious as a dog skull had to be a catch. Then one year the examiners, to their credit, double-bluffed and put in a real dog skull.'

So, marsupials and placental mammals can be startlingly similar even though they are separated by 65 million years of evolution. It's usually reckoned that the most recent common ancestor of marsupials and placental mammals lived just after the collision with the meteor that killed the dinosaurs. The thylacine itself is reckoned to have evolved about 4 million years ago, and it was found all over the Australian continent: that is to say, the mainland, New Guinea and Tasmania. But by the time the Europeans got there – Captain Cook arrived in 1770 – the thylacine was completely or very nearly extinct everywhere except Tasmania.

The reasons for the extinction are not clear, but it is likely that the establishment of human populations had something to do with it. It is a delicious and prevailing myth that, before the Europeans arrived, local populations of native humans across the world lived in perfect harmony with nature: across the Americas, in sub-Saharan Africa, in Australasia. Alas, it's not true: human hunting methods that included fire and driving herds over cliffs, and human alterations to the balance of species and the landscape, all made for conditions that not every species was able to survive. Australia has lost 90 per cent of its large terrestrial vertebrates in the last 40,000 years. Human habitation of Australia is normally reckoned to have begun 60,000–40,000 years ago. Not a coincidence.

Dead all right: studio portrait of Mr Weaver with thylacine taken in the late 1860s.

But the thylacine hung on in Tasmania: and yet very little is known about what it actually did. It was about wolf size, but no one is sure if it was a pursuit or an ambush predator. It was probably nocturnal or at least crepuscular (preferring dawn and dusk). Its main prey items can only be guessed at: some have suggested that it took only small prey, but there have been other suggestions that groups of thylacines could manage quite big mammals. A suggested list of target species has all the magic of Australia's apparent alternative creation: kangaroos, wallabies, wombats, potoroos and possums; and probably some birds – perhaps including the flightless Tasmanian emu, which was hunted to extinction – by humans, not thylacines – by 1850.

Thylacines probably hunted in small family groups; the extent to which these hunts were cooperative, with individuals taking different roles, is again not known. They seemed active and social animals, which normally makes for a high level of intelligence, but with thylacines we have no idea how smart they were. All we really know are a few anatomical details, like the fact that the males as well as the females had pouches, the males using theirs as a kind of scrotum-cosy.

The only clear thing about thylacines in the nineteenth and early twentieth centuries is that humans hated them. So they got rid of them. They established a deliberate policy of extinction, and it worked. In 1886, the Tasmanian government offered £1 per head (£128 in 2020 prices) for a dead thylacine and 10 shillings (half that amount) for a pup. In all there were 2184 bonuses paid, though it's reckoned that many more thylacines were shot than were claimed for.

The traditional view is that shooting did for the thylacines, but there were other probable causes. These include competition with feral dogs and a distemper-like disease. There were also radical changes in habitat as farmland was created, with a resulting decline in the thylacines' prey species. Thylacines were shot out; they were also crowded out.

There had been talk in 1901 about introducing legislation to protect the last thylacines, but talk it remained. The last known wild thylacine was shot in 1930. Another, sometimes remembered as Benjamin, was trapped in the wild and kept at Hobart Zoo in Tasmania, where it died three years later in 1936, of exposure after it had been left outside with no cover on a bitter night. The zoo authorities' response was to allocate £30 to purchase a replacement; the money was never spent. You can view footage of Benjamin in his cage on YouTube: moving jerkily, but looking dog-like and, as is the way of dogs, quite incomplete on his own. He looks like an unhappy dog with stripes.

Since the loss of the thylacine there have been tremendous efforts to find them again. An expedition to Tasmania of 1960 claimed the finding of scats (droppings) and footprints, and to have heard sounds like descriptions of thylacine vocalizations. But an expedition of 1972, equipped with camera traps, found nothing, and the species was declared extinct by the IUCN in 1982.

Since then there have been literally hundreds of unconfirmed sightings, many of them on mainland Australia. There are wobbly enigmatic videos available online, so you can make up your own mind. The only certain thing is that the thylacine is something that people very much want to see. The thylacine genome was sequenced in 2009, and there have been talks of cloning the animal back into existence, perhaps using eggs of the related – though not closely – Tasmanian devil. The thylacine survives – on the beer label, the Tasmanian coat of arms, Tasmanian vehicle licence plates – and in a million regrets.

There have been five great extinction events in the course of evolutionary history, of which the most famous is the Cretaceous-Tertiary Extinction, the one that finished the dinosaurs. The others are, going backwards from this, the Permian-Triassic, the Late Devonian, the Ordovician-Silurian and the Ordovician Extinctions. It is now widely accepted that we are in the middle of the sixth extinction. The thylacine was part of that process: others preceded it, like the marsupial lion, and others have followed, like the baiji, an animal we shall meet in Chapter 74.

FORTY-NINE

CROCODILE

'How doth the little crocodile
Improve his shining tail,
And pour the waters of the Nile
On every golden scale!'

Lewis Carroll, *Alice's Adventures in Wonderland*

The thylacines stand for extinction; crocodiles stand for endurance. Crocodilians are older than humanity. They evolved with the dinosaurs but, unlike them, they survived the meteor collision and the Cretaceous-Tertiary extinction event of 65 million years ago. Crocodiles are not merely as old as time: they are as old as Deep Time. For humans they have always represented a timeless menace: one that operates on a timescale quite alien to our own. I have many times sat above the Luangwa River in Zambia, looking down at gatherings of a hundred crocodiles, sometimes more than that, and always thinking that the same sight had been available to observers during the Cretaceous era.

I have also, on two or three occasions, seen crocodiles kill large mammals. It seems that they have only two speeds: stop and warp. In a lagoon full of floating logs with eyes, there was a sudden splash and a puku antelope that had been enjoying a somewhat nervous drink would worry no more. (For similar experience with a baby elephant see Chapter 54.)

Crocodiles are found on every continent in the tropics. The order Crocodilia contains three separate families; the gharials, with their long thin snouts, are the easiest to distinguish. The difference between alligators and crocodiles is more subtle, but if, when the mouth is closed, you can see teeth from both the upper and the lower jaw, you can be confident that you are looking at a crocodile. There are normally reckoned to be fourteen species of these.

One of the reasons for their success is their ability to catch and consume a wide variety of prey including fish, molluscs, crustaceans and birds. But what captures human imagination is their taste for large mammals. Crocodiles are semiaquatic animals, but they are adept at feeding on land-dwelling mammals, primarily because most species of mammal must come to the water to drink. Most notably, there are occasions when mammals must cross rivers: the wildebeest crossing of the Mara River in Kenya on their annual migration is one of the most famous

Annual feast: Nile crocodile at the Mara River crossing in Kenya.

wildlife spectacles in the world. The wildebeest always succeed in getting across, but they always leave a few of their number behind.

The Mara River crocs have the opportunity to grow huge. They do so by feeding only once a year, in any demanding sense of the term, but then hugely: a different rhythm of existence. The slow metabolism of ectothermic creatures, who borrow their heat from elsewhere rather than generating their own, makes their style of life possible. They are not cold-blooded in any literal sense: the sun will often heat their blood to temperatures above our own. It's the source of the heat that matters, not the heat itself. Crocodiles take their heat from the sun and so their need for energy is less hurried. A croc can wait, in a manner no human can empathize with.

Crocs will kill and eat a good few species of mammal – including humans. Humans must also find water in order to live: they also look for fish to eat and that brings them to the water for long periods. In some areas this means living with crocodiles, creatures that specialize in striking unseen at mammalian intruders into the water. The two largest species are the most lethal to humans: the Nile crocodiles of Africa and the saltwater crocodiles of eastern India, Southeast Asia and northern Australia. It's impossible to offer accurate figures for annual deaths, since few of these are reported, but we are talking in hundreds. We humans have our origin on the savannahs, where hunting and gathering certainly included a fair amount of fishing as well as a daily need for water. It's certain that, along with lions, crocodiles were early and efficient predators of our ancestors. No doubt that explains why we have such a deep response to them.

They kill us, but, as with snakes (see Chapter 28), they're not creatures we can easily relate to. In a sense we see ourselves as fair game for a lion, which is a fellow mammal, even a fellow hunter – but crocodiles, well, with their unreadable faces, their endless day-after-day stillness and their explosive turn of speed, are beyond easy understanding.

Crocodiles bite. They are probably the best biters that ever lived. If crocs had a motto, it would be: I bite therefore I am. Once a croc takes hold, it doesn't let go. The bite has been measured. A human chomping into a piece of steak can generate a bite of 0.1 tonne-force (890N). A hyena bites at 0.5 tonne (4450N) and a lion probably a bit less. A medium to large saltwater crocodile has been measured at 1.7 tonnes (16,400N). So when you scale that up to a 20ft (6m) croc – and Nile and saltwater crocodiles can both make that extreme of size – you are talking about 3.5 tonnes (34,250N).

It's all in the muscles of the jaw, which are rock hard and immensely powerful… and yet the muscles for opening the jaws are comparatively weak. Crocs trussed for transport (usually for crocodile farms, where they are kept for the leather that can be made from their skins) can be made safe with a couple of rubber bands, improvised from the inner tubes of bicycle tyres, strapped around the jaws. Of course, putting them on in the first place is the challenge…

Crocodiles are highly skilled ambush predators, with good eyesight, including excellent nocturnal vision, and strong senses of smell and hearing. All these senses operate well both in and out of the water. Crocodiles can be highly social, gathering in big numbers, apparently for the pleasure of each other's company; after all, with a slow metabolism there's not great competition to find the next meal. Nile crocodiles often share their rivers with hippos; when these die in the water, from age, wounds from fighting among themselves and when depleted by the slim pickings at the end of the dry season, the crocs will devour them at leisure and in numbers. Their bite is for holding, rather than cutting; they lack

god of gangga

Gunga

the carnassial or shearing teeth of mammals in the order Carnivora (and also some marsupials, as we have seen). So when they take a mouthful, they twist their entire body to wrench the morsel free, spinning their belly up through 360 degrees. They are equipped with a palatal flap, which keeps the water out of their throats when they bite below the surface. They can replace each of their sixty-four teeth as many as fifty times in the course of a lifetime, which can be up to seventy-five years.

They go in for maternal care: a mother will tenderly hold babies in her mouth, protecting them, on occasion feeding them. Crocs are vocal: up to twenty sounds have been recorded, each with a different meaning: so crocodiles have the beginnings of language.

They are creatures that have inspired awe across the centuries, and it is not surprising that crocodiles have made their way into the religious life of humans. Sobek is the ancient Egyptian crocodile god, sometimes shown as a man with a crocodile's head; among his many attributes he could protect you from the dangers of the River Nile. In the Hindu pantheon, Varuna rides a creature who is half crocodile, while his consort Varuni rides an unambiguous croc. Ganga is both river and god, and is often portrayed riding a crocodile.

The notion that crocodiles weep hypocritical tears of grief as they kill and devour humans has been traced to the *Bibliotheca* of Photios I of Constantinople, a ninth-century patriarch. It is found later in *The Travels of Sir John Mandeville* in the fourteenth century: 'These serpents slay men, and they eat them weeping.' Shakespeare liked the image and used it four times. Othello says:

If that the earth could teem with women's tears,
Each drop should prove a crocodile.

Rudyard Kipling also wrote about crocodiles. In *The Second Jungle Book* a jackal, an adjutant stork and a vast crocodile discuss life, and, this being their livelihood, death. The story is called 'The Undertakers'. In *The Elephant's Child*, perhaps the best known of the *Just So Stories*, the crocodile tempts the elephant's child into the water and seizes him by the nose: by the time he has been rescued (by the bicoloured python-rock-snake) the nose has stretched into a trunk.

But the most dismaying crocodile in literature appears in J. M. Barrie's *Peter Pan*. Having consumed the hand of Captain Hook, the crocodile likes the taste so much that he follows Hook for ever. He does so while ticking loudly, having swallowed a clock. It's a Freudian treat: a castrating crocodile is also warning us about the passing of time (which in the end diminishes us all). Our fear of crocodiles is a soul-deep thing.

Companion of gods: Ganga riding a crocodile, nineteenth century (artist unknown).

FIFTY

HORSE

'A horse! a horse! my kingdom for a horse!'

Shakespeare, *Richard III*

Humans learnt to consume, to ride and to drive horses – and horses drove the process of civilization. Human history, up to the beginning of the twentieth century, is the history of the domesticated horse. It's been suggested that rather than measure human history by means of the material we worked with – stone, iron, bronze – it should be measured by the uses to which we put the horse. In chronological order, dating from the starting point of each – though many categories overlap and some still continue – these constitute the ages of consumption, utilization and status, herding, chariot, cavalry, agriculture, carriage and leisure.

Horses have been so important to us that horsey terms are buried in the English language like the fossils: so when we've done the groundwork we'll take a breather, and, if he's not too long in the tooth, and doesn't get the wind up, we'll put him through his paces and then we'll be home and dry. I'm sure he'll be full of beans, as soon as he's got the bit between his teeth. I'd offer you a few more examples, but no sense in flogging a dead horse…

Odd to think that the horse evolved from Eohippus, a many-toed beast about the size of a Labrador dog. But this was the line that gave rise to the horse (a process that was once considered a classic lesson in evolution, second only to humans, demonstrating once again the incorrect but delightful notion of evolution's drive for perfection). Like kangaroos and some species of antelope, horses evolved to become grazers of the open plains, reliant for their safety on a sharp set of senses and a neat turn of speed. Horses have some of the largest eyes of any vertebrate, set high on their heads so they can still see all about them while grazing with their heads down. They also have very good hearing from loftily placed ears. Modern horses are more wary and spooky than usual when the wind is blowing, not because the wind itself is frightening but because it compromises their hearing. They evolved their long-limbed shape – effectively running on a single toenail – which combines a good takeoff speed with the ability to cover a lot of ground at an economical pace, so they can travel from one food source to another in arid country. (The same traits evolved in a different form in kangaroos, see Chapter 47).

Onward: Napoleon Crossing the Alps *by Jacques-Louis David (1801). The horse is probably Marengo.*

Horses were first domesticated in Kazakhstan about 6000 years ago; the Botai culture domesticated only horses and dogs, ignoring all other species. Horses were killed and eaten long before that, and it is likely that humans first sought proximity to horses so they could eat them.

When were horses first used for transport? For that is what changed everything. Horse skulls that appear to show wear from a bit – the part of a bridle that is placed in a horse's mouth for steering and for brakes – have been found among

Botai remains dated 5500 years back. Horse bones in Botai graves can be even older, indicating a close – and presumably highly valued, even status-conferring – relationship between humans and horses.

Scholars argue about whether horses were ridden first, or used to pull vehicles. It looks as if the answer is pulling, because chariot remains have been found that predate any riding equipment. I'm not a scholar but I'm certainly a horseman, and I'd plump for riding every time. You don't need any equipment at all to get onto the back of a horse: it's not even difficult, if the horse is small, unlike the big breeds that, by means of artificial selection, we produce today. If you lived close to half-tamed horses, were used to having them around for their meat, and if you were young and spirited, it would be the work of a second to vault onto a horse's back. It would be the work of a second to come off as well: but no one's claiming this was an overnight process. Finding a way to turn a horse's head to steer and otherwise exert control, and so inventing the bit, was not an exotic breakthrough, and using your legs to push a horse in the desired direction is instinctive. Riding was surely a very natural development.

The process of domestication and, to this day, the methods for preparing a horse for ridden and driven work are made possible by the social nature of the horse. Horses live in herds and they love to communicate with each other, by means of body language and through sound and touch. They have a deep respect for hierarchy, so taming a horse is about the human achieving a position of dominance. This can be done the shouting, bullying way, or with decency and consideration. You can choose whether you want to be the sort of leader a horse has to follow, or the one he wants to follow.

Once humans had horses to command, it was like the acquisition of superpowers. A human can walk at 4mph (6.5km/h) and so can a horse. But a horse can move into an economical trot and sustain it for long periods: travelling at 8mph (13km/h) is no big deal for a horse. A fit trained horse, given appropriate breaks along with food and drink, can cover 100 miles (160km) in a day. A man can carry a pack of around 50lb (23kg), a pack horse can carry 200lb (90kg). And a horse pulling a wheeled vehicle over decent going can pull more than twice his own weight: say a tonne.

The horse offered great distance and load-carrying ability and radically extended the human range of possibilities. Humans now had a greater range in their search for food. They could expand the time-honoured method of driving herds of animals over a cliff or into traps; if you can do this on horseback you can be a great deal more effective. Horses also made it possible for humans to make contact with a greater range of fellow humans. Horses opened up much greater possibilities of trade – and, of course, war. And horses revolutionized warfare, offering speed, mobility and terror.

For many centuries, warfare with horses was conducted by chariot. At Westminster Bridge in London there is a statue of Boadicea (or Boudicca), a local ruler in Britain who led an uprising against the occupying Romans. She is there in her chariot of war, with two dramatically rearing horses and a couple of daughters. People were well accustomed to riding horses for many purposes, but when going into battle they preferred a chariot. You can read all this up in the classic work, Xenophon's *The Art of Horsemanship* of 360 BC, in which he discusses the chariot, the ridden horse, the care and the psychology of the horse. It is still good sense today.

The reason for this preference for chariots is straightforward: the stirrup wasn't invented until the Early Han dynasty in China, that is to say, around 2000 years ago. Stirrups were widespread in China by the fifth century, reaching Europe in the six and seventh centuries.

Riding without stirrups is one of the first things you are taught, if you seek to ride in the classical tradition (in America called English saddle). It's an essential exercise, but in such circumstances you are never wholly secure: vulnerable to sudden changes of speed and direction. Were you to ride a horse into battle in such circumstances, you would not be safe, and you could easily be dragged off. To make things worse, you couldn't strike a solid blow; as any boxer will tell you, you need a firm base from which to make a meaningful strike.

The failure to invent stirrups is not a stupid oversight. You can't use stirrups unless you have a saddle with a tree: that is to say, a means of spreading the weight of the rider over a wide area of the horse's back. If your weight is acting over a small area, you will soon cause discomfort, bruising and chafing, and will rapidly lose the horse's cooperation. A saddletree is a relatively complex invention, and it changed the nature of warfare. It created an elite class of mounted warriors, and in Europe they were called knights. In the UK to this day, high-achieving men are rewarded with the title of knight: they are as important as a horse-powered warrior.

The notion of the horse-borne warriors as superior to all other kinds persists in the cavalry regiments of the British Army, even though they now use tanks rather than horses. The needs of equestrian warfare are part of classical riding to this day: you are taught always to mount from the left, though seldom informed that your reason for doing so is to avoid getting tangled up in your own sword.

Horses were also bred for less glamorous work, and were increasingly used to power agriculture from around 1000 years ago. Big strong horses with biddable dispositions were bred to pull ploughs and carts: faster than oxen, with no need to take a break to lie down and chew the cud. Following this, horses were used for transport: goods and people, and also information (by post, or mail). Such items could be shifted prodigious distances in a very short time by horse transport. This required smooth roads over which wheels could run: human communications

Lively old morning: mounted soldiers from the Parthenon Frieze (223–437 BC).

were revolutionized. When canals were constructed in the eighteenth and nineteenth centuries, a single horse could pull 30 tonnes – ten times more than was possible in a cart.

Towards the end of the nineteenth century, human civilization was still the civilization of the horse. Cities were powered by horses: private carriages, carriages for hire or rent, the public omnibus. People were employed as crossing-sweepers, to keep the roadway clear of droppings so that people might walk from one side to the other. Our wars, our agriculture, our communications, our trade: all these things were powered and made possible by horses. Horses were the world. The symbiosis between humans and horses was what made the world possible.

And then we invented the internal combustion engine.

No one inventor can claim the credit for it: it was a protracted development, fuelled by a series of new and better ideas. Towards the end of the process, you find names commemorated in the history of the automobile: Gottlieb Daimler, Karl Benz and Rudolf Diesel. In 1903 more than 30,000 automobiles were produced in France. The 6000-year history of the horse as the driver of human civilization was over. Now we could move goods, people and information at incomparably greater speed and greater convenience. We could go to war with self-propelled guns that could take a thousand bullets and not die. We still measure the power of our engines by their horsepower, but horses no longer power our lives.

Why, then, have horses not gone extinct? Why are there still horses in our lives? Horses serve very few practical purposes any more, certainly not in the developed world. Why do we still have them, at considerable expense and greater inconvenience?

We have them, it seems, because we like them. We have them for sport; some of these sports are based on the training of military horses, including dressage, show jumping and eventing. Racing one horse against another has surely been part of human life since we first tamed them, and that still amuses us. American traditions of riding have given rise to sports based around ranching for cattle. But more than that, people ride horses, hire horses and keep horses because they enjoy the horsey life. Being around horses is perhaps a nostalgia for the wilder lives we left behind a few generations back; it remains a fact that, of all the species we have domesticated, horses are the only one in which a degree of wildness is actively encouraged. The speed, agility and athleticism of the horse is an inescapable part of the fun of the horsey life.

This has led to a revolution in the way we train horses. Since we no longer have horses because we need them, but because we want them, there is an urge in many places to treat them in a different way: as partners rather than slaves; by means of persuasion rather than violence; by a shift of weight rather than a whip. These various techniques are usually referred to collectively as natural horsemanship, with pioneers like Monty Roberts and Pat Parelli.

In 2013 there was a scandal in the UK and Republic of Ireland when a range of frozen meat products was found to include horsemeat, labelled as beef. From the shock and horror that greeted these revelations, you'd have thought that people had been forced into cannibalism. The incident revealed the deep and powerful feelings we still have about horses. A practice acceptable in many countries is in others an abomination.

It is hard if not impossible to find historical or fictional horses who are not wholly admirable. Alexander the Great had Bucephalus, the horse that (of course) none but he was able to tame. General Robert E. Lee rode Traveller, Napoleon Bonaparte had Marengo, shown in one of the most famous equestrian portraits of all time, and the Duke of Wellington had Copenhagen. Noble horses abound in fiction: Black Beauty, Gandalf's Shadowfax in *The Lord of the Rings*, the Houyhnhnms in *Gulliver's Travels*, Bree and Fledge in *The Chronicles of Narnia*.

Human history is the history of the horse. We no longer need them, but we still have them in our lives. We may be tame modern humans but we have wild human ancestors, and many people find their lives incomparably richer and wilder for the presence of horses. I shall now bring this chapter to a close; I must go and ride my horse.

FIFTY-ONE
OWL

'*I am a brother to dragons, and a companion to owls.*'

Job 30:29

The planet Earth, seen from space. Seen from space *at night*. You can find images of this on the internet, and they are more startling even than the famous 'Blue Marble' picture of the Earth in daylight, taken from the Apollo 17 mission in 1972. There the planet looks innocent, vulnerable, a thing to be cherished.

The Earth at night shows us a different planet entirely: one not vulnerable but utterly conquered by its dominant species. It's not the darkness that shocks: it's the light. Cities close to cities surrounded by towns: and all of them pouring their light into space, sending a beacon out into the blackness: we're here! We've done it! We're in control!

Humanity's long war with darkness has been won. The fear that accompanied us throughout our history, from the days when our ancestors walked with lions, is now redundant. True, many people still feel an atavistic terror of night, and it's a souvenir of our ancestors – but in the developed world, all a modern human has to do for reassurance is to reach out and turn on the switch.

Light. The first thing that God made in Genesis, the symbol of wisdom and knowledge: God was the light that shineth in the darkness and the darkness comprehended it not, as it says in the first chapter of the Gospel according to John. Now light has conquered the Earth, to the extent that we now sometimes talk about light pollution, too much of a good thing. Look at the planet at night from space and no one can be in any doubt that the epoch of humanity, the Anthropocene, has been established.

The picture has become something of a glass half-full, glass half-empty personality test. Where do you want to be on that planet? Do you need to bask in the light? Or do the places of darkness and mystery hold a fascination? Or, more likely, both?

If you can find a place of darkness on the Earth, it's always worthwhile pausing and listening. Chances are that the darkness will find a voice. It's heard in every horror film: a voice that tells the audience about night, fear, secret enemies, a graveyard, perhaps a monster on the prowl – and then, echoing from the blackness, the voice of an owl. The owl stands for our hatred of the darkness, our soul-deep terror and our equation of darkness and evil.

The dark side: Escena de Brujas (*Witches' scene*) *by Goya (1746–1828).*

There are around 200 species of owl and they are found on every continent apart from Antarctica. Most of them are creatures of the night, or of the margins of dawn and dusk. They are comfortable in the dark as we are not: and that gives them their sinister quality. Along with a reputation for wisdom in some cultures: owls know dark secrets that we creatures of the day cannot penetrate.

They have the talons and hooked beaks of hawks and eagles but are not closely related: this equipment – the first for killing and the second for butchering – is another convergence. They have an upright stance and a wide face with two-eyed vision: so they look a bit like us, and that sort of resemblance always compels our attention. They have disproportionately large eyes; these are fixed in their sockets, as they are with many birds, but, whereas most familiar birds have something close to 360-degree vision with eyes set in the sides of their heads,

Athena's bird: also symbol of Athens. Silver tetradrachm (four-drachma coin) from the fifth century BC.

owls have a narrow angle of binocular vision. When they wish to see beyond this narrow angle they must turn their heads and they can do so with some elasticity: they are capable of turning through 270 degrees. Their wide dished faces gather and focus sound, as we humans can do by cupping our hands behind our ears. This face adds to the owl's human appearance.

There's a huge difference in size between owls: the smallest, the elf owl, is 5in (13cm) long, 1oz (28g); while a female Eurasian eagle owl (females are larger than the males in many species of owls) can reach 28in (70cm) long, 9lb (4kg). Most species are equipped with asymmetrical ears: the minute difference in the time the sound takes to reach the two ears allows the owl to get a very accurate crossbearing on something he can't see: useful when striking from the cover of darkness.

But perhaps the most important trait the owls have evolved is silence. I have occasionally startled owls from my barn and, each time, the owl has flown over my head within touching distance, but without making a sound. You hear a whir from the passing of much smaller birds, but not from an owl. The complex feathering of the wing makes for silent flight. This is useful, because the prey can't hear an owl coming. But it is also essential because an owl in flight can't hear the sound of his own progress: there's no interference when he is sound-fixing a prey item. That makes for accuracy, and that's what allows owls to make a living. Owls that specialize in catching fish, like Pel's fishing owl in Africa, a giant tangerine-coloured bird, don't have silencers fitted, there being no need; fish can't hear them underwater and the owls can't hear the fish. They hunt purely by sight.

Owls mostly live in the night, but like everyone else they need to communicate with each other. They can't do it by sight, and so they do so by sound, and that is why owls have far-carrying voices. The graveyard scene with the hooting owl is a cliché of the horror film: but owls really do make their presence felt at night. If you live near owls, you are frequently reminded of their presence. They call to communicate with each other – I'm here, where are you? – and they also call to announce, claim and maintain a territory. The tawny owl of Europe will mostly make the famous 'to-whoo' sound territorially; the 'to-whit' is generally a contact call. So the owls don't go 'to-whit-to-whoo'; they go 'to-whit' *and* they go 'to-whoo'.

In most civilizations owls are birds of ill-omen: the worst. The Aztecs and the Mayans saw them as birds of death and destruction. In Zambia if an owl calls from the top of a hut you destroy the hut, because if you don't someone will die in it. Farther north, the Kikuyu people see them as birds of death and bringers of bad luck. Ovid, in his great work *Metamorphoses*, writes of the transformation of Ascalaphus, custodian of the orchards of Hades: 'So he became the vilest bird, a messenger of grief, the lazy owl, sad omen to mankind.' Pliny the Elder wrote that the owl was 'the very monster of the night… when it appears, it foretells nothing but evil'. Shakespeare liked owl symbolism; in *Macbeth* the witches put a 'howlet's wing' into their hell-brew; and, just before his murder, King Henry VI tells Richard of Gloucester: 'The owl shrieked at thy birth, an evil sign.'

There is some let-up to this: owls are lucky in Japan, and even stupid in some parts of India. Athena, goddess of wisdom, has an owl for her symbol: the little owl to this day has the scientific name *Athene noctua*. Perhaps the point is that the owl, comfortable in darkness and with something of a human face, knows not just about darkness but about death itself, going beyond mere human wisdom.

The owl himself is seldom seen as evil. It is what he foretells or embodies that troubles us. He is darkness, he is fear, he is death: the oldest fears of humankind, fears we have felt since our earliest ancestors walked the savannahs. To combat that fear we have made a planet of light, in which true darkness is now genuinely hard to find. It's not so easy to find owls, either. We have banished darkness: but we can never banish fear. It is as much a part of the human condition as music and love.

FIFTY-TWO
SEAL

'Where billow meets billow, there soft be thy pillow,
Ah, weary wee flipperling, curl at thy ease!
The storm shall not wake thee, nor shark overtake thee,
Asleep in the arms of the slow-swinging seas!'

Rudyard Kipling, 'Seal Lullaby'

It's one of those for-all-time images: a baby seal, in its white fluffy coat, a face of 24-carat cuteness looking up in bewilderment at the human towering above him, the human wielding a special club called a hakapik, his face filled with the dispassionate eye-on-the-ball concentration of a top baseball player or cricketer. It was taken by Kent Gavin in 1968, and, like the rhino in Chapter 39, it made the front page of the British *Daily Mirror*. The headline: 'The Price of a Sealskin Coat'.

Violent interaction between human and non-human animals is usually acceptable so long as we don't have to see it: the modern slaughterhouse is only the most obvious example. In the 1960s, with increasing means of communication, better travel and better equipment, we could now view images of seal hunting when sitting at our breakfast tables. In that extraordinary decade, when there was music in the cafés at night and revolution in the air, the world was full of people trying to find a better way of living. This image was a universal cry: surely, we can do better than this?

Seals have been hunted for centuries and they are hunted now. They have been hunted for meat, for fur, for leather and for the oil that can be gained from their carcasses. This has been used for soap, treating leather and, most importantly, for lighting: and in the previous chapter we looked (perhaps by artificial light) at the importance of light for humans.

There are usually reckoned to be thirty-three living species in the subgroup known as Pinnipeds, which belongs to the order of Carnivora. These include walruses, who bottom-feed on the molluscs they dig up with their tusks; sea lions, which have visible external ears and can reverse their hind flippers to move on land with comparative agility; and 'true' seals, with no external ears who must shuffle or hump themselves along when on land. The Pinnipeds range in size from the Baikal seal, 3¼ft (1m) long and weighing 100lb (45kg), to the southern

Saviour of the seals: illustration for Rudyard Kipling's 'The White Seal' from The Jungle Book *(published in 1894).*

elephant seal, up to 16ft (5m) in length and weighing 7100lb (3200kg), the world's largest living member of the order Carnivora.

Most seals generally feed on fish, but the leopard seal specializes in penguins and other species of seal. The seals' adaptations for long and deep diving include fully collapsible lungs, insulating layers of blubber and, sometimes, fur: it follows that, though some species can cope with warm conditions, they are essentially creatures of the cold. You can see a colony hauled out on a beach and basking as if they were sun-worshippers at a Mediterranean resort, but in temperatures that would kill an ill-equipped human.

Humans have mostly liked seals, or at least enjoyed watching them. Many seaside resorts offer trips to the seals: in Norfolk, in England, several hundred people a day set off on a flotilla of boats to visit the seal colony on Blakeney Point. Seals were kept in ancient Rome: Pliny the Elder remarks that they are amenable to training. Roman citizens could watch seals salute with their flippers and bark in response to certain commands. They could also watch seals in animal shows at the circuses: there are records of polar bears being set loose on seals for the entertainment of thousands. Probably these were mostly Mediterranean monk seals, now an Endangered species.

The trainability of seals is a result of their sociable disposition, and their need to please someone higher up in the dominance hierarchy. They have been trained to amuse humans for at least 2000 years: so much so that they define a human who responds unthinkingly to the requirements of others. In the Clint Eastwood film *The Dead Pool*, a character talks of the difficulties of getting a drug-addicted singer to do his stuff: 'Don't worry. Johnny's like a trained seal. Throw him a fish and he'll perform.'

And while in recent years there have been increasing ethical concerns about performing seals in circuses and as public entertainments, this is nothing compared to the concern generated about seal hunting in Canada after the picture that shocked the world. The harp seal, the principal target, had declined to 2 million individuals by 1960, and it's estimated that they were killed at 290,000 a year between 1952 and 1970.

Many people spoke out against seal hunting, including Paul and Linda McCartney and Brigitte Bardot. There were responses from governments involved at either end of the trade. The Canadian government introduced quotas in 1971, and later made it illegal to kill 'whitecoats': that is to say, harp seals under two weeks old and still in possession of that white coat – in other words, you can't kill them until they've stopped looking cute. In 1983 the EU banned the trade in whitecoat pelts. The USA passed a ban on the hunting of all marine mammals, making some exceptions for hunting by indigenous people.

Commercial seal hunting continues today in Canada, Greenland, Namibia, Norway and Russia. The harp seal population was estimated at 5.5 million in 2005. In other words, objections to the hunting of harp seals cannot be made on the grounds of their likely imminent extinction. But there are still objections, passionate and committed. People object, not because we are likely to lose a species of seal but because they dislike the perceived cruelty of the business.

In 2005 the conservation organization WWF reported that opposition to seal hunting was based on 'emotion, and on visual images that are often difficult even for experienced observers to interpret with certainty. While a hakapik strike on the skull of a seal appears brutal, it is humane if it achieves rapid, irreversible loss of consciousness leading to death.'

It all depends on your definition of humanity, presumably. In 1978 the great marine conservationist Jacques Cousteau said: 'The harp seal question is entirely emotional. We have to be logical. We have to aim our activity first to the endangered species. Those who are moved by the plight of the harp seal could also be moved by the plight of the pig – the way they are slaughtered is horrible.'

There are two points arising here. The first is that there is an emotional dimension to the notion of saving any species from extinction. The second is that emotional responses are not by definition invalid, as anyone who has ever married is in a position to point out.

There is a difference between conservation and welfare. We have looked at the success of charities that raise money to prevent cruelty to domestic donkeys (see Chapter 20), something that sometimes irritates people trying to raise money for the conservation of rainforest. You can attract donors by showing moving images of starving children; you can do the same with images of seals being slaughtered with clubs – and it's much easier than trying to raise money by talking about biodiversity and carbon sequestration. Organizations like the International Fund for Animal Welfare look for donations so they can try to prevent perceived instances of inhumane treatment of non-human animals. Seals highlight this conservation/welfare divide with immense clarity. Few in conservation would say that animal welfare is a bad idea: few in welfare would argue against conservation. But they're not the same thing.

Hunting was probably what did for the Caribbean monk seal, which was hunted by Columbus in 1494. The species was thought extinct in the middle of the nineteenth century, but a small colony was found off the Yucatán Peninsula in 1896 – and at once the hunting began again. The last individual was seen in 1952. The last Japanese sea lion was seen in 1974.

Despite the slaughter, harp seals are still surviving in numbers. The danger to their future is probably not the men with the hakapiks; it's climate change and the increasing problems they face as the polar ice melts. That's an issue we will meet again in these pages.

FIFTY-THREE

BOWERBIRD

'It is not the language of painters but the language
of nature which one should listen to.'

Vincent van Gogh

Yes, but is it art? A question a million students debate every day, in seriousness and in jest. Must it be beautiful, need it involve craft, is one form of art higher than another – and now let's consider John Cage's *4'33"* and the *Fountain* of Marcel Duchamp.

Or we could take this a step farther than a silent musical work and the appearance of a urinal in an art gallery and turn our attention to the bowerbird. For if art is the creation of a work intended to have an effect on someone else – particularly if it is the deliberate creation of something beautiful intended to move a person other than its creator – the bowerbird is an artist.

There are twenty – some say twenty-eight – species of bowerbird, and they live in Australia and New Guinea. They can be found across a range of habitats: rainforest, eucalyptus woodland, acacia and scrub. All but three species build bowers: structures that have no function other than pleasing females. They are built to attract a female, to encourage a visit, to give the male a chance to display and show off his mate-worthiness and, if all goes well, to mate. In such circumstances the female will then fly off, build a nest, raise a brood of chicks and do so all by herself.

The male's job, then, is creativity. Nothing less and nothing else. He will construct a bower and make it beautiful. Bowers fall into two main types. Some species build a maypole bower: a conical structure built around a good upright, usually a sapling. Others make an avenue bower: twin lines of twigs set vertically into the ground, like the nave of a church. These structures are not used for nesting; for which they would be useless. They are entirely about aesthetic pleasure: the pleasure of the female who will select her mate on the basis of what the bower offers, and – we can perhaps speculate – the agony and the ecstasy of the creative artist.

These bowers would be remarkable enough in themselves if that was the whole story, but of course it is not. The decoration of these small enclosed spaces is still more extraordinary. The male collects objects, not at random but carefully chosen

Portrait of the artist: satin bowerbird by Elizabeth Gould (1804–41), for John Gould's The Birds of Australia.

– and very cunningly arranged. These objects are collected in their hundreds, for bower-building is the only job the male has to do, apart from surviving and copulating. The list of objects chosen to decorate these places is endless: shells, leaves, flowers, feathers, stones, berries, caterpillar faeces, fungus – and plenty of manmade objects including strips of plastic, coins, nails, rifle shells, pieces of glass. There are outlier examples of decoration, including a toy soldier and a plastic elephant. Male bowerbirds will kill beetles and use the bodies solely for decoration: as the scientist and author Jared Diamond said, they are 'the most intriguingly human of birds'.

The maintenance of these bowers is an all-consuming task: new objects are brought in, fading objects replaced. And then there is burglary and vandalism: both to inflict and to suffer. The cockbirds routinely raid each other's bowers, trash them a little and steal objects that catch their eye. In this way, the maintenance of a beautiful bower is a fairly honest reflection of dominance.

How did these structures come about? Why did the cocks start making them; why do females find them appealing? This is a subject of much debate: one intriguing suggestion is that the nature of the enclosed space makes a forced copulation impossible. A female inside a bower is safe from rape. In other words, she can make her own choice on her own terms. It's been observed that the vigorous, borderline-violent displays of the male will sometimes visibly intimidate a visiting female, which is the last thing the cockbird wants. So he will rein himself in, and perform in a more restrained and decorous fashion, and with luck and good judgement the female will stay where she is.

Females take their time, and visit a number of bowers before making their choice. Studies show that bowers are decorated to match female preferences: knowing what another individual wants is an essential skill for a male bowerbird, and a fairly advanced one. Often many females will select the same male, the male with the top bower, leaving all rivals frustrated, vowing to do better next time round. Age and experience matter.

It's not just about amassing a collection of objects. The arrangement is critical. Males will spend hours of every day tending to them, getting the display just right. Human experimenters have rearranged the objects when the bird is away: his response on return is invariably to set about restoring the bower to the way he wants it. Random and chance-driven effects are not to the bowerbirds' taste: they are conscious and deliberate artists. They seek and achieve control over their works.

One way some species do this is by creating what's called a forced perspective: creating an illusion that makes the bower look immense. They arrange the objects, often pebbles, so that the largest are at the front and the smallest at the back, creating an optical illusion of extensiveness. Again, if a human interferes and moves a pebble to spoil this, the returning bird will at once set things aright.

Observers have established that the bowers with the steepest size gradient – the most obvious and dramatic perspective – are the most attractive to females. They stay longer at such bowers, and the presiding males have greater mating success.

Some species are partial to certain colours. The satin bowerbirds have a thing for blue, their own colour; it's been suggested that this makes the cock look bluer and therefore better. Bowerbirds tend to favour objects that reflect ultraviolet light: like many species of birds (but unlike us) they can see into the ultraviolet parts of the spectrum. As a result, the tail feathers of the common parrot species crimson rosella and a certain type of plastic bottle top are both coveted objects.

The opportunity to operate a system based on so many hours of artistry comes from the richness of the environment in which bowerbirds live: it wouldn't work unless it was relatively straightforward for a female to bring up a brood unaided. When a species has the opportunity to consider tasks beyond surviving and becoming an ancestor, strange unexpected things happen. Bowerbirds are also remarkably long-lived for passerines, the great group of perching birds that includes sparrows, crows, thrushes, tits and chickadees. That expanse of time allows male birds to gather experience and knowledge – of what females want and how best to supply this – so that if all goes well after barren early years they will be, in their maturity, a success.

Darwin was most intrigued by bowerbirds and wrote about them in his work of 1871, *The Descent of Man, and Selection in Relation to Sex*. We will look at some of his conclusions in Chapter 79 on peafowl.

And that leaves us asking all kinds of questions about art and human uniqueness. Do birds have an aesthetic sense? Is what they do really art? The common-sense answer to both questions is yes, but others prefer to skew the definitions to assert the old notions of human uniqueness. As skylarks and nightingales sing (see Chapter 40) and create music, so bowerbirds create visual art. In all these examples, the males create and the females judge: and on that judgement she makes her choice about who shall father her offspring.

David Rothenberg wrote in *Survival of the Beautiful*: 'Bowerbirds, say biologists, are unique. There is perhaps no other species besides human beings that is known to create things so beautiful beyond their function, structures that we have a hard time calling anything else but art.'

FIFTY-FOUR

ELEPHANT

'Nature's great masterpiece, an elephant,
The only harmless great thing…'

John Donne, 'The Elephant'

When you're walking on the savannahs of Africa you don't want to come across a pride of lions unexpectedly. Lions are about fear: they are creatures that actively wish you harm. Elephants are a quite different matter. They are no problem at all so long as you give them respect. You must wait on their convenience, that's all. If they're in your way, you either find another way, or you wait until they have moved on. You must defer to them. Deference to a non-human animal seems incongruous in the twenty-first century, but it was essential to the survival of our ancestors. The notion that elephants deserve not fear but respect survives in our modern minds.

There are three recognized species: the Asian elephant, the African bush or savannah elephant, and the African forest elephant. All three use their noses as hands: two fingers on the trunks of the two African species, and one on the Asian. The trunk is a fusion of the nose and the upper lip. It contains no bone, and 150,000 muscle fascicles. It is used for breathing, smelling, touching and creating sound; an elephant's sense of smell is reckoned to be four times better than that of a bloodhound, which is specially bred for its scenting powers. A big bull elephant can lift 770lb (350kg) with its trunk, and can also wipe his own eye with perfect delicacy. He can crack a peanut with his trunk without breaking the seed inside. A trunk can hold 15 pints (8.5 litres) of water.

The big ears of the Asian elephant and the enormous ears of the African species are for thermoregulation: the ear flapping cools the blood passing through them. Their incisors grow into tusks, which can be used as weapons and as tools. (Most female Asian elephants lack them.) If you look at an elephant's tusks, you will see that one tusk, generally the right, is more worn than the other: elephants are, if you like, either right-handed or left-handed. The male African savannah elephant is the largest living land animal on Earth: they can stand 10ft (3m) at the shoulder and weigh 13,200lb (6000kg). The immense size compromises their

With added smile: poster of Jumbo the elephant while at London Zoo (c.1865).

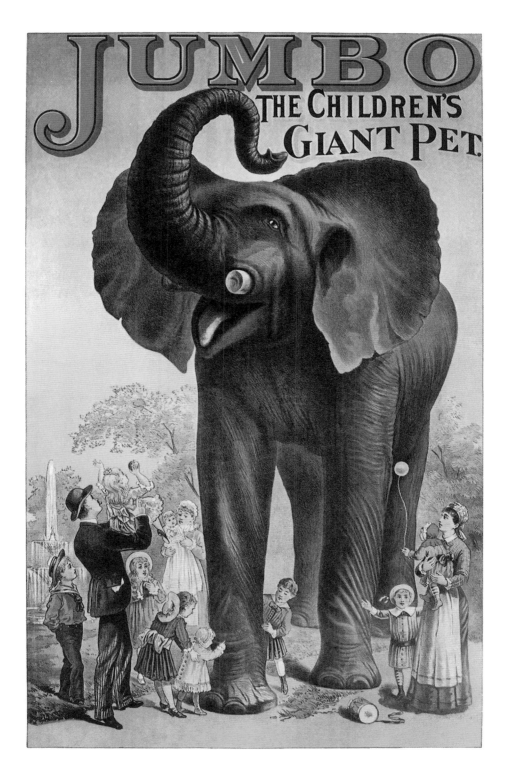

athleticism: they can never, even for a second, leave the Earth. Elephants can't jump, nor can they perform any gait with a moment of suspension: no trot, no gallop. That doesn't mean they can't move with some urgency when they need to: they can reach 16mph (25km/h) in a sort of Groucho Marx power-walk.

The human relationships with elephants changed with the invention of agriculture. When humans operated from a fixed place, it was no longer so easy to move out of the way when elephants were around. Crop-raiding by elephants has been going on for at least 10,000 years and remains a problem to this day.

Elephants are too enormous and too demanding to be considered fully domesticated animals, but tame elephants have been used for heavy work and for war. Tame elephants first worked in the Indus Valley 6000 years ago. There are still places where elephants have advantages over other forms of transport: when it comes to the steepness and softness of the going they can travel where no vehicle can. In war they have worn armour and their tusks have been equipped with points of iron or brass. They have been trained to impale the opposition, or to pass them up to the men on the elephant's back for slaughter. There are war elephants in the *Mahabharata*, which was written more than 2000 years ago. They were primarily a weapon of terror, most effective when surprise was also a factor. Countermeasures – elephants are not terribly manoeuvrable – eventually made them more of a liability than a strength.

The Carthaginian general Hannibal is famously associated with elephants. He set out to wage war on Imperial Rome, crossing the Alps to do so with thirty-eight elephants, along with 38,000 infantry and 8000 cavalry. He got the elephants as far as the River Po in Italy, but here the elephants succumbed to disease and many died. Hannibal, by this time blind in one eye, rode the last surviving elephant in triumph into Capua.

Elephants have always fascinated humans. They were kept in menageries in ancient Egypt, China, Greece and Rome. Pliny the Elder wrote about them with great enthusiasm: 'The elephant is the largest land animal and is closest to man as regards intelligence, because it understands the language of its native land, is obedient to commands, remembers the duties that it has been taught, and has a desire for affection and honour. Indeed, the elephant has qualities rarely apparent even in man, namely honesty, good sense, justice, and also respect for the stars, sun and moon.'

The elephant-headed Ganesh is perhaps the best-loved god in the Hindu pantheon: the lord of beginnings and the remover of obstacles. There is a story that the Buddha was a white elephant reincarnated as an exemplary human. Again, the pattern of human relationships with elephants combines both affection and respect. They are creatures we admire: creatures we would like to be, not just for their size and power, but for their grave demeanour and apparent wisdom.

They were must-have items in zoos: it seemed that all the wonder of non-human life could be summed up in the spectacle of an elephant. To gaze on an elephant was partly to experience a certain sense of human insignificance, along with a far greater feeling of human power: even beasts as great as this must, in the end, do what the ruling species wanted of them. Jumbo was the most famous of these elephants, coming to London Zoo in 1865, and was later sold to Barnum and Bailey's Circus. Two elephants, Castor and Pollux, were kept in Paris Zoo; they were killed and eaten during the siege of Paris in 1870. There are increasing ethical concerns about zoo elephants. They have been known to demonstrate what's called stereotypical behaviour: such things as mad swaying, compulsive pacing or endless fruitless fidgeting, all these things a response to boredom, stress and isolation.

Humans have been trading in the tusks of elephants for at least 3500 years. The ivory trade was so important that an entire country – Ivory Coast or Cote d'Ivoire – was named for it. Ivory was a major trading item, in many places closely associated with slavery; slaves carried the tusks. The elephants of northern Africa were hunted out 1000 years ago. In the nineteenth century, ivory was widely used for comparatively homely items: knife handles, balls for snooker and billiards, and piano keys. In Japan ivory is traditionally used for stamps for the elite to sign their name with. With increasing prosperity by the 1970s, everybody wanted one; Japan accounted for 40 per cent of the global trade in ivory.

The declining elephant population in India can be explained by habitat loss, at least to some extent, but rather less so in Africa. Eventually CITES made the trade illegal in 1990. The subsequent history of the ivory trade is complex, racked by politicking and corruption. Many African countries, notably Zimbabwe, South Africa and Botswana, have consistently argued for a legal trade, and there have been various amendments and backtracking and complex agreements and counter-agreements.

There was a legal sale agreed in 2008, when a great deal of stockpiled ivory was sold, mostly to Chinese and Japanese dealers. A few people got very rich very quickly. It's probably fair to say that the trade in ivory is completely banned and that ban is wholly effective, apart from cases where people have ivory to sell and other people have money to buy it. People argue that they should be able to make money from legally killed elephants; however about 80 per cent of the trade is in illegally killed elephants. Some say that the answer is a well-regulated trade in ivory; others that any trade in ivory creates demand and heats the market. The price of ivory continues to rise.

It has become an intensely polarized debate. The Zimbabwean position is that all wildlife should be exploited sustainably: it is the country's right and anything else is an imposition and unjust. The opposing argument is that the trade as it stands is neither sustainable nor moral.

The issue is made more complex by the discoveries about the nature of elephant life, mostly by Cynthia Moss. She revealed the intense nature of elephants' social life, built around small family groups led by a dominant female, the matriarch. She has written of the immensely strong emotional bonds that keep elephants together, and the complexities of their minds. Among her many revelations is the fact that elephants seem to understand the concept of death, another of those things that was previously believed to be unique to humans. She wrote of one incident: 'By nightfall they had nearly buried her with branches and earth. Then they stood vigil over her for most of the night, and only as dawn was approaching did they reluctantly walk away.'

I witnessed a disturbing scene in Zambia when a baby elephant was taken by a crocodile. The mother's distress – I assume it was the mother – was a terrible thing to see: she walked round and round through the herd screaming.

With increasing human populations there are increasing problems of conflict and of habitat destruction. Sometimes these can be controlled by electric fencing, and by fencing with chilli-impregnated string. Elephants can be shot with disintegrating balls of chilli juice and so turned away. A less confrontational approach is to encourage people to grow chillies as a cash-crop: one crop the elephants will never raid.

These problems of human conflict with elephants are more acute in Asia. The Wildlife Trust of India has been working to establish elephant corridors that link two good bits of remaining habitat. Sometimes people who live in these corridors – frequently traditional migration routes – can be persuaded to move out of these areas to a safer place, one free of elephants. At one such place I met a woman who had been inside her house while an elephant destroyed it. 'I kept very still and very quiet.'

Elephants are very large and very demanding of food and space. They also carry a valuable commodity on their person. It is, then, remarkable that we should still have decent populations of elephants, even if these are drastically reduced. In the 1930s the African population was 5–10 million, today it's less than half a million; a century ago there were around 200,000 Asian elephants, today there are fewer than 40,000.

There are still elephants in the world because people like them. People pay good money to come and see them, especially in Africa, and they play a significant part in the local and national economy; in India they are part of the religious life of the country. Drivers of trains who have been involved in accidental collisions with elephants are given help and counselling, for they have killed god.

The elephant that demanded respect from our ancestors now depends on the whims of humanity for its future. Like practically every other species on Earth – but in a particularly dramatic and generally enormous way.

Weapon of terror: The Battle of Zama, 202 BC, *by Giulio Romano (1492–1546)*.

The idea that an elephant never forgets is very dear to us; perhaps it comes from their slow and apparently thoughtful approach to life. But there is plenty of evidence to suggest that elephants possess good, long-term memories. Two elephants, thought to be strangers, greeted each other ecstatically when they were put together at an elephant sanctuary in the USA. It turned out that they had been in the same circus for a few months twenty-three years earlier. The elephant ethologist Iain Douglas-Hamilton was welcomed joyfully by an elephant he hadn't seen for four years. There is real survival benefit in memory. In Tanzania it has been recorded that, in times of drought, the older herd leaders – matriarchs – led their groups to distant, still-fertile places, which younger matriarchs had not experienced.

FIFTY-FIVE

PIRANHA

'You will see that my piranha fish get very hungry.
They can strip a man to the bone in thirty seconds.'

Blofeld, dialogue from the film *You Only Live Twice*

The British newspaper the *Sun* has never been noted for its restraint. The day I began the research for this chapter they ran a piece headlined: 'Mystery as man-eating piranhas found in lake in Doncaster'. We can all see the scene that's conjured up in those few words: the solid innocent citizens of this Yorkshire town walking proudly round their beloved lake: a sweet innocent child paddles, the water boils around her, she screams, an anguished parent leaps in to save her, and the water is churned into new frenzies by the ravening hunger of the most dangerous fish on Earth.

There's a scene in the James Bond film *You Only Live Twice* when Blofeld, stroking his white Persian cat, presses the toe switch that sends his assistant Helga Brandt – 'this organization does not tolerate failure' – into the water where she is devoured alive by her master's pet piranhas. It is an image we are all deeply familiar with: the waters that devour. There's a graphic description of such a death in the Bond novel of the same title: 'A mass of small fish were struggling to get at the man, particularly the naked hands and face… Once the man raised his head and let out a single, terrible scream and Bond saw that his face was encrusted with pendent fish… James Bond wiped the cold sweat off his face. Piranha!'

And while all this is very thrilling, it's not actually what piranhas do. The legend of the many-headed, multi-toothed shoals of cooperatively hunting piranhas reflects our ancient fear of the wild world, of what we can't see, of what might happen to us should we dare to step into an alien environment. But if you really did walk into a river full of piranhas you'd be unlucky to suffer anything worse than a bitten toe.

Piranhas – correctly pronounced pi-ran-yas – are freshwater fish found in South American rivers, lakes and floodplains. The term is usually used for four genera of fish, comprising around sixty different species. These include the pacus, which eat plant matter and have teeth uncannily similar to humans. But the rest – what we informally consider real piranhas – have highly specialized teeth. They have a single row of teeth in each jaw, and these are tightly packed and

Well armed: piranha species Pygopristis denticulata *by W. H. Lizars after an illustration by James Stewart from Robert Schomburgk's* The Natural History of the Fishes of Guiana *(c.1850).*

interlocking, flat and blade-like. They are effective at making a rapid puncture and shearing away a good mouthful; they have, pound for pound, the most powerful bite of all the ray-finned fish (which doesn't include sharks, see Chapter 14). The black piranha bite has been measured as, pound for pound, one of the most forceful bites of any vertebrate. We are right to be impressed by piranhas, but not in the traditional way.

Most species are confined to a single river catchment, like the Amazon or the Orinoco. The largest is, as it were, the default piranha, the red-bellied piranha, up to 20in (50cm) long. All species of piranha are omnivorous: they take plant matter like fruit, seeds and leaves. They also eat mammals, birds, reptiles, insects and crustaceans, but their main food item is fish. That's around 80 per cent of the diet in some species; with others the split is more even. They are eager scavengers when they get the opportunity.

The teeth of piranhas have been used as tools by humans: for carving wood, cutting hair and sharpening darts for blowpipes. The fish are not popular among local people, even though they are a useful food item: they steal bait, damage fishing gear, eat fish that have been caught and, when caught themselves, they bite the catcher.

The fascination with piranhas lies in their multitude: the way we believe that they gather in shoals for cooperative hunting, so that the small many can entrap the enormous one. The notion of death, not by one big monster but by many small things with the ability and the desire to work together to act like a single monster, is extraordinarily compelling.

And humans have been killed by piranhas. These incidents usually occur when the piranhas are stressed by hunger, by damage to their habitat, by human activity. Splashing attracts them, a possible explanation for the fact that children are the most frequent victims of these attacks.

In a lake in Palmas in Brazil there were 190 attacks by piranhas in six months – every one of them a single bite to the feet. That's jolly painful, but it's not death. Piranhas are not by nature insatiable; and nor – as the last record amply demonstrates – are they driven to frenzies as soon as there is blood in the water. They are pretty indifferent to blood. Nor do they form shoals for the purpose of cooperative hunting: observation has demonstrated that the main purpose of shoaling in piranhas is to avoid predation by cormorants, caimans and river dolphins (see Chapter 74).

The myth of the piranha can be traced to Theodore Roosevelt (see Chapter 34 on bears). He was an ex-president of the USA by the time he made an extended journey to the Amazon in 1913 for hunting, of course, and also for exploration. Naturally, everywhere he went people wanted to amuse and impress him. He was a man people wanted to please, and so, for his delectation, they invented the myth of the insatiable piranha.

They took considerable trouble over it, netting off a 100-yard (100m) stretch of river and, in the weeks before his visit, filling it with piranhas. They kept catching piranhas and, whenever they did, they chucked them into the netted-off sector. Naturallym they didn't feed them, so by the time Roosevelt arrived they had a river packed with an unnaturally high concentration of fish, all highly stressed and intensely hungry. The locals then regaled Roosevelt with all kinds of advice: don't, whatever you do, put so much as a foot or finger in that water... and, when Roosevelt cast doubt on this, they drove a poor old cow, sick and ancient, into the water – where she was, indeed, torn apart by the fish. Roosevelt was had. He was caught hook, line and sinker. And he wrote about the experience with profound awe in his 1914 book *Through the Brazilian Wilderness*: 'The head, with its short muzzle, staring malignant eyes and gaping, cruelly armed jaws, is the embodiment of evil ferocity. And the actions of the fish exactly match its looks… They are the most ferocious fish in the world… They will snap off a hand incautiously trailed in the water, they mutilate swimmers… they will rend and devour alive any wounded man or beast; for blood in the water incites them to madness.'

Many tall tales of ferocious and implacable foes of humankind have been told across the years. Many of them, like the sea serpent and Abominable Snowman, have been discarded in the face of evidence, or lack of it. But some of these persist notwithstanding all kinds of revealed truth. In the film *The Man Who Shot Liberty Valance*, the closing words are: 'This is the West, sir. When the legend becomes fact, print the legend.'

We have been printing the legend of the piranha for more than a century. We don't want the truth: it seems we need the legend of the boiling water, the swarming fish, the feeding frenzy, perhaps also the notion that those blade-like teeth are heading first for the victim's crotch. We need the illusion that nature is at war with humankind: perhaps we need that illusion more than ever, because it protects us from the horror of the truth: the situation is the exact other way round and in many places the war is as good as over.

Unwitting dupe: Theodore Roosevelt photographed in 1913 during the Roosevelt–Rondon Scientific Expedition to Brazil.

FIFTY-SIX
TITS AND CHICKADEES

'Feed the birds, tuppence a bag…'

R. M. Sherman and R. B. Sherman, from the film
Mary Poppins (1964)

Tits, titmice and chickadees are all in the same large family of Paridae: small birds built in the unusual fashion of having their feet at the top and their heads at the bottom. That, at any rate, is how we usually see them: inverted, helping themselves to food we have given them ourselves and almost near enough to touch. This is astonishing on two fronts: first because small birds usually fly away from humans, and second because we humans traditionally want everything for ourselves.

In chapter after chapter, this book logs the changing of human attitudes to individual species and to nature. The family Paridae demonstrates an almost ubiquitous human desire to reach out to non-human creatures and help them and, what's more, to have non-human creatures reach back to us in response.

No doubt we have always fed the birds. Done so deliberately, I mean, not just by creating opportunities with spilt grain and unwanted food. On the savannahs of Africa, I have tossed the odd crumb to see if a bird would come and pick it up, just for the fun of the thing. It's impossible to believe that our most ancient ancestors didn't do the same thing. As suggested earlier, it's likely that deliberate feeding was the precursor to the domestication of chickens and pigeons (see Chapters 29 and 22).

But in more recent times – and no doubt in earlier times as well – the payoff for deliberate feeding is a less tangible thing: pleasure. The pleasure of looking at birds close up, the pleasure of their movements, the pleasure of their negotiations and squabbles around a major food resource, and the pleasing illusion of their gratitude.

Of all the birds that come to garden feeding stations, the family Paridae is the most numerous and most gratifying. They are all pretty similar birds in shape: stocky and equipped with a short stout beak. They range in size from small to very small: between 4in (10cm) and 8in (20cm) in length. They are found right across the northern hemisphere, in the USA, the UK, across Europe and Asia into Japan, and in Africa, right down to the Cape. They are adaptable in both habit

Blue tit bonanza: A Feast of Strawberries *by Eloise Harriet Stannard (c.1900).*

and diet, like most species that have adapted well to human environments, though they are not urban scavengers but suburban exploiters of gardens, which provide a kind of ersatz, edge-of-woodland habitat. Broadly speaking, these birds feed on invertebrates in warm weather and seeds in the cold: this adaptability means they can stay in the same place all year round. Insect specialists must migrate from north to south, from cold to warm, during the winter: tits can hang around, changing not location but diet.

They are also natural problem-solvers, always eager to apply their sharp minds to exploit new sources of food. Researchers have created ingenious puzzles that require tits to perform a series of manoeuvres – like taking away a succession of rods, eventually allowing a tasty morsel to fall within beak range. The birds have consistently performed such tasks to an unexpectedly high level.

Until recent times, milk was delivered daily to households in the UK, in glass bottles with a lid of foil. Tits learnt to puncture these lids and drink the cream beneath. This behaviour spread rapidly across the country, demonstrating the birds' very high learning ability, as well as their capacity to live alongside humans. It was widespread for fifty years, dying out in the 1980s when people, now more health-conscious, no longer had full-cream milk delivered. It was cream that floated to the top, fat- and energy-rich, that the tits went for. Mere milk is no good for them; they can't even digest the stuff.

All the Paridae are naturally birds of the canopy, usually out of sight in woodland, though they tend to be very vocal. The garden feeding station is like a spell that conjures them into being: cavorting before your eyes and performing astonishing acrobatics in order to reach the delights offered to them. There is a strange intimacy in this relationship: a bird that makes itself visible just for you, the kind supplier of food.

They behave in a similar way across their range: forming noisy sociable flocks in the winter, splintering into loud territorial pairs during the breeding season. In natural circumstances each species will tend to forage in a different way: for example, the European great tit will work closer to the trunk than the smaller and more agile blue tit, which is more normally found on the tips of the outermost twigs: species side by side exploiting slightly different ecological niches. It's only when they come to the super-resource of the garden feeder that they come into competition. Here the smaller birds must defer to the larger, and the food-provider can witness competition between species, and between individuals of the same species.

The notion of feeding birds for pleasure has been with us a long time. It was recommended by the romantic Henry David Thoreau, in the American classic of 1854 *Walden; or, Life in the Woods*: 'To many creatures there is in this sense but one necessary of life, Food.' So why not help them out – and reap the pleasure of their presence and the pleasure of helping them out?

Bird genius: blue tit feeding on cream from a milk bottle.

In the desperately hard European winter of 1889–90, newspapers urged people to put food out for the birds: bread and kitchen scraps. In 1910 the British magazine *Punch* noted that feeding the birds was 'a national pastime'. In 1926 there was commercially available food to give to wild birds. These days food for wild birds is an international industry, nothing less. One British firm offers, among many other treats, mini suet pellets with berries, muesli with mealworms, peanut butter with mealworms, sunflower sticks, robin blend, duck and swan sprinkle food and birdcakes in the shapes of snowflakes, snowmen, hearts and Christmas trees.

It's reckoned that half the adults in the UK feed the birds. In the United States the figure is one-third of the adults or 55 million people, spending an annual US$3 billion. Feeding the birds is the most popular hobby in the United States, after gardening. As a species we are destroying nature at a frightening rate: at an individual level we can't get enough of the stuff.

And mostly it comes down to the Paridae, the birds that in North America will give an alarm call that sounds like 'chickadee-dee-dee'. They have, more than any other group, driven the urge to spend our hard-earned money on nature: for the direct consumption of wild creatures whose needs have traditionally been considered in direct opposition to our own. Here then, almost two centuries from the example of Thoreau, we humans feel an urgent need to cherish wild creatures: to take responsibility for them; to go out of our way to make the local environment equally suitable for human and non-human life. The bird table exists in a kind of no-man's land or perhaps an everybody's land: a little like the Christmas truce of 1914, when British and German soldiers held an unofficial ceasefire on Christmas Day as a break from trying to kill each other.

At the heart of our urge to feed the birds lies what the great American scientist and science writer Edward O. Wilson called biophilia: the tug exerted on us, the love impelled from us by non-human life: the feeling that without non-human life around us we are incomplete. Anyone who has patted a dog, stroked a purring cat or smelt a rose knows precisely what he means. The pleasure of seeing the Paridae descend from the trees to feed on stuff we have provided ourselves is biophilia in action.

All this seems to reveal a fundamental human goodwill towards the wild world: more than that, a fundamental need. When we look at a bird feeder, with three or four members of the family of Paridae hanging beneath, there seems to me to be a strange admission: maybe we've got it wrong in the way we run the planet. Maybe what we want in our lives is not less but more nature.

FIFTY-SEVEN
SPIDER

'Weaving spiders come not here.
Hence ye long-legged spinners, hence!'

Shakespeare, A Midsummer Night's Dream

We fear sharks to an irrational degree, but some species of shark can kill you (see Chapter 14 on sharks). Our fear of piranhas is based on an exaggerated conception of their potentialities, but they can certainly cause pain and, as we have seen, they have occasionally killed humans (see Chapter 55 on piranhas). We fear snakes, and some species can indeed kill you: we examined the figures for deaths by snakebite in Chapter 28.

We also know that very few spiders can kill a human and that relatively few species are even capable of causing pain. Of the 50,000 or so species that exist, only a handful – perhaps that's not quite the right expression – are worth the trouble of fear. And yet very many people do fear spiders, to the extent that they can't be considered in the human context without addressing that fear.

Spiders are in the phylum of arthropods, which includes insects. They come from the class Arachnida, which includes scorpions and sun spiders or solifugids: these are not spiders at all though superficially similar. The 'true' spiders are found in the order Araneae: eight legs and body divided into two parts rather than an insect's three. The front section is more than a head: the cephalothorax combines the front two parts of an insect's body all in one.

Spiders have fangs to inject venom and they bear spinnerets. These can produce seven different kinds of silk, which are used for at least ten different purposes. If we (some of us, anyway) weren't so worried about their effects on our own troubled minds we would revere all spiders for their life in silk. They can produce different kinds of silk for:

The outer rim of webs and for ballooning (see below)
Temporary scaffolding
Capturing lines of the web
Protecting egg sacs
Wrapping prey
Glue
Bonding threads together

Spinning a yarn: Arachne als Spinne *by Gustave Doré, for Dante's* Inferno *(1861).*

Spiders use silk for:

Capturing prey

Immobilizing prey

Making shelters for the purposes of reproduction

Ballooning – small spiders eject a long thread of silk, to catch the wind – they initially catch the electrical charge in the air – and take to the air for chance-driven migration

Food – silk is made from protein and some species eat the webs of other spiders; all web-making spiders will eat – reabsorb – their own silk when appropriative, for energy economy

Lining nests and wrapping eggs

Guidelines: they will follow a trail of silk

Emergency drop-lines: they can simply fall from their own webs when threatened, saved by their own silk

Alarm-lines: triggered by the vibrations of an approaching predator

Creating shelters to rest in

Trussing a female in preparation for sex

Wrapping nuptial gifts

Sex: a trail of pheromone-impregnated silk can be used to attract the opposite sex

Not all spiders use all those techniques, but they are all part of the great vocabulary of silk. Water spiders create a diving bell of silk, which they fill with air bubbles and use to rear their brood underwater. Bolas spiders use missiles of silk to capture prey. Other species carry nets of webbing and use them to ensnare moving prey; such is the elastic quality of spider silk that a spider can expand the net to ten times its size at the moment of striking. Trapdoor spiders build silk-lined tunnels with lids, and emerge from these to capture prey. Two spiders were taken into space on Skylab 3 in 1973 to see if they could make webs in zero gravity; they found it difficult at first but soon adapted. Spiders from the *Portia* genus demonstrate intelligence in their hunting techniques, seeking to outflank potential prey, showing the ability to learn and adapt when confronted with novel prey species.

Only one known species of spider is not a predator, and this, lovers of *The Jungle Book* will be delighted to learn, is *Bagheera kiplingi*, which feeds on acacia trees. Spiders range in size from the minute, *Palu digua* from Colombia with a body length of 0.15in (4mm), to some species of tarantula that have a body length of 3½in (9cm), with a leg span of up to 10in (25cm). (The camel spider, which became an internet sensation during the Iraq War of 2003, is a solifugid, not a true spider, and can have a leg span of 6in (15cm); more dramatic claims are based on exaggeration and distorted perspective in photographs.)

There were around a hundred reliable reports of death from spider bites in the course of the twentieth century. The recluse and widow spiders (which of course include the famous black widow) bite only from self-defence and when trapped. The notorious Australian funnel-web spider has caused thirteen deaths in the last fifty years. The redback spiders of Australia, equally famous, have caused only a single death since antivenom was developed in the 1950s; however, the bite is extremely painful, and since the redbacks have an affinity for human dwellings they are creatures to beware.

There has been research in spider venom, which includes attempts to make an insecticide that is less damaging than the more conventional kinds. Spider venom has been investigated for the possible treatment of cardiac arrhythmia, Alzheimer's, stroke and erectile dysfunction.

Spiders are part of human lives. Some species come into our homes and, once inside, make their own. A prevailing myth is that they enter by means of the water pipe, emerging into baths and washbasins from the plughole. They don't: the waste trap, a U-bend filled with water, makes such a means of entrance impossible. This myth is perhaps designed to protect us from the fact that spiders are permanent residents in our houses. They enter our wash places, sometimes by accident and sometimes in search of a drink. Once down, they are unable to scale the slippery sides of a bath; you can, if you have a generous heart, leave them a

On her tuffet: Little Miss Muffet, illustration from A Nursery Rhyme Picture Book *(c.1900).*

ladder of lavatory paper so they can climb out again and retreat to the secret places of your home where they live.

Their web-making skills have fascinated humans almost as much as the irrational terror they inspire. Uttu, the ancient Sumerian goddess of weaving, was depicted as a spider. Arachne, in Greek mythology, was foolish enough to challenge the goddess Athena (she of little owl fame, see Chapter 51) to a weaving competition, and compounded her folly by winning. Furious, Athena destroyed her rival's tapestry. Arachne then hanged herself, but Athena brought her back to life as the first spider.

A spider's web is the basis for the traditional illustration of the Buddhist notion of interconnectedness of everything: there is a jewel at every vertex – where lines and edges meet – and in each jewel every other jewel is reflected. The notion of the web of life exists less formally in Western thinking: it has been proposed that each strand of the great web represents a species, and you can't destroy a single strand of a web without weakening the entire structure. We use the concept of the spider's web every time we look something up on the internet: www, which comprises nine syllables, is an 'abbreviation' for World Wide Web, a more compact three.

A spider also represents patience: a web is a long-term plan that will bring a future reward. In Britain this is associated with the fourteenth-century Scottish hero Robert the Bruce, who, while taking refuge in a hut – some say cave – in Ireland, watched a spider try and fail to swing from one place to another. With the seventh swing the spider at last succeeded – inspiring Robert the Bruce not to accept defeat but instead to return to Scotland and wage war on the English once again, driving them from Scotland in the battle of Bannockburn in 1314.

Charlotte's Web (see Chapter 41 on pigs) is a rare example of a good spider in fiction; she inspires and saves Wilbur the pig. Most other spiders are straightforward horrors: Aragog in the *Harry Potter* series, Shelob in *The Lord of the Rings*. They are a staple of horror films: *Tarantula* (1955), *The Giant Spider Invasion* (1975), *Arachnophobia* (1990).

Phobias remain mysterious. Their origin is contentious. There are two main theories: one, that they represent an exaggerated and distorted survival mechanism, namely the avoidance of dangerous creatures; and, two, that their origin is purely cultural. Speaking as a sufferer from (comparatively mild) arachnophobia, I am inclined to reject both. I had no example of arachnophobia to follow; my mother had a phobia about mice (see Chapter 93). And the fact is that fear of spiders is a pretty useless survival mechanism: spiders really don't do us much damage. I am inclined to put it down to movement, for it is the movement of spiders that gives me that feeling of irrational horror; perhaps the same is true for those who fear snakes, and indeed mice. All these animals move, or seem to move, in a way that breaks the pattern of normal life: there is something *wrong* about it. And there's always the possibility that something that breaks the pattern may be dangerous: ask any horse (see Chapter 50); they will often spook at a plastic bag blown by the wind: it doesn't look right, perhaps it's dangerous. I put my arachnophobia down to a bit of faulty wiring in the brain: the association of strange and apparently unnatural movement with danger. In other words it is indeed a survival mechanism gone wrong, but it is a response to pattern-breaking movement rather than genuine danger.

FIFTY-EIGHT
SILKWORM

'The worms were hallow'd that did breed the silk…'

Shakespeare, *Othello*

The Empress Lei Zu, aged fourteen, was the wife of the Yellow Emperor. As she took tea beneath a tree one day, something fell into her cup. She looked closely: it was a cocoon: a pupa wrapped in a bundle of silk that it had created for itself. As the cocoon began to unravel in the hot liquid, the empress, intrigued, began to aid the process with her nimble fingers, unwinding the single strand of silk that made up the cocoon. In this fashion the great silk industry began: maker and breaker of powerful people, driver of trade, responsible more than anything else on Earth for creating contact between the civilizations of the East and the West. The trade also brought about the direct precursor of the instrument on which I am writing these words: the silk industry began the series of processes that led to the invention of the computer.

The story of the empress is mythological, as is the Yellow Emperor. It tells how Lei Zu was deservedly promoted to the rank of goddess for her discovery. The production of silk began 5000 years ago in China; throughout the many complexities and convolutions of history and industrial power since then, China is back as the world's leading silk producer today.

The domestic silkworm is the caterpillar of the moth *Bombyx mori*. The species is so thoroughly domesticated that the adults can't fly and they need human assistance to mate and produce eggs. They are descended from the wild silk moth *B. mandarina*; the two species can hybridize when given the opportunity. Domestication and selective breeding have increased the size of the cocoon, speeded up the caterpillar's growth rate and increased their tolerance of crowded conditions. The silk production of a single worm (larva) has increased ten times over 5000 years. The domestic species has been bred to lose all pigment. They have no fear of predators, and so no strategy for evading them. This species was created by humans by means of selective breeding (artificial selection in Darwin's terms) and it cannot live without humans; so far as silkworms are concerned, humans are god.

There is a story of an exchange between Mrs Horace Greeley, wife of the American newspaper editor, and someone called Miss Fuller. Mrs Greeley

commented on the kid gloves of Miss Fuller: 'Skin of a beast!' Miss Fuller asked what Mrs Greeley preferred: 'Silk, of course!' Miss Fuller replied: 'Entrails of a worm!'

Which is a fair point, even if the anatomy is dodgy. (Excrement of a worm is probably closer.) The details of silk production are not the sort of thing to encourage a love of a flowing cloth that refracts light in thrilling and unexpected ways. The eggs of the silk moth hatch after fourteen days, and then the caterpillar must eat, caterpillars being largely eating machines. They mostly eat the leaves of the white mulberry tree, though they can cope with some other species. After four moults they reach the final stage: each caterpillar, now at its maximum size, proceeds to weave itself a cocoon from the silk it excretes and, once inside, it metamorphoses into a pupa. If left to itself, the pupa will eventually metamorphose into an adult moth and emerge from the cocoon by partially destroying it with the enzymes it secretes. Silk production – sericulture – naturally requires good numbers of adults, so they can lay eggs and make more silk moths. The damaged cocoons are not altogether wasted: they are used, among other purposes, as stuffing for jackets.

But the great majority of cocoons never produce a moth. Instead they are boiled; as if in homage to Lei Zu's cup of tea. That kills the pupa inside and starts the process of unravelling the silk: a single cocoon can produce up to 1000 yards (900m) of silk in a single strand. It takes 2000–3000 cocoons to make 1lb (450g) of silk. It's been estimated that 159,648 tonnes of raw silk is produced across the world every year and it is produced by 10 billion cocoons.

Silk creates a low-density fabric that makes light, comfortable clothing. It keeps its shape, resisting deformation. It is a good insulator, warm in winter and cool in summer. It is the strongest of all natural fibres. It has good tensile strength; each thread is elastic and reluctant to break. It drapes sweetly over a body. It absorbs well, making it easy to dye. It shines without being skiddy and slimy, like some artificial fabrics. The threads reflect light at many different angles, creating a sheen. Silk is, then, quite remarkable stuff for clothing. It has also been used for parachutes; in the construction of tyres for bicycles and racing cars; for prosthetic arteries; in the treatment of wounds and burns; for medical sutures; and as insulation coils for telephones.

In China, silk was first used for women's clothing and was a badge of rank; the colour of the garment indicated the social class of the wearer. Silk remained a Chinese monopoly until about AD 500. There are many stories of how it eventually got out to Japan, Korea and India. One states that a Chinese princess, married to a Khotan prince (from what is now part of Iran), was so appalled by the thought of living without silk that she smuggled silkworms out in her hair.

How silk began: illustration from Myths and Legends of China *by Edward T. C. Werner (published in 1922).*

Another is that travellers hid silkworms in hollow sticks, to take back and sell to the Byzantine emperor. One more is that Japanese entrepreneurs stole silkworms and captured four young girls and forced them to teach the secrets of silk.

It was for centuries a fiercely protected trade, and women played an important part in the production. Secrecy surrounded the production of silk, and in the Roman Empire it was suggested that the stuff came from leaves. Readers will be unsurprised to learn that Pliny the Elder knew better, and wrote about silkworms with calmness and accuracy in his *Natural History*. He was an ascetic man, and so when it came to silk he let himself go rather more than usual: 'Even men have not been ashamed to adopt silk clothing in summer because of its lightness. Our habits have become so bizarre since the time we used to wear leather cuirasses that even a toga is considered an undue weight. However, we have left Assyrian silk dresses to the women – so far!'

The Silk Road was opened up around the second century, and operated by caravans of up to 500 people: moving villages. Though the route can be joined together to make a single whole, hardly anyone travelled its entire length. It led from one trading centre to another: linking cultures by means of a shared desire for the cloth that clings and shines in the light. By the Middle Ages, the Silk Road was in decline: sea routes, though still hazardous, were now considered less dangerous than the long journey across land. You could say what you like about the sea, but at least you didn't encounter so many potentially dangerous people.

Silk came to Europe in a second wave with the returning crusaders. Italy set up its own silk industry and became a major exporter to the rest of Europe, and France followed. With the Industrial Revolution in the mid-eighteenth century the pace hotted up, and France became the centre of the trade. The new spinning technologies were irrelevant, as silk comes straight from the worm in the form of a thread, but the growing complexity of looms changed the nature of the industry. The Jacquard loom operated on a punched-card system, which meant that you could programme a loom to perform certain tasks; it is, then, the direct precursor of the computer. Looms of this design didn't become obsolete until the 1970s.

Machines are all very well, but silkworms are still members of the animal kingdom, no matter how degenerate they have become by means of artificial selection. So it happened that France lost its supremacy in the silk business because of a series of epidemics that killed silkworms.

The difficulties of international trade during the Second World War put the silk trade on hold and accelerated the development of artificial fabrics for parachutes and stockings. This continuing development meant that silk was no longer considered an essential material: and so it became a luxury item once again. Japan dominated global silk manufacture in the early twentieth century, but China subsequently reclaimed the top spot.

Spinners of gold: making nests for silkworms, Japan (c.1905).

Silk requires the slaughter of insects on a massive scale. A pretty silk scarf is the result of several thousand deaths. There have been some ethical concerns about this. Mahatma Gandhi, who played the leading role in the independence of India from the British eventually achieved in 1947, constantly opposed the use of silk and promoted cotton as a worthy alternative; these days there are also concerns about cotton as a crop that requires heavy use of pesticides, herbicides and often scarce water.

These days it is conventional to proclaim the superiority of natural over synthetic fibres: and silk cannot be matched for beauty, practicality, lightness or warmth by anything manmade. By stressing the superiority of the natural we are once again proclaiming our ties with nature: and, with that, our reluctance to complete the divorce from nature.

FIFTY-NINE

FALCON

'Turning and turning in the widening gyre
The falcon cannot hear the falconer;
Things fall apart; the centre cannot hold;
Mere anarchy is loosed upon the world.'

W. B. Yeats, 'The Second Coming'

It would probably be wrong to claim that humans are the only animals that kill other species of animal purely for amusement – for fun – for sport. A fox that gets into a henhouse is likely to kill more than he can carry off, simply because the situation seems to demand it. I have witnessed a pride of lions, so full from last night's gorge they could hardly move, respond to the appearance among them of a herd of 500 buffalo by killing three of them. Their reaction to this unexpected and almost unwanted bonanza was more playful than anything else: one young lion dragged the foetus from a pregnant cow and performed a little dance with it, like a cat with a catnip toy. It was either horrific or hilarious depending on whether you happened to be a buffalo or a lion.

The philosopher Ludwig Wittgenstein famously remarked: 'If a lion could speak, we could not understand him.' This is speculation without data, which frees me up to do the same thing – and I reckon I could have a very decent conversation with a lion: the intimacies of pride life, the pleasures of food and sex, the profound satisfaction of hunting, and perhaps the still more profound satisfaction of cooperative hunting: teamwork; working as one to win a prize for all. That sportive triple-killing was an example of hunting for the pleasure of it: the reward of a meal was more or less incidental.

So humans are not the only animal that kills for the fun of killing, but, all the same, killing for sport is a practice that has played a major part in human history. Hunting without eating your quarry can perhaps be compared to sex without reproduction... or perhaps the point is that we mostly pursue both these things for their own sake, for their own pleasures and gratifications. I don't suppose any species, including our own, is invariably thinking about potential offspring while soliciting and performing sex; and perhaps it is also true that hunting animals think of the hunt first and deal with the nutritional aspects second.

Humans have hunted with falcons for around 4000 years; the practice developed in Mesopotamia and spread quickly across West Asia. Falconry is difficult. It takes time and patience and knowledge to train a bird and to keep one fit and healthy and willing to hunt. That means that using a bird of prey to hunt for the pot requires a great deal of commitment for a small and uncertain reward. Nevertheless it can be done: the traditional quarry for falconers in West Asia and North Africa is the houbara bustard and MacQueen's bustard: weighty birds. There is evidence that the Bedouin tribes trained falcons for more practical reasons than sport and status.

I have held a trained bird of prey on my gauntleted fist, and flown him and summoned him back. I have had the extraordinary sensation of holding such a bird for an extended period, its hooked beak a few inches from my left eye. After a while the bird began to feel like an extension of myself: as if I, sharp-winged and ferocious, could take to the sky and climb and stoop and kill. In short, I felt ever so slightly like God Almighty. That feeling is at the very heart of falconry, and perhaps of all forms of hunting for sport.

Falconry reached Europe around AD 400. A treatise on falconry, written in Arabic by Moamyn in the ninth century, was translated into Latin in the thirteenth century. Falconry was primarily a sport for the wealthy: for those who could employ a falconer to do the hard work of training and looking after the birds. But there is evidence that the same activity was practised at a lower and more surreptitious level by the less well-off; King John, who loved falconry, banned poor people from hunting with birds, all to increase his own bag.

It's been claimed that falconry was the favourite sport of every king of England from Alfred the Great to George III: all save James I, who was a natural exception to most rules; he was fascinated by the idea of training cormorants and ospreys for fishing.

Status, as this royal tradition shows, was always an important part of falconry. The better the bird you carried on your fist, the greater the person you were. This is spelt out by the famous table in the *Book of St Albans* of 1486, which assigns a different bird of prey to each social rank. It is somewhat fanciful: the idea that a king should go hawking with a vulture is absurd on two counts: first, a vulture is too heavy to carry on your fist for more than a few minutes and, second, vultures don't hunt, they look for carrion. But the notion that falconry was also about the nuances of social prestige is right on the money.

Many different species of birds of prey have been used for falconry. Falconry is an inexact term: many more species than falcons are used. There are about forty species in the genus *Falco*, which includes the tiny merlin, which can sit on your naked fist without causing the slightest discomfort with his talons, and burly birds like the saker falcon and the lanner falcon. The more famous – peregrine – is at the upper end of the size gradient, and has the honour of being the fastest of all flying birds: capable of powering downwards at speeds of 200mph (320km/h) and more. Its basic hunting technique is to drop from the air onto a flying bird, which is normally shocked to death on impact.

It's easy to see, then, that watching such a bird, identifying with such a bird and then having the bird return to you is a truly thrilling thing, particularly when the quarry is large. In Britain falcons were flown at herons and cranes: having your falcon – yours, the one that rides to the hunt on your own fist – bring down a bird that stands 5ft (1.5m) high and, as a bonus, makes the centrepiece of a banquet – well, if such a thing didn't actually make you feel like God Almighty, it would certainly make you feel like a king.

A few falconry terms lie fossilized in the English language. Hoodwinked describes a hawk wearing a blinding hood, which convinces him that it is night and sends him to sleep. A hawk that is fed up has eaten well and will not fly after a quarry. A lure is a specific term for a training device, one that lures a falcon into flight and attack. If a falcon is under your thumb, you are holding the jesses – leg-straps – with sufficient firmness to stop him flying off.

Feeling like God: miniature from Treatise on Falconry and Veneration *(France, fifteenth century).*

The Anglo-Saxon poem 'The Battle of Maldon' begins as one of the leaders lets his beloved hawk fly from him, a gesture of startling commitment to the battle: abandoning sport, fun and joy before setting his face sternly on the task ahead. Shakespeare loved a falconry reference. Hamlet says:

I am but mad north-north-west.
When the wind is southerly,
I can tell a hawk from a hansaw.

A hansaw is a young heron (contraction of heronshaw) and a prime target for falconry.

Othello, in despair and believing that his wife Desdemona is unfaithful to him, says:

If I do prove her haggard,
Though that her jesses were my dear heartstrings,
I'd whistle her off and let her down the wind
To play at fortune.

The falconry image is chosen to reveal the depths of Othello's anguish; today the simile also works the other way round, and tells us how greatly a favourite falcon was loved, if it could be compared to Othello's love for Desdemona.

Falcons are traditionally birds we love and admire, even when we don't own them. It was birds of prey that suffered most spectacularly when the uses of organochloride pesticides (like DDT, see Chapter 10 and elsewhere) were at their height. The poisons were designed for insects and killed them; the birds that ate the insects suffered and the birds that ate the insect-eaters also suffered as the poisons built up in their systems and caused their eggshells to become thin and unviable. Such revelations about agricultural chemicals were crucial to *Silent Spring*, the great work of 1962 (see Chapter 23 on mosquitoes, and elsewhere). Once these chemicals were slowly and reluctantly banned, birds of prey made a triumphant comeback in many places: so much so that peregrines are now city birds, with city pigeons a main prey item.

Falconry continues as a modern hobby or perhaps obsession: it has few followers but all of them deeply committed. It is a passion among the wealthy and well-bred in West Asia. In most countries their birds must be captive-bred; it is illegal to take chicks from a nest. The most popular bird for all-purpose falconry is the American species Harris's hawk, which in the wild hunts cooperatively and, no doubt for that reason, is social and cooperative with humans – so much so that it has been called 'the Labrador of falconry'. The flying display of birds of prey is more or less a statutory item at country shows in the UK. Zoos and bird-of-prey centres offer experiences in which members of the public can spend a day with the hawks, and fly them for themselves. For a fee, you too can be king for a day.

SIXTY

PHEASANT

'For a few minutes he almost choked with pleasure at the
prospect of so great a carnage. Then he sprang from his
horse, rolled up his sleeves and began to aim.'

Gustave Flaubert, 'The Legend of St Julian the Hospitaller'

From about the middle of the seventeenth century falcons ceased to be beloved best friends and started to become ferocious enemies. Creatures that were once cultivated and cherished were now killed whenever possible. There was a serious attempt to wipe out all birds of prey: and it was pretty successful.

The reason for this was the invention of the shotgun. It not only made the distant killing of birds of prey possible, it also made their killing desirable. It was now possible to kill birds without the help of another bird. It was easier and cheaper and you could kill a great many more birds. In many places, the prime target for the shotgun was the pheasant. That made birds of prey potential rivals for the same quarry: and so the long campaign against them began. In many places it still continues.

There are around fifty species in the family Phasianidae, but one of them we love so much that we breed them and release them into the wild in vast numbers. This is usually referred to as just pheasant, or sometimes common pheasant, *Phasianus colchicus*. Never has a bird prospered so much from its ability to die. An Asian bird that in natural circumstances would live no nearer Britain than the Black Sea is released into that country at a rate of 40 million a year. Pheasants

Jolly good sport: painting by Ange-Louis Janet (nineteenth century).

were introduced into the United States in 1773, and there, too, they are cherished because people enjoy killing them.

'The history of the pheasant has its origins 2000 years ago,' announced the website for the British shooting publication *The Field*, at a sweep eradicating several million years of evolution and existence before pheasants were introduced to Britain. It's generally reckoned that pheasants were brought to Europe by the Romans for the combination of their decorative and nutritional qualities. Pheasants were also cherished by the Normans, who protected them from killing by unimportant people. In 1170 St Thomas à Becket dined on pheasant the night before he was murdered in Canterbury Cathedral.

But with the invention of the shotgun, it became possible to fire a tight group of lethal projectiles at a bird from a distance. Early shotguns were difficult to use: the longfowler was some 7ft (2m) long. But, with improving technology, barrels became shorter, and by the beginning of the eighteenth century it was possible to take birds on the wing. Shooting birds was now a measure of the skill and composure of the shooter. By the late eighteenth century you could shoot with barrels less than 3¼ft (1m) in length; some were even shorter and designed for shooting in wooded country. The side-by-side, double-barrelled gun was developed, giving every shooter a second chance and the best a second bird. Henry Nock's patented breech was introduced in 1787; it used less powder and provided a much shorter gap between the trigger and the bang. The shotgun was becoming an affordable and accessible weapon, designed for taking up arms against birds.

Like falconry, carrying a gun is an aspect of feeling a little bit like God Almighty, with the power of life and death over anything you come across. Its use requires some skill, though there is a fair margin for error. I once travelled across country in a shooting-break towed by a tractor, a sick-making, juddering experience. 'How does anyone shoot straight after travelling in one of those?' I asked. I was told: 'Very few of them can shoot straight to begin with.'

Pheasant-shooting became more and more intense with the British Enclosure Acts of the eighteenth and nineteenth centuries. It was now possible for a squire to keep the hoi polloi out of the woods, and therefore to use them to rear pheasants by means of artificial feeding. He could then kill them in due season.

Pheasants must here stand for all the other target species of bird whose deaths are sought so eagerly. In mainland Europe the British obsession with pheasants is often the subject of incredulous laughter: the shooters there prefer to kill native fauna. In many rural areas British roads are paved with dead pheasants: the young released pheasants have little idea of how to survive and are routinely mown down by passing traffic.

Pheasants are strongly sexually dimorphic: that is to say, the males and the females look very different. They are omnivorous, taking grains and other seeds,

King of the wood: painted by a member of the circle of Tobias Stranover (1684–1756).

insects, other invertebrates and also small mammals, reptiles and amphibians when they can get them. They are ground birds that fly reluctantly, as a last resort. Like the red jungle fowl, to which they are related, they forage on the ground and roost at night on perches. They are well-muscled and heavy – a lot of meat on them – which makes them able runners but clumsy fliers.

They have a double strategy for avoiding predation. The first is to lie doggo and hope that danger passes. The second, put into action when the predator gets close, is to adopt a startle ploy: to leap into the air while making a loud noise, actions calculated to freeze the predator for the vital second. So in the nineteenth century the French came up with the idea of the driven shoot, in which beaters walk through the pheasant-thronged countryside, whistling, hollering and whacking at the vegetation. The pheasants abandon the lying-low notion, take to the air with a great din and fly: into the waiting guns.

In the United States pheasant-shooting is particularly popular on the Great Plains, where the pheasants have settled on natural grassland and farmland; it's been calculated that in South Dakota every year 200,000 shooters shoot more than 1 million pheasants.

In theory the birds are shot for the table, but in practice many are dumped and left to rot. In Britain most years there are scandals when people come across such dumping sites; shooting interests routinely say that these are birds too greatly damaged to be acceptable to the commercial market. 'I shot a 500-bird day,' one enthusiast wrote on a shooting website: no doubt he ate them all. It should be noted that killing a pheasant with a shotgun is not that difficult. A flying pheasant represents a large, slow-moving target, with no skills at all in evasive manoeuvres; when driven directly towards the guns, the problems of full deflection shooting

– shooting at a target moving at 90 degrees to the shooter – are taken out of the equation. What's more, a shotgun throws the pellets in spread pattern: to say that the odds are on the shooter's side is a considerable understatement.

Conservationists are routinely concerned about the effects of alien species on an environment, as we have seen with the question of rats on islands (see Chapter 25 on rats and Chapter 43 on the albatross); alien species potentially cause the extinction of natives. There is, then, inevitable concern that the annual addition of 40 million pheasants to a single small nation is having an effect on native species, particularly on amphibians and reptiles.

There is a record of George V of England shooting 1000 pheasants over six days in 1913, in competition with a friend; the friend won. It is clear, then, that the pleasure of the activity lies in numbers: a good day is one with a large bag.

And so, throughout the nineteenth century, landowners who possessed a pheasant shoot fought an unforgiving war on birds of prey. In the UK by the beginning of the twentieth century, five species of diurnal birds of prey had been shot to extinction: goshawk, honey buzzard, marsh harrier, white-tailed eagle and osprey, while five more were down to fewer than 100 pairs: Montagu's harrier, golden eagle, hobby, hen harrier and red kite.

The two world wars, in which there were more urgent uses for men and ammunition, slowed down the rate of killing, allowing birds of prey to make some kind of recovery. In 1954, the UK government passed the Protection of Birds Act, which forbade the killing of birds of prey apart from sparrowhawks. These finally got protection nine years later.

Birds of prey are still killed illegally by the shooting industry. Figures from the Department for Environment, Food and Rural Affairs (Defra), which cover England and Wales, demonstrate that there is room for 300 pairs of breeding marsh harriers in the uplands of England: in 2013 not a single pair managed to breed. Hen harriers nest on moors favoured by grouse: and the grouse-shooting industry is very protective of its profits. Hen harriers have been fitted with satellite tags: these routinely cease to function when the birds fly over land managed for the shooting of grouse; it can never be proved that they have been shot and their tags destroyed, but it is the obvious assumption.

In other words, a great deal of land and a great deal of wildlife are manipulated so that people can shoot pheasants and other birds. The fact is that people enjoy killing birds very much. People will pay a good deal to do so: about £50 a pheasant in the UK, for example; a great deal more for grouse. And this mounts up since huge bags are rather the point. Birds are shot for pleasure all over the world. We have shaped large areas of the world so that this pleasure can be pursued. It is a pleasure particularly associated with the elite. Those who suggest that the practice is damaging routinely suffer persecution and vilification, as I know from personal experience.

SIXTY-ONE

BARNACLE

*'I must call on you, and report for my own satisfaction, a really
(I think) curious point I have made out in my beloved Barnacles.'*

Charles Darwin, letter to Richard Owen

Perhaps barnacles are the most surprising creatures in this book. They are surprising not just because they aren't related to the limpets that you find anchored on the rocks alongside them, being more closely allied to the butterfly that passes haphazardly over the beach. They are also surprising because they are an important reason for the rising maritime might of the UK during the eighteenth and early nineteenth centuries, paving the way for the establishment of the British Empire, and because they played a crucial role in the revolution in human thought that came with Darwin's brutal and beautiful revelations.

Barnacles are not molluscs, like limpets, like the former owners of most of the shells you find on a beach. They are arthropods: in the same group as insects, spiders and crustaceans (crabs and lobsters). As such, they are a comparatively rare example of creatures that Linnaeus got wrong. Georges Cuvier got them wrong as well. It was only in the mid-nineteenth century that scientists became aware that barnacles go through a larval stage – two, actually – and needed reassigning. The man to do that was waiting for just such a challenge to rise to.

Curly feet: illustration of gooseneck barnacle from Encyclopaedia Britannica.

What adult barnacles do better than practically any other living animal is stick. They stick to rocks, they stick to marine structures, they stick to the bottoms of ships, they stick to great whales. They are most commonly seen by humans in the intertidal zone: waiting for the twice-daily soaking and the bounty of food the waters will bring them. Once immersed they waft small items of nutrition from the seas with their feet: they are technically cirripedes, or curly feet, and there are getting on for 1500 species.

Some species of barnacles perch on an elegant stalk and look rather goose-necked, a similarity that created the wacky notion that geese are hatched from them; the barnacle goose is named for that reason. It was a convenient thing to believe because it meant that geese were in fact fish – so you could eat them during Lent and on other days of fasting (see also Chapter 91 on beavers).

But, for the most part, barnacles have been an unmitigated nuisance to humans ever since we attempted to make our way out to sea, and they remain so today. You would hardly have thought that creatures so small could cause anything more than minor inconvenience to the monstrous works of humankind, but that's not the case. A ship that's thickly encrusted with barnacles will compromise its performance by as much as 40 per cent. It's a matter of fluid dynamics: you want water flowing smoothly and easily over the hull of your ship: barnacles make the flow complex and turbulent. If you've got barnacles all over your bottom you might as well be driving with the handbrake on.

If you're a sailing ship, an excess of barnacles means you exploit the wind with a good deal less efficiency, making you slower and less manoeuvrable. If you are powered by fuels and propellers, you will require up to 40 per cent more fuel for the same speed. What's more, all shipowners must take their boats from the water regularly to have the barnacles – they call it biofouling – removed.

It follows that dealing with barnacles has been an issue for centuries. In the eighteenth century the British came up with an answer: covering the bottom of the boat with copper. The metal reacts with seawater and creates a film that the barnacles don't like. So a copper-bottomed British ship could outperform boats from other nations: a notion that survives in expressions like 'a copper-bottomed guarantee'.

But barnacles have no problem adhering to steel craft, and you can't fit a copper bottom to a steel craft because copper makes the steel corrode. There has been a constant search for the perfect anti-barnacle paint; paints have been developed with copper, arsenic, mercury, strychnine, cyanide and tin. These have been variously effective but consistently efficient at polluting the sea; while the necessity to use more fuel to combat biofouling results in increased emissions. The longer a craft is stationary, the more barnacles you will get. Navies, unlike commercial cargo vessels, are required to have comparatively long periods at rest; it's been estimated that barnacles cost the US Navy US$500 million a year because of maintenance and higher fuel use.

The great botanist Joseph Hooker was a good friend of Charles Darwin; they corresponded throughout their lives and their exchange of ideas and knowledge was crucial to both their careers. At a comparatively early stage in their relationship, they were discussing a new book full of bright ideas. Hooker was inclined to be judiciously dismissive: 'I am not inclined to take much for granted from anyone [who] treats the subject in his way and who does not know what it is to be a specific Naturalist himself.'

It was a sentence that stabbed Darwin through the heart: 'How painfully (to me) true is your remark that no one has hardly a right to examine the question of species who has not minutely described many.'

Hooker was distressed in his turn: he never meant any of this personally. But Darwin chose to take it personally: as an implicit calling to order. He had been working on his great idea for some years: he now realized that no matter how he presented it, no one in science would take it seriously. He had not minutely described many species. He was a gentleman amateur: an adventurer who had written a best-seller, *The Voyage of the Beagle*. He was not a pro. So he resolved to make himself one: and do some serious and intense work on classification. The group he chose was barnacles.

He spent the next eight years devoted to the study of barnacles, a period that included two years of sickness. Not that a study of this depth was the original plan: he was going to cover just a few species, enough to get his scientific street-cred – but, Darwin being Darwin, once he got going he had to do the whole thing righteously and from top to bottom. The work was soon all-consuming: one of his children, visiting another house, asked, 'Where does your daddy keep his barnacles?'

Darwin produced a series of monographs that revolutionized the human understanding of barnacles: meticulous, detailed, precise, accurate. He wrote about his 'beloved barnacles', observing small but crucial differences which show that two similar specimens are in fact different species, incapable of breeding together. He observed how, in the right circumstances, a group of species will develop around one effective idea: what's now called an adaptive radiation. He noticed in delight that part of a barnacle's head was analogous to the head of a crab but 'wonderfully modified'. The oviduct – egg-laying implement – of a crab has in barnacles become 'modified and glandular and secretes a cement'. The larva uses the cement to stick to a rock – or a ship or whale – when it embarks on the stationary adult life: in other words, an organ can change its function from one related species to another, another central truth of the theory of evolution by natural selection.

Thomas Huxley called the barnacle monographs: 'One of the most beautiful and complete anatomical and zoological monographs which has appeared in our time.' By the time the work was finished (1854), the brilliant amateur was no more: a hard-bitten professional had taken his place and the process for doing so

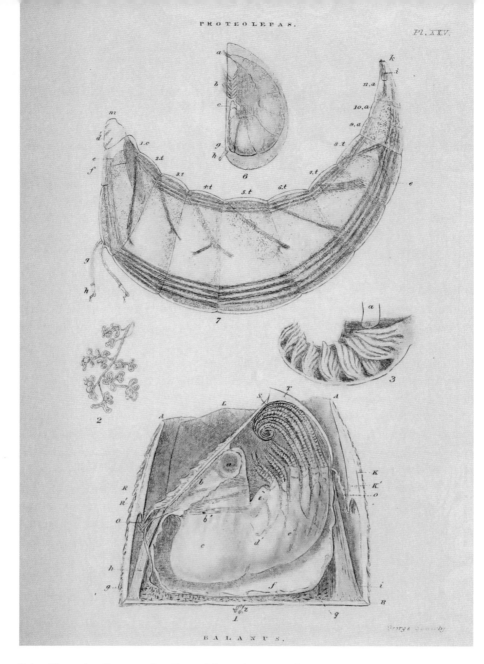

BALANUS.

Beloved barnacles: illustration from Darwin's barnacle monographs.

had added to his knowledge and understanding about species – and therefore about evolution.

It was also a process in which Darwin's love for barnacles eventually suffered: 'I hate a Barnacle as no man ever did before, not even a sailor in a slow-moving ship.' It was time to move on. The year was 1855, and Darwin began to put together *The Origin*, which would be published four years later.

SIXTY-TWO

HEAD LOUSE

'Sir, there is no settling the point of precedency
between a louse and a flea.'

Dr Samuel Johnson

Humans really don't like the idea of being part of an ecosystem. As we have seen already in relation to lions (see Chapter 1), sharks (see Chapter 14), crocodiles (see Chapter 49) and piranhas (see Chapter 55), we are deeply uncomfortable with the idea of being prey. But we are even more resentful of the idea of being habitat. To a head louse, a human is not the crown of creation, built in the image of God: it is somewhere to live and feed and copulate and reproduce.

A human body can be host to a number of parasites. Some of these live with us on a permanent basis, rather than dropping in for an occasional meal. These include the tiny *Demodex* mites that live in our eyelashes (see Foreword), and intestinal worms that can grow up to 100ft (30m) in length. Such truths are damaging to our notion of uniqueness. If other animals can feed on us and live on us, how can we be angels? There are species of lice that affect most orders of mammals, and every order of birds; the head louse evolved specially to parasitize humans. There is a closely related species that lives on chimpanzees, but the human louse is quite separate and its scientific name spells this out pedantically. The head louse and the body louse are subspecies: *Pediculus humanus humanus* for the body and *Pediculus humanus capitis* for head louse.

I remember the horror – and the humour – that attended an outbreak of head lice among the hippy-student community I lived in years ago. It seemed to question and challenge our values and our self-esteem. It wasn't helped by the fact that one of our number had decided that he was a Buddhist who must live in peaceful coexistence with his lice, meaning that we all had to do the same, because he kept reinfecting us. (Best not ask the route.)

Head lice are irritating and itchy but harmless, more damaging to pride than health. The only possible danger is from over-scratching the bites and opening the way for infection. It follows that the problems of dealing with them are more cosmetic than medical. Head lice are insects, up to ⅛in (3mm) long, wingless creatures that spend their entire lives, from egg to corpse, on a human head. They lay eggs and glue them (like their fellow arthropods the barnacles, they are adept

How to cope with lice: Louis XIV in a wig, by Charles Lebrun (seventeenth century).

at glue) to the base of a single hair shaft. These hatch and produce what are inappropriately termed nymphs. They are pretty much like the adults, but they have a few moults to go through before they reach sexual maturity. They do so by feeding on blood just as the adults do: nibbling the skin of the scalp and sucking.

They are not great at moving: they are designed to stay put on a fast-moving, active animal prone to head-scratching and head-combing – with fingers and with actual comb. But they can travel from one person to another, mostly by means of head-to-head contact, clambering from one hair to the next. They can't leap: so the two heads need to be good and close for the lice to make the transition.

Body lice don't travel with us in the same way. They prefer to keep in places where they are likely to come into contact with humans, most often in clothes and bedding. They bite most often in the places where seams touch the skin: shoulders, neck, waist and groin. They are most often associated with places where humans live in crowded and unsanitary conditions: and are therefore a problem in refugee camps and shelters for the homeless. They are best treated by dealing with their places of refuge: treat clothing to a hot wash and a hot dry. Unlike head lice they can pass on diseases: typhus, French fever and relapsing fever. There have been suggestions that they, rather than the oriental rat flea, were the true vectors for the plague (see Chapter 6 on oriental flea rats). They have probably been with us since we started wearing clothes 30,000–100,000 years ago; and so they diverged into the subspecies that lurks on clothing and the subspecies that lives with us full-time on our heads.

As just about every parent knows, head lice crop up. Their prevalence in middle-class households has discredited the idea that they exist only on dirty people from dirty homes: they are insects well adapted to their niche. There is a confusion about what constitutes nits: sometimes the eggs, sometimes only the old eggs from which a nymph has emerged, sometimes the nymphs themselves. Head lice are a problem most particularly associated with pre-teen children. It's been estimated that 6–12 million people a year are treated for head lice in the USA; in the UK perhaps two-thirds of all children have a go with head lice before leaving primary school. Infestations are routinely treated with pesticides, though this inevitably is followed by strains with increasing resistance. It's as if their job was to live with humans and provide them with a constant nagging reminder: you are a member of the kingdom Animalia whether you like it or not.

Head lice are irritating, humiliating and remarkably hard to get rid of. As a result, they shaped the way people of the West chose to present themselves to the public for damn near two centuries. The most effective way of getting rid of head lice is to have no hair: without hair to cling to, the lice can't stay attached to a human, and so are unable to take their regular meals of blood. So humans started to shave their heads and wear wigs.

Early pioneers of wig-wearing included Elizabeth I of England, who wore a tight cap of red curls, but perhaps the crucial figure in the history of the wig was Louis XIII, who took to wearing a wig when he went prematurely bald. Louis XIV, his son and successor, took this notion to another level and, with his lead, wigs became huge and recklessly extravagant; it's been said that he kept forty wig-makers in business at the Palace of Versailles. This fashion came to England when Charles II returned there after his exile in France. These huge wigs were heavy, uncomfortable and expensive, but they conferred status: and as we have seen many times over in this book, most recently in Chapter 58 on silkworms and Chapter 59 on falcons, when it comes to the question of status we are inclined to put up with a very great deal.

Head-scratching: head louse engraving, from Esperienze intorno alla Generazione degl'Insetti *by Francesco Redi (1688).*

Wigs became more or less obligatory in Britain for all those who aspired to a certain level of society. Wigs made from human hair were the best, but also the most expensive. Those who wished to claim status on the cheap could wear wigs made from the hair of horses and goats.

Samuel Pepys took an unselfconscious pleasure in dressing up. At one point he wrote in his diary: 'I did go to the Swan; and there sent for Jervas, my old periwig maker, and he did bring me a periwig, but it was full of nits, so as I was troubled to see it (it being his old fault) and did send him to make it clean.' The fashion was ended in Britain by a stroke of law-making genius: in 1795 a tax was levied on the powder used for hair and wigs – finely ground starch, scented with orange flower, orris root or lavender. Suddenly wigs and powdered hair became unattractive.

By this time wigs had been largely abandoned in post-revolutionary France and America. George Washington never wore a wig, though his long, carefully arranged hair often looked like one. The next four presidents have all been painted in wigs: John Adams, Thomas Jefferson, James Madison and James Monroe, but none of the others.

Wigs are still worn by judges and barristers in the British law courts. In the film *Withnail and I*, a character reports a court scene, with the defendant dressed in caftan and bells and the judge in his wig. 'This was more like a long white hat. And he looks at the Coalman and says, "What's all this? This is a court, man, this ain't fancy dress." So the Coalman looks at him, and he says, "You think you look normal, your honour?" '

So if you ever fall foul of a British judge, remember to ask him if his wig helps him to keep the head lice at bay, because that's the reason he's wearing it. Head lice, who care nothing for human status, take another sip of human blood and hold tight to the hairs of their chosen head. You can be a judge, a king, a great writer and social commentator: but so far as the head louse in concerned you're just a habitat waiting to be exploited.

SIXTY-THREE

CROW

'When crow cried his mother's ear
Scorched to a stump.'

Ted Hughes, *Crow*

*Too clever by three-quarters: illustration by
Richard Heighway from* The Fables of Aesop, *published in the 1890s.*

Mostly we respect intelligence when we find it in non-human animals. We are more distressed by the circumscribed lives of laboratory chimpanzees than we are about rats or pigeons; there are British laws to treat octopuses, who are intelligent (see Chapter 37 on octopuses), with more consideration than any other invertebrate; and we are inclined to revere the idea of an ideal aquatic community of mindful dolphins (see Chapter 38). Crows are as intelligent as any of these species: and yet they are routinely shot as pests and very little objection is made. In England, in 2019, a cessation of the universal right to kill crows on sight brought fury and dismay.

There are about 120 species in the family of Corvidae, and forty-six of them in the genus *Corvus*. The family includes crows, rooks, ravens, jackdaws, jays, magpies, treepies, choughs and nutcrackers. They are passerines, so related to songbirds and most of the birds that visit garden bird feeders, but they tend to be larger and more robust than most of these, strong-legged and for the most part stout-billed. They are imposing, often bold, frequently noisy and they force themselves on the attention of humans.

They are smart and they are omnivorous, traits that have allowed many species to adapt to human habitations. We are seldom entirely relaxed about animals that live alongside us on their own terms: crows have surely done this for as long as humans had even temporary settlements. All kinds of edible, human-generated rubbish are suitable food for crows: twenty-first-century crows will feed eagerly in places where people abandon food from McDonald's: the bun, the fries and the burgers are all good in their eyes. Crows tend to be enthusiastic eaters of carrion and are not fussy about which species provides the bonanza of protein. The remains of human meat meals, dead livestock and, for that matter, fallen humans have been traditional additions to the crow's diet. It follows that they are particularly associated with battlefields. The scene of the dead may be honour and victory, disaster and dishonour or glory for ever to the humans involved: for the crows it's all protein. They are, then, black birds associated with death and horror.

Crows have powerful beaks, though there are limits to what they can get from a fresh carcass. But they will feed opportunely on a carcass when they find one, heading first to a vulnerable unprotected part of the dead (or dying or otherwise helpless) mammal. The eyes. A human corpse left unattended in a wild place will, in many countries, be routinely found with its eyes removed and consumed. The same is true of vulnerable livestock: crows will, on occasions, peck the eyes from living lambs, which causes horror and distress among humans (who were raising the animal so that they could eat it themselves). There is also the question of financial loss.

Aesop put several crows into his fables. Sometimes the crow is foolish: a fox suggests that a crow, carrying a piece of cheese in her beak, is the queen of all birds, and longs to hear her sing. So she opens her beak to croak: letting fall the cheese, which becomes the fox's breakfast – so never listen to flatterers. Another of Aesop's crows wanted to be respected like the raven as a bird of ill-omen. So he perched ahead of a group of human travellers and cawed lustily. Some of the travellers were concerned, but the wiser ones among them said that it was only an old crow, pay it no mind – so never pretend to be something you are not, you'll end up looking silly.

But another of Aesop's crows is intelligent. Desperate for a drink and encountering a jug of water, he finds the liquid just beyond the reach of his beak. So he drops a series of pebbles into the jug until the level of the liquid is high enough for him to sup: a moral about being smart and thoughtful and looking for solutions. I have seen this tale replicated by a jay – a species of crow – at Cambridge University, where Professor Nicola Clayton (also a great adept of the Argentinian tango) had been running a series of experiments on the intelligence of crows. She didn't make her birds so thirsty that they were desperate for water; this would not be an ethical experiment. Instead she floated in a narrow vessel a juicy waxworm, just out of reach. The jays solved the problem exactly in the manner of Aesop's crow. When offered dummy pebbles made from Styrofoam, which floats, the jays rejected them contemptuously; didn't even bother trying them. Clayton says that anything a chimpanzee can do, a crow can do as well. Crows have a brain-to-body mass ratio in the same league as chimpanzees and dolphins.

Studies have yet to find the limits of corvid intelligence. A jay is the only bird species known to recognize himself in a mirror. Many species use tools, and even make them; New Caledonian crows will create a hook to fetch the larvae of insects from tight crevices. They will also use a tool to fetch another tool that is out of reach; the resulting find allows them to capture insects. Some anecdotal evidence has been still more spectacular. Hooded crows have been observed scattering bread on the surface of a pond and then catching the fish that came to eat the bread. Japanese crows have been found deliberately placing nuts on the road surface, so they would be broken open by passing cars. They refined this method by exploiting a pedestrian crossing for the nut-cracking; the crows would then wait for the traffic to stop before picking up the exposed kernels.

Many crow species lead complex and intense social lives and (as we have already seen in Chapter 38 on dolphins and Chapter 54 on elephants) that often leads to intelligence. Konrad Lorenz, the pioneer ethologist, observed the complex hierarchy among jackdaws, and their tight pair bond. He also noted that each half of a pair has the same status, and that a female who becomes the mate of a higher-ranked bird acquires her partner's rank. Young birds spend an extended period with adults: learnt behaviour is an important part of the way they live.

Clever as any ape: tool-using New Caledonian crow.

They have excellent memory. They recognize individuals of their species and are aware of social positions. There is anecdotal evidence of crows grieving after the death of a mate. Ravens have been observed consoling the loser of a dispute.

But sympathy for crows is hard to come by. Crows were seen attending the place of execution in the film of *Harry Potter and the Prisoner of Azkaban*. Noah sent a raven from the ark, but it failed to return and tell him that the flood had abated; this raven is sometimes seen as selfish, not civic-minded like the loyal dove (see Chapter 22) who was also released. (Perhaps the raven was feeding on animals drowned by the flood.) In the *Mahabharata*, the messengers of death are compared to crows. Lady Macbeth has one of the most chilling lines in a chilling play: 'The raven himself is hoarse that croaks the fatal entrance of Duncan under my battlements.' Duncan is murdered by the Macbeths that night.

The unsavoury reputation of crows continues into the twenty-first century. In the UK, crow species, particularly magpies, are blamed for the loss of songbirds; and it's true that many species of crow will raid a nest of smaller birds if they have the opportunity. Magpies, loud and conspicuous, are traditionally birds that bring bad luck, at least when seen on their own: 'One for sorrow, two for joy'. Magpies are as keen as any corvid on the pair bond and are usually seen in twos, so the superstition is rigged.

But crows continue to bear the blame for the decline of British songbirds: and so their killing is supported by the most tender-hearted. There have been many pieces of research on corvid predation of small birds, and they uniformly come up with the conclusion that the threat corvids pose is small to negligible. An American study came to that conclusion; a British study of twenty-three songbird species concluded that crows pose no threat; a review in the scientific journal *Ibis* examined forty-two studies reporting 326 interactions between corvids and prey species and concluded: 'in 81 per cent of cases, corvids did not have a discernible impact on potential prey'. They added that, on occasions, the presence of corvids was actively beneficial to songbirds.

Songbirds are certainly declining fast in all developing countries. Habitat destruction and the intense use of herbicides and pesticides are the principal and obvious reasons for this, but, as we have observed, facts about non-human animals have no chance when they are faced with a satisfying piece of mythology. We can prove that crows are harmless, even beneficial: but there's no point. At a deeper level we believe – know – they are evil.

The UK's *Daily Mail* in 2019, reporting on the suspension of licensing of the killing of crows, put up the headline: 'The savage cruelty of a law that lets crows torture and kill lambs'. Crows are a visible enemy. They are perceived as a threat to the human way of life. For that crime they can expect, and they receive, no mercy.

SIXTY-FOUR

BAT

'The bat that flits at close of eve
Has left the brain that won't believe.'

William Blake, 'Auguries of Innocence'

If an animal makes the roll call of ingredients in the witches' brew in *Macbeth*, you can be certain that it is associated with saucy doubts and fears. And not rational ones, either:

> Wool of bat and tongue of dog,
> Adder's fork and blind-worm's sting…

Humans dressed up as bats represent two of the world's most instantly recognizable images: and both Batman and Count Dracula are associated with darkness and death. And yet, in China, bats are symbols of happiness and good fortune: a bat image on your silken garments is not sinister in the least: it is as joyous an image as a robin or a kitten or a swallow would be in the West.

Bats inevitably cross paths with humans because there are so many of them, and so many different species. What's more, they are adaptable: even in London I have seen bats circling streetlamps to catch insects attracted by the glare. Bats seek out day roosts that are safe and quiet, and often choose abandoned human dwellings.

Night lovers: short-nosed fruit bat from Wellesley Albums *(illustrator unknown, 1798–1805).*

Like us they are mammals but, unlike us, they are creatures of the night. They are equipped with senses we find hard to imagine, finding their way in the true pitch-black of a deep cave as easily as you and I can walk across Liverpool Street Station. And, unlike us, they can fly.

There are a fair number of vertebrates capable of gliding: flying fish, flying frogs, snakes and lizards, flying squirrels, colugos and flying phalangers. In all these cases, the 'flying' part of the name is misleading: they are incapable of powered flight. As their speed runs down after takeoff, they rely on gravity to supply airspeed. This form of gliding is more or less falling with grace. In the entire animal kingdom, as we have seen, only four groups have evolved powered flight: insects, birds, the extinct pterosaurs (including pterodactyls) and bats.

About one-fifth of all mammal species are bats: there are around 1400 of them, making them the largest order of mammals after rodents. Of these, only three species feed on the blood of mammals, and yet we associate vampirism with the entire group. Bats are traditionally divided into two groups: megachiroptera, which include the great fruit bats and which don't echolocate (there is a single exception, *Rousettus aegyptiacus*); and the rest, which are mostly (but by no means entirely, see below) insectivorous and use echolocation. This classification has been challenged in recent years, but it will do for now. There is a considerable range in size: the bumblebee bat, not inaptly named, can weigh as little as 0.09oz (2.6g), with a body not much more than 1in (2.5cm) long, to whoppers like the giant golden-crowned flying fox, weighing 4lb (1.8kg), with a wingspan of 5½ft (1.7m).

Bats love the night. They fly, and what's more – perhaps what's worse – they fly in the night. They are at home in places, in a world, that we can never aspire to. They own the air of the night as birds own the air of the daylight hours. Some bats are more crepuscular than nocturnal, being most active at dawn and dusk. An African bird of prey, the bat hawk, specializes on that very particular half-hour of the day when the insects are swarming and bats are about, but when a sight-orientated bird of prey can still operate. The two brief frenzies at either end of the day – or night – must be fully exploited, for that narrow belt of time is life and death for the bat hawk as well as the bats.

The darkness is the first sinister thing, of course: as we have already seen with owls (see Chapter 51), we associate creatures of darkness with fear and, above all, with death. The second sinister thing is the wing: a piece of leather stretched over a skeletal framework queasily reminiscent of a hand. I'm not sure why we humans find that troubling: we have exposed skin ourselves, rather than feathers or fur. Conservationists who help people with bats on their premises tell stories of deeply alarmed house owners who, on being shown one of their home-sharers in hand, find the furry body and the bright mammalian eyes deeply reassuring, even touching: converts at a stroke.

But perhaps, as with spiders and snakes (see Chapters 57 and 28), there is something in the way they move that causes unease. Bats are not such powerful fliers as birds, and not so efficient over the long haul, but they are more manoeuvrable: uncannily so, it sometimes seems, when you see the bat caught for an instant in a beam of artificial light and jinking in the air in a way that looks aerodynamically impossible.

The micro bats find their way in the dark by means of echolocation: the sound of their own voices comes echoing back from stationary and moving objects, and that allows them to build a sound picture of the world they travel through. They aren't deafened by their own voices because they shout above the limits of their own hearing; they can hear only the return sound, and do so because they exploit the Doppler shift, which concerns the change in frequency of a sound if you are moving relative to the sound source (or vice versa, if you are attending a motor-racing event).

The echolocating ability of bats was suggested in 1920 by Hamilton Hartridge and demonstrated beyond doubt in 1938 by Donald Griffin. Radar was in the process of being invented at the same time and is based on the bouncing of radio rather than sonic waves; which is why the process was so-named (a contraction of Radio Detection and Ranging). If one discovery was an influence on the other, it has never been made clear: either way, the bats got there first by a few million years.

Some bats are fruit specialists; most others specialize on insects. But some subsist on other kinds of food: some take fish in low swoops over the water, others take small mammals, birds and amphibians. And three species of bats drink blood, often. It is pretty tiny: body length 3½in (9cm), wingspan 7in (18cm): not the dimensions you expect for something that strikes more or less worldwide terror.

Common vampires are agile crawlers, landing close to large mammals at night and making their way over ground to make their bite and lap. These days they will frequently take the blood of domestic stock like horses and cattle; they can spread trypanosomiasis and rabies, so they are never welcome visitors. They will, on occasions, bite the exposed flesh of a sleeping human. They go in for reciprocal altruism: a bat who fails in a night's blood-hunting will solicit a warm meal from a roost mate, who is not necessarily related, so we should admire the generous, civilized and genuinely altruistic home life of the vampire.

The Chinese have never found any difficulty in admiring bats. There is a traditional Chinese love for puns and homophones: many apparently identical words take different meanings from the tone (like a musical note) in which they are spoken. So *fu* can mean both 'bat' and 'good fortune' in Mandarin. Accordingly, a bat represents the five blessings: long life, wealth, health, love of virtue and a peaceful death. They have a reputation for living longer than other mammals of a similar size; for the Chinese, bats also symbolize longevity.

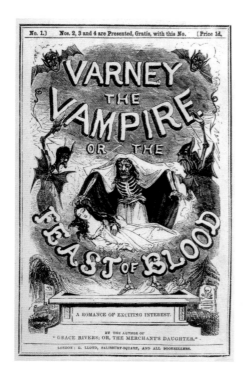

Penny dreadful: title page of
Varney the Vampire.

Which is the opposite of death and a contradiction of the symbolism of the bat in Western culture. *Dracula*, the novel by Bram Stoker published in 1897, established the bat-like Transylvanian count and his need for fresh human blood. (He had a precursor in another titled blood-sucker, who appeared in *Varney the Vampire*, a series of penny dreadfuls – cheap sensational magazines – between 1845 and 1847.) I remember reading *Dracula* as a teenager, and the terrible thrill as Jonathan Harker, the narrator, watches the count leave the castle from his window and descend the wall – with his head downwards.

The huge success of the novel seemed to fill a vacancy in the Western psyche: here, at last, was the monster we had always feared and never known, at last made undead flesh before us. The fear of darkness and death was added to some curious sexual neuroses and excitements: and an industry of vampire films was established.

Batman is another figure hagridden by death. He first appeared in comic-book form in 1939. The wealthy playboy and philanthropist Bruce Wayne is forever scarred by witnessing his parents' murder when he was a child, so he resolves to fight for justice and establishes his alter ego of the man in the bat suit. He has no superpowers and must rely on his own wit, training and resolve. The series of *Batman* films has been increasingly dark, both tonally and to an extent morally: neither Batman nor Bruce Wayne is at peace with the world.

It seems we need bats and bats in human form to come to terms with the darkness, in a form that is in turn sexy, ridiculous, comforting and inspiring. Dracula and Batman are both figures of death who have conquered death: one to do evil and the other to do good. Bats represent the night and the ability to navigate it: to be at home at times and in places forever inimical to light-loving humans.

And that is the way that bats live their lives, untroubled by human fears: but troubled greatly by declining numbers of insects as we humans get ever more efficient with our chemicals and our destruction of habitat where insects can flourish.

SIXTY-FIVE

BUMBLEBEE

'If you took away insects the entire thing would collapse.
It would be a catastrophe.'

Edward O. Wilson

Bumblebees must here stand for all the different species of animals that help plants to have sex: to make possible the process of pollination, to transfer the potent clinging dust from one flower to another. Humans have been in debt to wild pollinators ever since we began to eat food: and that debt continues to this day. It's reckoned that, for every third mouthful of food, we must thank our pollinators, as said before. There are many mind-numbing figures concerning the importance of pollinating insects: in 2015 the Intergovernmental Science-Policy Platform on Biodiversity and Ecosystem Services stated: 'The annual market value of additional services is estimated at US\$237–577 billion worldwide.'

We have already discussed the part in pollination played by domestic honeybees (see Chapter 12). We must now look at the still greater part played by wild pollinators, and the desperate disservice we do ourselves by causing their decline and extinction.

Busy bees: white-tailed bumblebee nest by Bob Bampton.

Bumblebees are highly visible creatures, active in the warmer months in the northern hemisphere; they have also been introduced to New Zealand and Tasmania (to pollinate clover, which was introduced as a fodder crop). They are large, slow-moving, noisy and brightly coloured: it is impossible not to be aware of them. There are more than 250 species in the genus *Bombus*. Like honeybees they are social, but on a much smaller scale. They don't overwinter as a colony – they do in the form of a fertilized queen – so they have no need for winter stores and don't make honey. Instead the queens hibernate, emerge in the spring and set about forming a colony of around fifty individuals.

Bumblebees are furry in the body, which is a help in making an early start on the cold bright days at the beginning of spring, and their colouring, mostly black and yellow, or black and orange, advertises the fact that they are dangerous. (Dangerous creatures tend to resemble each other, thereby offering a universal warning, a phenomenon called Müllerian mimicry.) They can sting repeatedly, having no barb on their sting in the manner of honeybees, but, as most of us know, they are unaggressive creatures with a strong faith that their warning will be enough. They will operate the sting, particularly in defence of the hive, but, unlike those late broods of wasps, they are too busy for needless aggression. Their bustling need to get on with the job in hand has made them creatures that traditionally arouse affection and even admiration from the humans they share their spaces with.

Before we move on, one thing to clear up is their flight. There is the glorious legend that, according to science, the bumblebee is unable to fly: a parable about the vanity of the human quest for knowledge. The origin of the story is much disputed: a back-of-the-menu calculation by an aerodynamic engineer in conversation with an entomologist at a banquet, who 'proved' that the bumblebee's wings are insufficient to keep it aloft. In fact, science is perfectly capable of explaining the bumblebee's flight in considerable detail: the rough-and-ready calculation presumably missed out the phenomenon of dynamic stall, which is a separation of the airflow that creates a vortex above the wing, thereby providing lift. Or to put that more simply, a bumblebee flies more like a helicopter than a fixed-wing aircraft.

There are a few crops out there that cannot produce fruit or seed without pollination. Some examples: okra, kiwi, potatoes, onions, cashew, celery, star fruit, Brazil nuts, beets of all kinds including sugar beet, mustard, rape, broccoli, cabbage and all the other brassicas, peas, all kinds of beans, chilli peppers, papaya, caraway, melons, oranges, lemons and all other citrus fruits, coconuts, coffee, tobacco, cucumbers, squash, carrots, strawberries, cotton, sunflowers, flax, lychee, apples, mangoes, alfalfa, avocados, apricots, cherries, almonds, pomegranates, pears, raspberries, aubergines, clover, blueberries, tomatoes and grapes. That list is by no means exhaustive.

A study from Cornell University in the United States concluded: 'Without insect pollinators roughly a third of the world's crops would flower only to fade and then lie barren. Unmanaged native bees provide crucial pollination services to many important crops.'

As seen in Chapter 12 (on honeybees), we have attempted to make good the loss of the free pollinating services of wild insects and other pollinators by employing the services of commercially managed honeybees. But they cannot provide the whole answer. A study based at the National University in Argentina examined forty crops in 600 fields on every populated continent of the world and came to the conclusion that wild pollinators are twice as effective as domestic honeybees.

Cluttering the corridor: illustration for The Tale of Mrs Tittlemouse by Beatrix Potter (first published in 1910).

There is a further problem: the more you rely on a single species, the greater the risk you put yourself under. A single disaster can rip through a species and devastate the numbers. We have seen this in the UK, when Dutch elm disease wiped out the elm as a British species – the plant can be found in hedges, but if you don't cut the hedge and the elm reaches 10ft (3m) or so in height it dies. The UK hasn't been reduced to a treeless waste because there is more than one species of tree in the country, and the problems that affect the elms didn't reach other species. The biodiversity of the British countryside, such as it is, protected it from devastation.

So if honeybees start to encounter problems and we rely on them exclusively for pollination, we are in trouble. As we have already seen, honeybees are in trouble with the problems of colony collapse disorder. Over 1 million hives are required to pollinate the almonds in California, and it costs the growers a fortune. In Sichuan in China, workers are employed to pollinate the fruit trees by hand, using chicken feathers and the filter tips from cigarettes.

Jason Tylianakis of the University of Canterbury in New Zealand summed up the study on wild pollinators already mentioned: 'The studies show conclusively that biodiversity has a direct measurable value for food production and that a few managed species cannot compensate for the biodiversity on which we depend.'

In these pages, then, bumblebees stand for a great range of wild pollinators which, by living the way they do, make our world what it is and allow us to grow food for our own use – for wild plants also require the service of wild pollinators. Animals that provide pollination service include hunting wasps, bee flies, butterflies and moths, pollen beetles, midges, thrips, carrion flies, hoverflies, fruit flies, cockroaches, ants, flatworms, mosquitoes, bats, birds, monkeys, rodents and even some lizards.

The idea that we humans are dependent on wild creatures – a great suite of wild creatures – for our daily nutrition and for our survival is not one we can easily deal with. We traditionally see most non-human species as rivals, in direct competition with us, and across the millennia we have managed the planet on that basis. We have destroyed great swathes of habitat on which bumblebees (and other pollinators) can live and thrive and make more of their kind. We have broken up natural habitat, effectively marooning pollinating insects in a small patch of countryside. We have also gone out of our way to poison insects. After DDT was banned (see Chapter 10 on eagles, Chapter 23 on mosquitoes, and elsewhere) we simply turned to other kinds of insecticides. Among recent examples are neonicotinoids, which were seen as a most marvellous and effective way of growing high-value crops.

It has sometimes seemed to me that one of the problems we have in thinking about insects is the catch-all term 'bug', which has become a designation used for any insect, and practically any other small form of invertebrate life. The word is ancient, but was taken up in America as a blanket term. All insects are bugs and we keep them out with a bug screen and bug spray, because a bug is, more or less by definition, nasty. A bug in a computer programme is a bad thing. An infectious illness is blamed on a bug, an irritating person is someone who bugs me. All insects are bugs, therefore all insects are bad.

This is increasingly proving to be one of our bigger mistakes.

But it has triggered an outbreak of public concern. It's not hard to work out that a world without pollinating insects will be a sterile one. It is a truth we find easy to grasp intuitively. The problems of declining wild bees, along with the publicity given to colony collapse disorder, have created a new awareness of bees: and a new understanding that we need to do something about their decline.

Conservation organizations have established projects to encourage wildflower meadows. There are campaigns to get people to create bee-friendly gardens: with nectar-rich plants and no insecticides. Farmers are paid to leave flower-rich borders in their fields. There are strategies for creating bee-roads: roadside verges managed for flowering wild plants, bringing back bee habitat. B-Lines is a scheme in the UK that aims to join up remaining bee habitats, allowing the insects to travel. We are beginning to wake up to the idea that bumblebees (and all the other wild pollinators) are not only nice to have around, but they are also essential to the continued survival of the human species.

SIXTY-SIX

SALMON

'What a day! Two salmon this morning and the offer of the Exchequer this afternoon.'

Quoted in *Neville Chamberlain: A Biography* by Robert Self

The combination of their palatability and their heroic lifestyle – great journeys accompanied by glorious leaping – gives salmon a kudos no other fish can possess. The fact that most of the salmon consumed in North America and Europe have never seen a waterfall, still less leapt over one, has no effect at all on our notion of the salmon as the fish of glory.

I have sat at a fall and watched the salmon leaping it: a spectacle that forces an empathy from all those who look on, each successful leap bringing an involuntary cheer from your throat, as if you had witnessed the most wonderful goal at a football match. You identify with the fish: and, by the rule of you are what you eat, consuming a salmon gives you a share in that glory.

Salmon is not a precise term. There are nine commercially important species of salmon, which are for convenience divided into Atlantic and Pacific salmon. There are other species that are called salmon without being closely related to the *Salmo* genus: these include Australian salmon, Indian salmon, Hawaiian salmon and others. As seen before (see Chapter 14 on sharks) British fish-and-chip shops used to sell a delicacy known as rock salmon, which was, in fact, a cut from species of shark.

It is the heroic migratory lifestyle of the true salmon that appeals to the human imagination, though salmon that have found their own way or been introduced into lakes frequently evolve a non-migratory life. Migration, in all species and by whatever means they travel, is a high-risk strategy. Folklore has always claimed that salmon return to the self-same spot in which they were spawned, and modern research has shown that this is, by and large, the truth. The Chinook and sockeye salmon of Idaho make a return journey upstream from the Pacific of 900 miles (1450km), involving a climb of 7000ft (2100m).

These journeys end in spawning and death for most of the fish involved: and that creates an annual goldmine of food for any species capable of exploiting it. These famously include bears and eagles, and of course humans. Wild salmon are keystone species in the ecosystem to which they return, for as they do so they transfer the protein of the ocean to the forest.

Fish as hero: leaping landlocked Atlantic salmon on the line of a fly fisherman, by W. Blackwood Law (1950).

Most salmon species are anadromous: they make a transition from fresh to salt water, and then back again. The journey through the brackish water of tidal rivers and estuaries helps them to make that adjustment. They hatch from eggs and are then known as fry, and pass through a series of stages as parr and then as smolt. In this stage their body chemistry changes to allow the fish to make the transition to salt water, where, depending on the species, it will spend between one and five years as it grows towards sexual maturity. Some species acquire a hump, or canine-like teeth, or a curvature of the jaw: often these mature salmon have a grotesque, gargoyle-like appearance.

Everyone who has ever paddled a boat knows that, in most circumstances, the best plan is to start against the current while you are fresh, and to come back with the stream, when you are weary. Salmon take the opposite view, making upstream journeys that can be hundreds of miles long, taking on rapids and waterfalls as they go. It is certainly a serious test of the quality of a fish: only the best can make it to the spawning grounds. All this puts salmon among the most spectacular animals on Earth today. They catch the human imagination as few others do: and they are considered supremely delicious. What's more, food science has revealed that there are great nutritional benefits in eating salmon: they are oily fish with high protein along with omega-3 fatty acid: a diet high in such acids leads to a lower risk of heart disease: thus allaying the fears that come from eating too much beef (see Chapter 7 on cattle).

Salmon also thrill recreational fishers, who find combat with wild salmon while armed with a rod and line a supremely satisfying occupation. The ritual photographs of the successful angler threatening the cameras with a dead fish invariably show a person with a smile that simply can't be suppressed: meeting at the back of the neck: experiencing something close to supreme happiness. People will pay for such pleasures, and in many rivers the owners make far more money from recreational fishing than they could from commercial fishery. Salmon have been introduced to the Great Lakes of America, and they're all there for recreational fishing.

Salmon fishing is part of the pattern of human ownership of nature: the ancient need to take it on and win, emerging with a prize. It is a process that shows not only the prestige of humans as a species but also the prestige of the individual human concerned. Such great prizes are beyond the reach of most. They represent an economic as well as a physical achievement.

As we have seen with the cod (see Chapter 17), the problem with commercial fishing is that, while it is in everybody's interest to fish sustainably, it is in the interest of no single individual out on the fishing grounds. As a result of this simple truth, global fisheries are routinely overfished; it's been estimated that 75 per cent of the world fisheries are worked at a level near or beyond the point of sustainability.

Portrait of the artist as a fisherman: Sir Edwin Landseer fishing for salmon with a ghillie, painted by John Frederick Lewis (1804–76).

This has recapitulated the situation that humans first encountered in the Fertile Crescent 12,000 years ago: to make food more accessible, you must domesticate the animals we eat. If we can't catch wild fish and eat them, we can set up a farm and eat tame fish instead. It is reckoned that about half the marine fish (salmon included) that we eat now come from fish farms, about 62 per cent of them in China. These farms can take the form of suspended cages or concrete tanks.

They don't provide a long-term solution to the problems of oceanic pollution, overfishing or of feeding a growing human population. Carnivorous fish such as salmon are usually fed on fish lower down the food chain, which are wild-caught and fed to the captive salmon; it takes 5lb (2kg) of small fish to create one pound of salmon. Farmed fish have also been fed on plant material, like soy, most of which has already been grown at a high environmental cost; again, the equation between food for the salmon and the resulting food on the table is unbalanced.

Farmed fish are kept in intensely crowded and therefore stressful conditions, which make them vulnerable to disease and to parasites like sea lice. The answer for the farmer is to treat them with pesticides and antibiotics; the resulting problems from such substances are found in Chapter 7 on cattle, and *passim*. There are also problems caused by pollution from the excrement of fish and from uneaten food. There are ethical concerns: as in battery chickens (see Chapter 29), these fish are being treated as plants to harvest rather than animals to nurture. The salmon we see on the plate – 99 per cent of all the Atlantic salmon that reaches the table is farmed – may not be the wild and free salmon of our imagination, leaping the falls on its way to glory, but instead a fish that has lived its entire life as confined as tightly as a person on a commuter train, a member of a shoal that spends every second of its existence swimming in its own excrement.

SIXTY-SEVEN

ORYX

*'Jill had, as you might say, quite fallen in love with the unicorn.
She thought – and she wasn't far wrong – that he was the
shiningest, delicatest, most graceful animal she had ever met...'*

C. S. Lewis, *The Last Battle*

L et's start this chapter with two very clear visual memories.
First: Namibia, in the desert. Perhaps an hour since I saw anything that wasn't sand or gravel. And there, just below the top of the long slope of a white slanting dune, a white horse that bore on its head a single horn... for a long moment it really was a unicorn, until the vehicle moved on and the perspective shifted and one horn became two. This was an oryx, one of the Hippotraginae or horse-like antelopes.

Almost a unicorn: Galloping Oryx *by Mark Adlington (2010).*

Second: a cartoon by Charles Addams, the American cartoonist with a dark strain of humour, whose work inspired *The Addams Family* television series and film. I saw this cartoon as a child – perhaps twelve – and never forgot it. It is not too much to say it has haunted my life. Two unicorns are standing together in the rain, a rising flood at their hooves. In the distance, Noah's Ark is slowly drifting away...

It's likely that the myth of the unicorn came from the reality of the oryx. Unicorns weren't part of Greek mythology; they were considered perfectly real. The physician Ctesias, visiting the Persian king Darius II in the fifth century BC, wrote about an animal whose single horn was used 'as a protection against deadly drugs'. Aristotle ran with this to an extent: 'We have never seen an animal with a solid hoof and two horns, and there are only a few that have a solid hoof and one horn, as the Indian ass and the oryx.' I suspect that the Indian ass was the Arabian oryx and the other the scimitar-horned (or just scimitar) oryx. There are four species in total, the others being the South African oryx, called in Afrikaans gemsbok, and the East African oryx. The Arabian oryx has the remarkable distinction of being the only animal that has ever been downgraded by the IUCN from Extinct in the Wild to merely Vulnerable: promoted by four entire categories, by way of Critically Endangered and Endangered. It seems that the ark came back for this one.

Oryx are dry-country antelopes. Their gaits are pleasingly horsey and in some attitudes they seem to be a light, rather airy-fairy kind of horse. They were once widespread in the arid, semi-desert parts of Africa and West Asia, adapted to go without water for extended periods and capable of enduring temperatures that would kill most other mammals. Among their adaptations for thermoregulation, they have blood vessels that cool the blood before it reaches the brain, the organ most vulnerable to overheating. They stand at around 3ft (1m) at the shoulder, though some species, notably the gemsbok, can be 1ft (30cm) taller. They all tend to stand tall with their high head carriage and long pointed horns: two, it should perhaps be stressed here, in number, though the brittle nature of these horns makes them prone to breaking, and single-horned (if asymmetrical) oryxes are far from rare. It was perhaps the trade in these horns – and the idea that a unicorn horn protects users from poison – that bumped up the role of the unicorn in mythology.

Unicorns were seen as creatures of rare, almost impossible purity and grace, other-worldly creatures midway between us and heaven. As such they are beyond the command of most humans, but could be tamed by a virgin: and thus the image of Mary with a unicorn in her lap represents the principle – as well as ocular proof – of her eternal virginity. By extension, unicorns represented chaste love and faithful marriage. Unicorns could make poisoned water drinkable and cure sickness: the horns of unicorns were procurable at a great price: the horns of narwhals and (no doubt) oryxes were there to play the part at need.

Vision of purity: Virgin and Unicorn *by Domenichino (c.1602).*

In the Bible an animal called the *re'em* (Hebrew) appears eight times, variously translated; it is usually considered to be reference to a wild ox or an oryx. The King James Bible, more gung ho, prefers unicorn: 'Save me from the lion's mouth… for thou hast heard me from the horns of the unicorns…' (Psalms 22:21).

Unicorns are a staple of magical fiction. A unicorn makes an appearance in *Through the Looking-Glass, and What Alice Found There*, fighting the lion all round the town. There is a frolicking herd of cute, pastel-shaded unicorns in Walt Disney's 1940 film *Fantasia*. Jewel the unicorn is the great friend of King Tirian in *The Last Battle*, the final book of *The Chronicles of Narnia*; in *Harry Potter and the Philosopher's Stone*, the first in the *Potter* sequence, the evil Lord Voldemort keeps himself going by drinking the blood of a unicorn: 'The blood of a unicorn will keep you alive, even if you are an inch from death, but at a terrible price. You have slain something pure and defenceless to save yourself and you will have but a half-life, a cursed life, from the moment the blood touches your lips.' Which is a long way from the rainbow-surrounded and unmysterious unicorns that have become modern toys, pink, glittery and girly.

The nearest a real oryx gets to popular fiction is in *The Story of Doctor Dolittle* by Hugh Lofting, and in two subsequent volumes. This takes the form of the pushmi-pullyu, a creature with two heads, one at each end of its body. This was shown as a double-headed llama in the 1967 film starring Rex Harrison, but Lofting draws this quite clearly as a double-headed oryx. This, again, is something of a magical creature, rare and fine and sensitive.

Humans have always liked oryxes, though they have been exhaustively hunted, for sport and for nutrition, in regions short of ready sources of meat. There have been declines in all species, but the Arabian oryx was shot to extinction. There were, however, a few individuals left in captivity. In 1962 a programme of captive breeding was started. It was based at Phoenix Zoo, Arizona, which is close in climate to the natural habitat of the species. It involved the Fauna Preservation Society, now Fauna and Flora International, and the World Wildlife Fund, now just WWF. They started with nine individuals and after 240 successful births, groups were sent to other zoos, where they too bred. Populations were released into the wild and can now be found in Oman, Israel and Saudi Arabia; in the last country they can be found in Mahazat as-Sayd Protected Area, which covers 770sq. miles (2000sq. km).

This can be savoured as an example of humanity doing right by a non-human species. It is now recognized that, with most species (the African rhino species being exceptions, see Chapter 39), the best way of ensuring their future is not necessarily by means of a holding population in zoos, but instead by looking after their habitat and preventing persecution. In 2007 Oman's Arabian Oryx Sanctuary was opened up for oil prospecting. The oryx population dropped from 450 to a bare handful, and is a rare example of a place being stripped of its status as a United Nations Educational, Scientific and Cultural Organization (UNESCO) World Heritage Site.

The scimitar-horned oryx of northern Africa was declared Extinct in the Wild in 2000. There are a number of captive populations, with an estimated 11,000 in Texas and a further 4000 in the Gulf States. There are schemes that have established various semiwild populations in fenced reserves: so yet again, in the case of this group, which humans have always liked and which adapt well to confinement, the dream continues: the hope that in at least some cases we really can put the toothpaste back into the tube.

SIXTY-EIGHT
SHEEP

‘ *"Worthy the lamb that died," they cry, "to be exalted thus!"*
"Worthy the lamb," our hearts reply, "for he was slain for us." '

Isaac Watts, 'Come Let Us Join Our Cheerful Songs'

Sheep are as important to human civilization as God: so much so that we identify God with sheep. In traditional Christian imagery, Jesus is not only the good shepherd (and we are members of his flock) but he is also the Lamb of God: the necessary sacrifice that saves us all from death.

Sheep were the first large mammals we domesticated for food. There was a good reason for this: sheep are easy. Comparatively, anyway. You are in less danger from a flock of sheep than from a herd of aurochs or a sounder of wild swine. Sheep accept confinement with reasonably good grace, and they can subsist on much less: where one cow grazes you can keep half-a-dozen sheep.

Lamb of God: stained-glass window of Christ the Lamb in St Matthias church, Budapest.

Sheep have a strong understanding of flock, of leadership – and, most significantly, of followship. They are also prey animals who will turn and run together, seeking safety in numbers. These facts make them fairly easy to move and control. I did this myself when fifty sheep broke out from the field a hundred yards away; alone and without a dog or any other such artificial aid, I gently persuaded them to return, anticipating their moves and making sure I didn't split the herd.

It is reckoned that sheep were first domesticated in Mesopotamia, perhaps as many as 13,000 years ago. There are six species of wild sheep; the moufflon *Ovis orientalis* was the ancestor of domestic sheep, which is classified as *O. aries*. Sheep provided clothing and warmth in the form of wool, and food in the form of meat and milk, which could be made into cheese for portable food that kept longer than meat. The minute you had a flock of sheep, you had something pretty close to civilization: a very great deal of what you needed to survive was now conveniently near, in a non-taxing and non-aggressive form.

Sheep abound in the Bible, most famously in Psalm 23: 'The Lord is my shepherd, I shall not want. He maketh me to lie down in green pastures; he leadeth me beside the still waters.' Those words still conjure up an Arcadian notion of peaceful living, in which easeful death is no great hardship.

The ancient Greeks were great shepherds: in the *Odyssey*, Odysseus and his men escape from Polyphemus by hiding under the sheep: the blinded Cyclops sees them out by touch and is convinced he has released only the sheep, but finds himself outwitted. The Romans were also keen on sheep and took them around their empire. Later European colonialists brought sheep with them to feed their new worlds: central and southern South America, Australia and New Zealand were filled with sheep. Other important sheep-raising nations include China, India, the USA, South Africa and Turkey.

The Bible also talks about the problems of distinguishing sheep from goats: a necessary skill in animal husbandry. Rugged, thin-fleeced breeds of sheep can look very similar to goats, but they are but distantly related and any unlikely offspring from their union will be infertile. Look for the beard of the goat and the split upper lip of the sheep (see also Chapter 77 on goats).

There are several hundred breeds of sheep, adapted for different forms of pasture and climate, and for different priorities (meat or wool or milk). Many breeds have dwindled in popularity, though some are kept going by enthusiasts. Often they will claim that such and such breed is 'rarer than the giant panda', which is not comparing like with like. Domestic sheep aren't wild animals, and a breed is not the same thing as a species – a rare Jacob's sheep can cross-breed with a merino sheep without any problem whatsoever.

Sheep thrive on rough grazing, though they won't take the woody stems that goats do. They are able to get the most out of this in the classic way of ruminants:

A beast to worship: The Adoration of the Mystic Lamb, *by brothers Hubert and Jan van Eyck, in St Bavo Cathedral, Ghent (1432).*

they bring up a bolus of pre-chewed and fermenting vegetation and chew it again, getting more from it the second time around. This technique of making comparatively little go a long way is the great edge that all ruminants possess: it helps them to survive in the wild and it has made them the most useful of all domestic mammals bred for food. You need less pasture and less winter fodder for a ruminant than you do for, say, a horse of comparable size. Sheep are still more obliging in that they don't need much of anything in the way of artificial shelter, even in hard weather, and they don't require artificial feeding, in the form of grain or soy, as cattle do in many parts of the world.

It wouldn't do to claim that the life of the shepherd is undemanding. All modern sheep farmers will tell you that, from birth, a sheep has a single ambition in life, and that is to die. They are prone to disease and to parasites. Most spectacularly, they are vulnerable to flystrike, in which flies lay eggs in the nice thick fleece, and the resulting maggots eat the sheep alive from the outside. This is hard to deal with (I have done so) and the best way is prevention: by means of highly toxic sheep-dips. The poisons involved get into the waterways and cause pollution downstream. Equally powerful sheep wormer is needed to deal with internal parasites, which also find their way into the ecosystem via the droppings.

Sheep have become an image of people who are timid and easily led, eager to abnegate any will of their own. People with a meek bashful demeanour are sheepish. The traditional way of getting to sleep is counting sheep: soothingly anonymous and unaggressive creatures to people a brain and to lull its owner back to sleep. But rams have more vigorous implications: the Los Angeles Rams play in the National Football League of the United States, while natives of the astrological sign of Aries are courageous, determined, honest and passionate. People born in the year of the sheep in the Chinese horoscope (sometimes interpreted as the year of the ram or goat) are gentle, courtly and broad-minded.

The historical roll call of shepherds includes some of the biggest names in human history: Abraham, Isaac, Jacob, Moses, King David and Muhammad. 'There was no prophet who was not a shepherd,' Muhammad said. Sheep have been important items for sacrifice in Hebraic, Arab, Greek and Roman cultures; which is why Jesus (sacrificed on the cross) is considered to be the Lamb of God. Lamb is the traditional Easter dish for Christians; traditional Jews eat lamb on Passover.

There are more sheep in traditional nursery rhymes than any other animals, showing their ancient importance: Mary had a little lamb, so did Little Bo-Peep, Baa-baa black sheep had plenty of wool. And the sheep in wolf's clothing is a stock image to this day.

In former times (and, in some places, to this day) the main problem with sheep was predation by wild animals. Shepherds were required not just to keep an eye on the flock and drive them back towards people at the appropriate time, but also to chase off predators, sometimes an unnerving business. Big fierce dogs were bred to help the humans out, the most famous being the German shepherd.

Despite the ancient part that sheep play in human civilization they have also appeared in some of the most startling developments of modern science. In 1995 two sheep, Megan and Moray, were cloned from differentiated cells, and a year later the more famous Dolly was created from an adult somatic cell.

The romantic feeling that people have for sheep and shepherds can run counter to logic. Upland areas of Britain are routinely stocked with large numbers of sheep, which are heavily subsidized by government. There is little economic justification for this, but this idea is indignantly and emotionally resisted. The fact is that, by overstocking high land with sheep, the upland blanket bogs have been destroyed and their ability to hold water in heavy rains has drastically diminished. The water runs straight off the hills and causes floods downstream, which cost a great deal of money to make good and/or to prevent with large earthworks. The government is effectively paying sheep farmers to cause flooding: this is not sensible or logical, but try telling British people that mountain shepherds are doing wrong. Sheep matter to us at a very profound level.

SIXTY-NINE

NENE OR HAWAIIAN GOOSE

'The most effective way to save the threatened and decimated natural world is to cause people to fall in love with it again, with its beauty and reality.'

Sir Peter Scott

T he start of the Environment Movement is generally put at 1962 and the publication of *Silent Spring* (see Chapter 23 on mosquitoes). I'm not convinced that's right. If we define the Environment Movement as a combination of two things: (1) the realization that something needs to be done to protect the environment and (2) a genuine attempt to do it, then we can perhaps advance the date of this double-dawning to 1950: the year that Peter Scott (see also Chapter 46 on ducks) started to breed the nene at what was then called the Severn Wildfowl Trust at Slimbridge, in Gloucestershire, in the west of England.

Visionary: Sir Peter Scott, founder of the Wildfowl and Wetland Trust, with the first British-raised nenes.

Or perhaps even earlier than that: Scott opened Slimbridge to the public in 1946, a place set up to research the needs of wildfowl – which is a shooter's catch-all term for ducks and geese. It was opened to the public, which was then an unusual step and showed a remarkable early grasp of the fact that increasing numbers of the population were increasingly nature-deprived, and that places such as Slimbridge could reconnect people and wildlife. This disconnection has been subsequently shown to be deeply damaging to human life as well as to conservation.

Islands are rum places (see Chapter 4 on the Galápagos mockingbirds and Chapter 19 on the dodo) and they are frequently where rum animals evolve. Speciation – the process of creating new species – tends to occur in two different ways: when different populations in the same place come to occupy different ecological niches and start to live their lives in different ways; and when populations of the same species are separated from each other: sympatric and allopatric speciation. The nene, like most species specific to islands, is an example of the latter: speciation by means of isolation from others of their kind.

The nene or Hawaiian goose is related to the Canada goose, but has chosen a very different way of living. They are waterbirds that don't like water very much. They do most of their feeding on the lava plains of Hawaii, along with scrubland, grassy areas and coastal dunes as well as manmade habitats like pasture and golf courses. Nenes don't even have completely webbed feet, an adaptation for the rocky terrain. Unusually for geese, they mate on land.

They were first known to Europeans on the voyage of Captain Cook, who arrived in Hawaii in 1778, and it is reckoned that there were around 2500 of them. By the middle of the twentieth century there were only a couple of dozen; estimates vary.

Hawaii is one of the world's extinction hot spots; New Zealand is another. Both these places had a large and diverse population of birds unique to their islands, many of them flightless. They were, then, ill-adapted to the arrival of humans and the animals that humans brought with them. But in both these isolated places the pace of destruction hotted up considerably with the subsequent arrival of colonists from Europe. Hawaiian species wiped out before 1778 include the moa-nalo, which is a group of four huge birds that evolved from dabbling ducks (like the mallard, see Chapter 46) and reached massive size, standing up to 4ft (1.2m) tall and weighing up to 16lb (7.2kg). They are known from subfossils, old bones found on the islands of the archipelago.

The state of Hawaii comprises 0.25 per cent of the landmass of the United States but contains more than 25 per cent of its threatened species. Hawaii has lost more species and contains more threatened species than any other state. That is the nature of islands. And it looked as if the nene was certain to go. There were thirty-two wild nenes left on three of the Hawaiian islands when Scott acquired a pair in 1950 and began to breed them at Slimbridge.

Triumph: Nenes on Mauna Loa *by Sir Peter Scott.*

Scott was a remarkable man, son of the Antarctic explorer Captain Robert Falcon Scott, whose doomed expedition to the South Pole of 1910–13 has haunted the British imagination ever since. In his last letter to his wife, Kathleen, which was found with his body in 1913, he wrote: 'Make the boy interested in natural history if you can; it is better than games – they encourage it at some schools.'

His son Peter's early interest in natural history was in killing: he became a keen wildfowler. He was also highly talented at drawing and painting, and some of his classic work of geese and ducks in flight became almost archetypal images, drawing-room favourites. I was given one myself when I was ten. And, like not a few shooters, Scott gave up shooting and became a conservationist, and a great one: Sir David Attenborough called him 'the patron saint of conservation'. The organization he founded, now the Wildfowl and Wetland Trust, is thriving and has nine reserves in Britain and Northern Ireland, including the startlingly lovely reserve in the middle of London, close to the River Thames, the London Wetland Centre.

The rescue of the nene was a classic good work. Reintroductions from Slimbridge added numbers and resilience to the Hawaiian population. But reintroduction wasn't a complete success on its own. Even as numbers increased, the birds were still facing problems from mongeese (a better plural than the more correct mongooses). European colonialists introduced mongeese to a number of islands to get rid of the snakes. But mongeese eat plenty of other stuff as well, and the eggs and chicks of ground-nesting birds suited them admirably. The population of nenes stabilized and became self-sufficient after campaigns to extirpate mongeese from Hawaiian islands.

There is something a little quaint and dated about the story of the saving of the nene. In these very early years of practical conservation, the problems of habitat destruction and introduced predators were not fully appreciated. The idea that we are running out of planet is a big idea to take on, after all. It was widely assumed that there was plenty of planet left: after all, there always had been, in all the millions of years of human existence, so why should things change? It required a colossal shift in human thinking to come to terms with this most disturbing truth.

Captive breeding, with the help of zoos, was then seen as a crucial part of conservation: an insurance population that could lead to reintroduction, as with the nene, as with the Arabian oryx (see Chapter 67). It was hard to envisage a situation in which there would be nowhere left to put a threatened species back. It was possible to come to terms with the notion of saving the gorilla, for example: much harder to grasp the notion of saving the rainforest. And, of course, much harder to raise funds for such an abstract idea.

The notion of saving a species is linked to biodiversity: the understanding that life operates by making lots and lots of different species, and that the destruction of biodiversity weakens the resilience of life on Earth. But the importance of biodiversity has fostered a wrong notion: the idea that our most important job in conservation is to maintain a token number of every species. Other, more terrible understandings of the loss of nature have developed with the growth of the Environment Movement.

Scott went on to become a founder member of what was then the World Wildlife Fund, now WWF. He designed the panda logo, a species that spells out the need (see Chapter 16 on pandas) to save habitats, and that saving habitats is at base the only way we have of saving ourselves. He also co-designed the Red List of Endangered Species, which is now operated by the IUCN, and which is referred to in chapter after chapter of this book. He became Sir Peter Scott and died in 1989, aged eighty.

SEVENTY

ORANG-UTAN

*'Destroying rainforest for economic gain is like burning
a Renaissance painting to cook a meal.'*

Edward O. Wilson

In 1984 the rainforests were revealed to the world as both a wonder and a crisis. Their destruction became a major focus and concern. The idea of saving the rainforest went mainstream. It was a drastic promotion, considering that, a year earlier, few people other than specialists knew what rainforests were or why they mattered. And as concern for the rainforests grew, we needed a single species to stand for them all. The orang-utan, though now found on only two islands in the world, has been most often chosen to play that role (but see also Chapter 96 on jaguars).

Sir David Attenborough opened the world's eyes to the rainforest in *The Living Planet*, a series about ecology, taking the planet habitat by habitat. The fourth episode is called 'Jungle' and it remains a powerful piece of television. Attenborough shows the bewildering diversity, the astonishing specialization and the dizzying camouflage. In a glorious coup de television, he dramatically reveals to us all that he is standing there right in the middle of the richest habitat on Earth. And as he goes on to show in the final episode, this habitat is under the greatest threat from human pressures. He concludes the series with these words: 'Immensely powerful though we are today, it's equally clear that we're going to be even more powerful tomorrow. And, what's more, there will be greater compulsion upon us to use our power as the number of human beings on Earth increases still further. Clearly we could devastate the world... As far as we know, the Earth is the only place in the universe where there is life. Its continued survival now rests in our hands.'

The orang-utan became a symbol of the rainforest: for its human-like qualities, of course, because it is one of us, one of the great apes, and also perhaps for its exoticism; it is far less often found in zoos than chimpanzees and gorillas. Also, there is an enigmatic quality in an orang-utan's facial expression. Orang-utans are less social than gorillas and chimpanzees, and so they have less need to communicate with their faces. For that reason, you can read anything into that level, golden-eyed stare: blame, reproach, suffering, patience, perseverance: above all, the fact that one of our own is in need of our care.

There are three species of orang-utan, all of them rainforest specialists: the Bornean, the Sumatran and the Tapanuli orang-utan. The first is confined to Borneo, an island that is part Malaysian, part Indonesian and also contains the independent kingdom of Brunei. The second two are both found on the Indonesian island of Sumatra; though by one of those quirks of biogeography, the Tapanuli orang-utan is more closely related to the Bornean than the Sumatran. They have been known in the West since the seventeenth century. The word orang-utan comes from the related local languages, *bahasa Malaysia* (or *Melayu*) and *bahasa Indonesia* (the first word means language). It is a mild bastardization of words that mean people of the forest. (The correct word for forest is *hutan*, and Hutan is the name of an excellent conservation NGO in Malaysian Borneo; see also Epilogue for more.)

The Dutch scientist Jacobus Bontius wrote about the orang-utan in 1631; he said that people believed that orang-utans could talk, but preferred not to 'lest he be compelled to labour'. There is also a curious link between orang-utans and Rembrandt, who painted one of the most chilling works of art ever hung in a gallery, *The Anatomy Lesson of Dr Nicolaes Tulp*. Tulp also dissected an orang-utan and wrote up his findings in his book *Observationes Medicae*, under the chapter title 'An Indian Satyr'.

Petrus Camper, another Dutch scientist (the Dutch colonized most of the orang-utan heartland), gave the first accurate description of the animal in the eighteenth century, and Linnaeus described the orang-utan in the 1760 edition of *Systema Naturae*, classifying it in the genus *Homo*. The creature makes the title page in the great work of 1869 by Alfred Russel Wallace. Wallace was a scientist and explorer; he was also co-discoverer, with Charles Darwin, of the principle of natural selection (he graciously acknowledged – in a refusal of priority almost unique in the history of science – that Darwin, who sat on his findings for many years, had got there first). Wallace's book was called *The Malay Archipelago: The Land of the Orang-utan and the Bird of Paradise*.

Darwin also encountered orang-utans, not wild but in the zoo at Regent's Park in London in 1838, where he had access to an individual called Lady Jane, usually addressed as Jenny. She was dressed in human clothes to amuse visitors and taught to drink tea: in other words, her human-like attributes were stressed and exaggerated. Darwin made lengthy notes of his time with Jenny, and they appear in what are called 'the transmutation notebooks': in other words, the most significant of all the many notebooks he filled. To him, the kinship with humans was obvious and by then non-negotiable. He records more than once that Jenny was 'like a child', and wrote: 'Let man visit Ourang-outang in domestication, hear expressive whine, see its intelligence when spoken [to], as if it understood every word said...'

Friend of Darwin: Jenny, the London Zoo orang-utan.

317

People of the forest: sub-adult male orang-utan in Lamandau Nature Reserve, Malaysian Borneo.

Wallace saw them in the wild in 1855 and, indeed, shot a good number of them. His closest relationship with a living orang-utan was a baby that survived for three months in his care after he had shot the mother.

At this time, all the non-human great apes were frequently referred to as orangs, or orang-utans. After the publication of *The Origin* in 1859, Charles Lyell – author of *Principles of Geology* and the man whose notion of Deep Time was essential for Darwin's (and Wallace's) theory – announced that the only way to accept it was 'to go the whole orang'. In other words, we had to accept human descent from non-human animals.

But rather like Wallace, orang-utans became an overlooked, half-forgotten part of the big story. They were very much the third of the non-human great apes, after gorillas and chimpanzees. Jane Goodall (see Chapter 42 on chimpanzees) and Dian Fossey (see Chapter 3 on gorillas) are comparatively well-known: both of them were brilliant primatologists who worked on their chosen subjects with great support from Louis Leakey. They are sometimes referred to as 'Leakey's angels' and there were three of them. The forgotten angel is Biruté Galdikas, a Canadian Lithuanian who studied orang-utans, again with Leakey's support.

Her work reveals creatures as complex as you would expect. They are the most arboreal of the great apes, seldom coming down to the forest floor; they live mostly on fruit and are solitary and social at the same time. Females have a prolonged association with their offspring, up to three years, and this is a period of education. Males in maturity develop exaggerated cheek ornaments called flanges. Each evening an orang-utan will construct a nest; I once watched one doing this, with the absorbed but slightly absent expression of a craftsman performing a task long-mastered. Infants learn this skill from their mothers, elaborate structures that can include a roof, a mattress, a pillow and blanket all made from vegetation.

They have been observed using and modifying tools to reach insects and to extract seeds. Not all populations behave in the same way: geographical variations in tool use imply that this form of behaviour represents a culture: unsurprisingly, the more social populations are more advanced culturally. In Atlanta Zoo orang-utans have been recorded playing games on touch-screen computers. Research in Leipzig Zoo has shown that orang-utans have mastered the principle of 'calculated reciprocity': that is to say, they can weigh up the benefits of gift exchange, the first non-humans to have been observed doing so.

All three orang-utan species are classified at Critically Endangered; the Bornean species has declined 60 per cent in sixty years, the Sumatran 80 per cent in seventy-five years. There are about 800 Tapanuli orang-utans left; they are Critically Endangered and the most vulnerable of all the great ape species. They have been poached for meat and for the illegal pet trade. Clint Eastwood made two films featuring a pet orang-utan called Clyde, both were critical failures and popular successes. As we have found again and again in this book, though, by far the greatest problem faced by orang-utans is habitat destruction. Sumatra and Borneo are both ideal places for growing oil palms, and this is a product with a thousand uses. Products made with palm oil include bread, crisps, margarine, soap, ice cream, pizza, instant noodles, shampoo, chocolate, detergent, lipstick, biscuits and biodiesel. It's been reckoned that each year everyone on Earth consumes 17lb (8kg) of palm oil.

Enormous areas of rainforest have been cut down to create space for oil palms. Some companies are more benign than others, and conservation organizations recognize the need to work with such companies rather than take a hopeless stance and fight the entire industry. The point is looking after such forest we have left, and the orang-utans that live in them, rather than fighting an idealistic war that can't be won.

The orang-utan is a rainforest specialist and, therefore, is caught in the tender trap. We are likely to have orang-utans for as long as we have rainforests. That fact has helped humans to focus their concerns. Rainforests are the most efficient places on Earth at storing carbon, rather than releasing it into the atmosphere, which makes them an essential part of the planet's cooling system. That is a crucial truth for us all, but it is hard to be passionate about air-conditioning. We have orang-utans to inspire passion instead.

No nonsense: blue and yellow macaw, lithograph by Edward Lear (1832).

MACROCERCUS ARARAUNA.

Blue & Yellow Maccaw

⅓ Nat Size.

SEVENTY-ONE
PARROT

*'Live in such a way that you would not be ashamed
to sell your parrot to the town gossip.'*

Will Rogers

Parrots sit upright on a perch like a man on a chair. They cock their heads in an irresistibly human manner, and the light of intelligence in their eyes is unmistakable. They will use a zygodactyl foot as a hand, eating a nut in the manner of a child with an ice-cream cornet. Oh – and they talk. They talk our language. They are birds that remind us of ourselves: charming pretend humans in clownish colours, doomed to be forever comical, no matter how serious the story.

Pretty Polly! Pieces of eight! Give us a kiss! Parrots have spoken the punchlines in our stories again and again. A recent story, claimed as entirely true, featured a couple who owned a parrot. The name Gary was heard on television, whereupon the parrot made kissing noises. When attention was drawn to this behaviour, the parrot said, in the exact accents of the woman: 'I love you, Gary.' Gary was the name of a third party; the woman confessed all. Once again – in irresistibly comic circumstances – the parrot had brought us the punchline.

The parrot has become part of the furniture of human narrative. He's there in Aesop, making a wise rather than a comic point: the parrot is asked why he has become such a favourite and replies that back where he comes from he is common enough, but out here he is a rare thing.

There are getting on for 400 species in the superfamily Psittaciformes, which can be divided into New Zealand parrots, cockatoos and the rest – the 'true' parrots or Psittacoidea. They are all characterized by a hooked beak and those zygodactyl feet – with two toes pointing forwards and two back in the manner of chameleons and woodpeckers. They range in size from the buff-tailed pygmy parrot at ⅓oz (10g) and 3in (8cm) in length, to the hyacinth macaw, 3¼ft (1m) long, and the kakapo, the flightless parrot of New Zealand, weighing nearly 9lb (4kg).

They are found in the tropics and subtropics across most of the world, though some species operate in temperate climates. Some species have been inadvertently introduced in colder places, including Britain, where they have been found as far north as Edinburgh. More on these in a moment.

It's the beak that defines the parrot: very large and very powerful. The upper mandible is not fused to the skull but moves independently, a trait that permits parrots to administer a bite of considerable power; a hyacinth macaw can bite as powerfully as a large dog. The beak has touch receptors that give the bird a great deal of manipulative skill. This is backed up by a strong tongue in most of the seed-eating parrots.

One more important physiological trait is their brain-to-body ratio, which is similar to that of crows, dolphins and the non-human great apes. It would be fascinating if someone equivalent to Leakey's angels (see Chapter 70 on orang-utans) could make extended field studies on parrots, but the practicalities of this are insuperable. The first problem is that many of them like forest canopy, which in some ways is harder to observe than the bottom of the sea. The second is that conventional ways of marking individual birds don't work on a parrot: if you attach a leg ring (band) or a wing tag they will remove it with that powerful and dextrous beak. So most of what we know of parrot intelligence we know from tame birds.

Grey parrots have been observed to associate words with their meaning. The most famous of these is Alex, who was trained in language – what else? – by Dr Irene Pepperberg. He could use words to identify objects, he could describe them and he could count. He could answer complex questions – how many red squares? – with an accuracy greater than 80 per cent. Another parrot called N'Kisi was claimed to have a vocabulary of 1000 words. The parrot species kea have been observed using tools and they can solve puzzles.

These are all striking examples of animal intelligence, which takes us once again into the badlands that lie between humanity and the rest of the animal kingdom: a place in which we are forever uneasy, forever curious, forever inclined to make exaggerated claims, and equally inclined to pooh-pooh. We use the word parrot as a verb, meaning that somebody is repeating something not fully understood – much as I used to parrot Newton's laws of motion for my physics exam. But parrots are capable of doing more than parroting.

Most parrot species are monogamous, and the pair bond is the central thing in their lives, celebrated with much mutual preening, and almost constant companionship. This, combined with their intelligence, makes them very poor pets. Deprived of a mate, they focus instead on their owner or keeper, and when deprived of this person they go into a state of extreme distress and scream. They need a good deal of stimulus in their lives: bored parrots go into a decline and exhibit stereotypical behaviour, such as plucking out their own feathers. For these reasons, pet parrots, seen as one of the great glamour acquisitions, often lead disastrous lives, handed from one person to another before they die or are taken in by a welfare charity.

And yet the demand for their company as pets has helped to bring about the decline of many species. Habitat destruction has also been a major factor – many parrots are rainforest dwellers – but the pet trade, and more recently the illegal pet trade, has pushed many species into danger. Again, more of that in a moment.

Parrots can be difficult to keep, and in the wrong conditions will suffer from psittacosis, a disease that affects many species of bird and is often fatal. It is presumably the prevalence of this disease that brought about the folk-expression 'sick as a parrot', much favoured and much mocked in the 1960s as an expression of disappointment among footballing people. An elaboration on this – 'sick as a parrot with a rubber beak' – has been used for comic effect.

Parrots are still much-desired pets. In 1996 Tony Silva, former director of Tenerife's Loro Parque (parrot park), was convicted for smuggling hyacinth macaws into the United States; he was given an eighty-two-month sentence and fined US$100,000. The *Straits Times* of Singapore in 2016 reported the sale of a hyacinth macaw for £28,000.

Parrots appear again and again in fiction: perhaps most famously on the shoulder of Long John Silver in *Treasure Island*, the pirate fantasy by Robert Louis Stevenson; no historical pirate is associated with a parrot. The parrot's irresistible association with the comic muse is best seen in the 'Dead Parrot Sketch' in the British television programme *Monty Python's Flying Circus*, in which John Cleese plays Mr Praline, a customer in a pet shop complaining that the parrot he had purchased is dead.

I always liked Polynesia, Doctor Dolittle's parrot, who teaches him the language of non-human animals: 'After a while, with the parrot's help, the Doctor got to learn the language of the animals so well that he could talk to them himself and understood everything they said.'

But perhaps the greatest parrot in literature appears in Gustave Flaubert's short story 'A Simple Heart', in which the servant, Félicité, comes to conflate the Holy Spirit and her mistress's parrot Lou-Lou. The great story ends: 'And as she breathed her last she thought she could see, in the opening heavens, a gigantic parrot hovering above her head.' Comedy and language are two of the most important things in human life: and again and again we have revelled in bringing the two together by means of the parrot.

We have also brought a measure of comedy, some of it light, some of it distinctly dark, into the lives of wild parrots. These days, parks, particularly in southeast England, are full of ring-necked parakeets, descendants of pet birds that escaped or were released when their owners tired of them. There are two origin myths for this population: the first is that they escaped during the filming of *The African Queen*, a production of 1951 that starred Katharine Hepburn and Humphrey Bogart. Some of the scenes – most of the action takes place on a small boat –

Loved to extinction: Spix's macaw, pet-trade favourite.

were shot at Isleworth Studios in west London, on the Thames.

The second, still more picturesque, is that all the birds are descended from two birds released from confinement by Jimi Hendrix, the great innovative guitarist (see also Chapter 40 on nightingales). An elaboration on the myth is that he was in the middle of an acid trip at the time. The prosaic likelihood is that the parakeet population was the result of many releases, stoned or unstoned, and the population has grown and prospered over the past half-century, until the birds have become a feature of London liked by some, but by others bitterly resented, perhaps not so much on ecological grounds as a socially acceptable form of xenophobia.

But most parrots and most parrot species inspire affection and laughter. This was part of the undoing of Spix's macaw, which is – or was – a specialist that lived in the Bahia area of Brazil, where it fed only on silver trumpet trees (*Tabebuia aurea*). The bird suffered from the destruction of this habitat, and, being a specialist, was caught in the tender trap and unable to switch to another species of tree. It also suffered from its attractiveness to the pet trade, one that increased as the bird grew rarer. By 1990 the wild population was down to a single male, which had paired up with a female blue-winged macaw, with which it could not breed. There were and are still captive Spixes. In 1995 a female Spix was released on the same site as the mispaired male. The male preferred to stick with his chosen partner; the female was killed by flying into a power line.

In 2013 Spix's macaw was declared Extinct in the Wild. In 2018 there were 160 Spixes in captivity, 146 of them at a project in Germany. There are plans to attempt another reintroduction programme in 2021. Perhaps they are not, after all, ex-parrots.

SEVENTY-TWO
COLORADO BEETLE

'You like potato and I like pot-ah-to.'

George and Ira Gershwin, *'Let's Call the Whole Thing Off'*

The Colorado beetle evolved in Mexico as a specialist feeder on flowering plants from the family Solanaceae, particularly of the genus *Solanum*. That meant they were mostly to be found on plants like buffalo bur and various other species of nightshade. But they have no great objections to other plants in the same genus.

The scientific name of potato is *Solanum tuberosum*; of aubergine *Solanum melongena*, of tomato *Solanum lycopersicum*. In other words, humans have gone out of their way to create an ideal world for Colorado beetles, which are also known as ten-striped spearman, ten-lined potato beetle and potato bug. The adults are around ½in (10mm) long, and are bright yellow with bold brown stripes.

They evolved to eat the leaves of *Solanum* plants, and eating the leaves of *Solanum* plants is what they do. There are rather more such plants around than when the Colorado beetle first evolved, for we humans have covered vast areas of the Earth in *Solanum*. So far as this beetle is concerned, it's as if the whole world had turned into ice cream. What beneficent, God-like creature had made a world so perfect for Colorado beetles?

Hungry as ever: Colorado potato beetle by Steve Roberts.

They were first observed in 1811 and formally described in 1824. In 1840 they had adopted potatoes as a host plant and rapidly became the potato's most significant defoliant. The adults and the larvae both feed on the leaves; they can cause total loss of the crop if they get there before the tubers have formed. An adult can eat 1½sq. in (10sq. cm) of foliage in a day, so you need a lot of them to damage a crop. They can breed quickly: they are good at becoming a lot. In 1859 they were found west of Omaha in Nebraska, and from there they spread out in all directions at a speed of 87 miles (140km) a year. By 1874 they had reached the Atlantic coast. They can now be found in every US state apart from California, Nevada, Hawaii and Alaska. In 1875 Germany, Belgium, France and Holland all banned the import of potatoes from the USA, but the beetles still reached Europe, and did so almost at once. They were found in the Liverpool docks in 1877, but didn't get established. There have been further outbreaks: it's been estimated that the Colorado beetle has been eradicated from the UK 163 times, most recently in 1976.

I have chosen Colorado beetles to represent all invertebrate consumers of human crops: all those competitors for whom we create ideal conditions and then do our damnedest to eradicate: such species as cotton worm (sometimes called cotton leafworm) which infests cotton, red flour beetle (wheat), fall armyworm (maize) and brown planthopper (rice). I have gone for the Colorado beetle because it is imprinted on the minds of British people of a certain age: there used to be warnings about them in every post office, with pictures of its bright and beautiful self, along with instructions about the drastic steps that should be taken if such a dreadful creature should be sighted. It was part of the traditional British fear of invasion that dictates decision-making to this day: it's been estimated that the cost of measures to exclude the Colorado beetle from the UK is far higher than the likely cost of control. But it's not beetles that the British traditionally fear: it's foreigners.

Colorado beetles established themselves around American army bases in Bordeaux during the First World War; by the beginning of the Second World War they were common in Spain, Belgium and Holland. They increased after the Second World War and can now be found on most of the European mainland. During the Cold War, half the potato fields in East Germany were affected, and it was believed that the beetles had been deliberately dropped onto the countryside by planes from the United States. It's a nice story of national and political paranoia, but the beetles are perfectly capable of making their own way, given helpful conditions.

Humans have fought back, and the results have been devastating, if not necessarily for the beetles. Most of the advances in control on crop-eating species were tried first on the Colorado beetles; the copper-based insecticide Paris Green was used on potato plants in 1860. More firsts: Colorado beetles were the target for the first hand-operated compression sprayers, the first wheel-drawn sprayers, the first traction-operated sprayers, the first engine-operated sprayers, the first air-blast

Warning to all: poster issued in the Second World War as part of the Ministry of Information's wartime campaign. Painting by Dorothy Fitchew.

sprayers and the first aeroplane sprayers. Despite all this, the beetles are still here: of all the competing species (those that humans call pests) the species most resilient to chemical treatment is the Colorado beetle.

We have dealt with this issue already in Chapter 23 on the mosquito, but here's a refresher: an insecticide kills the individuals susceptible to such treatment, but the immune individuals survive, breed and pass on their immunity to subsequent generations; Colorado beetles can go through three generations in a year. Colorado beetles have evolved immunity to fifty-six different chemical insecticides. Chemicals that worked initially failed with subsequent generations; Colorado beetles were immune to DDT by 1952 and to dieldrin in 1958.

Other methods of controlling the beetles have been tried, including bacterial insecticides which can target the very fresh larvae. The plants themselves have been sprayed anti-feedant, but that has the disadvantage of affecting the plant.

The most effective, if rough-and-ready solution to Colorado beetles is crop rotation: if you try to grow potatoes on a field that wasn't used for potatoes the previous year, you will be ahead of the beetles: reducing the problems by 96 per cent. The beetles overwinter as adults and when they emerge they can't fly. They need to regenerate their flight muscles, and they can't lay eggs until they have fed. So long as the new crop is planted 650–1300ft (200–400m) from the old, you should be reasonably clear of beetles. The beetles can be foiled by digging pitfall traps, exploiting the fact that adults can't fly straight after emergence. Flame-throwers have also been used. Other possible solutions include genetically modified strains of potatoes. Such a strain was introduced by Monsanto, later withdrawn in 2001 for commercial reasons.

Colorado beetles also affect aubergine and tomato crops, but it is as potato pests that they have made their name. We deal with them linguistically by calling such competing animals pests, bugs, blights, scourge, plague, as if it was the duty of the world to please us. We have waged war on such creatures, using weapons that not only destroy their target but many others as well, a process that continues in different forms to this day. So let us finish this chapter with Sir David Attenborough, closing his series on terrestrial invertebrates of 2005, *Life in the Undergrowth*, with the following words:

> If we and the rest of the backboned animals were to disappear overnight, the rest of the world would get on pretty well. But if [the invertebrates] were to disappear, the land's ecosystems would collapse. The soil would lose its fertility. Many of the plants would no longer be pollinated. Lots of animals, amphibians, reptiles, birds, mammals would have nothing to eat. And our fields and pastures would be covered with dung and carrion. These small creatures are within a few inches of our feet, wherever we go on land – but often, they're disregarded. We would do very well to remember them.

SEVENTY-THREE
LOCUST

'That which the palmerworm hath left hath the locust eaten.'

Joel 1:4

Humans have traditionally made up stories about shapeshifters: usually people who can turn from benignity to uncontrollable hostility in the blink of an eye. We have the older myths of werewolves and were-tigers; the more modern story of Dr Jekyll and Mr Hyde, and current versions like the Incredible Hulk. They are stories about ourselves, about our own ability to shift from civilized humanity to primeval fury: stories that tell us that, for all the great achievements of humanity, we can never stop being animals.

Such a shapeshifter exists in the class of insects. It is mostly found as a solitary creature, in form and behaviour indistinguishable to all save experts from any other short-horned grasshopper. When numbers are low, they pose no threat to humanity whatsoever. But in certain circumstances they change. They change in shape, they change in colour, they change in behaviour. They change from charming grasshoppers to the locust swarms of the Bible: and once they have formed they are devastating and almost impossible to control. The process begins with overcrowding. The presence of many others of their own kind charges their brains with serotonin: they eat a great deal more, they breed far more easily and they are extraordinarily attracted to each other's company. To use the technical terms, an outbreak becomes an upsurge becomes a plague.

There are examples in this book of human terror at the wild world that are unjustified by performance: piranha (see Chapter 55) and shark (see Chapter 14). That is not the case with locusts. A locust swarm can comprise billions of insects over thousands of square miles: there can be 200 million insects in a square mile (more than 80 million in a square kilometre). Each locust can eat its own weight in plant matter in a day: it follows that they can devastate crops grown by humans.

Their creation of plagues is associated with changes of the wind or weather and is a response to overcrowding. Plagues have been reported in North and South America, in Asia and Australia, as well as the traditional areas of Africa and in West Asia. In China there are records of 173 outbreaks in less than 2000 years.

They are most famously associated in Western minds with the plague reported in the Book of Exodus. Here the God of Israel inflicts ten plagues on Egypt, of

which the eighth is locusts: 'Else, if thou refuse to let my people go, behold, tomorrow will I bring the locusts into thy coast: And they shall cover the face of the earth, that one cannot be able to see the earth: and they shall eat the residue of that which is escaped, which remaineth unto you from the hail, and shall eat every tree which groweth for you out of the field...'

Perhaps this biblical reference gives locusts a sense of distance, reducing them almost to mythology. But that is not appropriate: plagues of locusts have punctuated the history of the world for as long as humans have gone in for agriculture: and such visitations can only have felt like the wrath of God. Locust images have been found in the tombs of ancient Egypt almost 5000 years old, predating the Exodus plague by a good millennium. Aristotle studied locusts and made accurate observations about them. John the Baptist is recorded as eating locusts and wild honey; this has been interpreted as an acceptance of the troubles of this world, embodied in the locusts, along with a promise of salvation in the world to come, in the honey. The Torah prohibits many foods to the faithful, but some locusts are permitted: no doubt this is because in times of plague the only food available is locusts. The Quran reports that the Prophet Muhammad ate locusts during a military raid.

In AD 311 a plague of locusts in the northwest provinces of China was followed by an outbreak of plague (as in the Black Death, see Chapter 6); it is likely that rats came to eat the millions of fallen locusts, and their accompanying fleas passed on the disease. The twin disaster wiped out 98 per cent of the population.

Plagues have been recorded throughout human history and continue today. In 1747 a plague of locusts hit Damascus; a barber named Almad al-Budayri wrote: 'They came like a black cloud. They covered everything: the trees and the crops. May God Almighty save us.'

A roaming plague affected Africa between 1966 and 1969. In 2003 there were plagues in West Africa, which followed a period of unusually heavy rain; the locusts reached Egypt, Jordan and, for the first time in fifty years, got to Israel. In 2013 a plague covered half of Madagascar. In 1954 a swarm from northwest Africa reached England; in 1988 a swarm from West Africa got as far as the Caribbean. In 2019 locust swarms formed in Kenya.

Once a plague has been established, it is very difficult to deal with. The answer, so far as an answer exists, is to get to the swarms before they have formed: in other words, it's about monitoring and early detection. Such things require organization, finance and political stability. In other words, the poorer your country, the more vulnerable you are to locusts: but a plague can establish itself in poor places and then move on to more wealthy neighbours. Subsistence farmers are the most

Uncountable: A Swarm of Locusts, *lithograph from the German scientific textbook* Brehms Tierleben *(1860s).*

God's will: plague of locusts, from the Nuremberg Bible (coloured woodcut, fifteenth century).

vulnerable: it has been estimated that one-tenth of the world's population lives under the threat of locusts.

At the beginning of the twentieth century, people started using pre-emptive methods of dealing with swarms. These involved cultivating the soil where locusts lay eggs, working the lands with rollers, trapping them in ditches and using flame-throwers. In the 1950s dieldrin was used widely and was effective, but that has caused huge environmental problems that outweighed the benefits and the substance is now banned.

I should point out here that locusts should not be confused with periodic cicadas, which are loosely referred to as locusts because they also appear in large numbers. These insects hatch out in numbers after a long period of dormancy: one species appears every seventeen years. They are very noisy but don't pose a threat to human crops. These are the insects referred to in the Bob Dylan song 'Day of the Locusts': 'And the locusts sang – and they were singing for me.'

There is a curiosity in the history of locusts, and that is the extinction of the Rocky Mountain locust. This was the species that created what is known as Albert's swarm of 1875, which hit the western United States. Albert Child estimated that it covered 198,000sq. miles (510,000sq. km); it was reckoned that the swarm contained 3.5 trillion individuals and weighed 27.5 million tonnes. The last Rocky Mountain locust was seen alive in Canada in 1902; it is possible that agriculture destroyed their breeding grounds.

SEVENTY-FOUR

BAIJI OR CHINESE RIVER DOLPHIN

'In the middle of one of the biggest, longest, noisiest,
dirtiest thoroughfares in the world lives the reincarnation
of a drowned princess.'

Douglas Adams, *Last Chance to See*

Once upon a time there was a princess. She was, of course, beautiful. Or maybe she was just a beautiful girl: but no matter, she had an evil stepfather who wanted to sell her. Then, as they were in a boat together, travelling along the river on their way to market, the stepfather was overcome with desire and made a pounce at her. To save her virtue, the girl leapt into the river and was drowned. But she was reincarnated as a dolphin, a goddess and therefore immortal, forever swimming the great River Yangtze, a perpetual reproach to all that is evil, a perpetual celebration of all that is beautiful.

That is the story of the baiji or Yangtze river dolphin. Funky things, river dolphins: four species evolved for four of the great river systems of the world, but they did so quite separately, without a close common ancestor, a classic example of convergent evolution (see also Chapter 41 on pigs, Chapter 48 on thylacines and Chapter 51 on owls). Dolphins evolved to live in the Ganges, the Amazon, the Plate and the Yangtze. In 2006 the baiji was declared 'functionally extinct'. It was the first megafauna extinction in more than three decades: in the 1950s we saw the last of the Caribbean monk seal. And in the 1970s we lost the Japanese sea lion.

Megafauna is a catch-all term for any large animal – usually a big mammal and the sort of thing that's quite easy to notice. Many species have already been lost since humans evolved: we are in the midst of what has been called the sixth extinction (see also Chapter 48 on thylacines).

In recent years we have lost many species of invertebrates, and good numbers from our own phylum of chordates. Many of these are island species: we have already looked at their vulnerability as well as their singularity (see Chapter 4 on Galápagos mockingbirds, Chapter 19 on the dodo, Chapter 97 on the pink pigeon).

Another time: baiji by Else Bostelmann (twentieth century).

But many of these are not creatures that we readily notice. We talk about megafauna because they are creatures that we actually miss.

The baijis were a decent size all right, females up to 8ft (2.5m) long, males a little less. They were a pale bluey-grey on top and plain white below. They had a long beak that pointed slightly upwards, and a low triangular dorsal fin that was said to look like a flag. They had a top speed of 37mph (60km/h), when trying to escape from a threat, and they routinely cruised at about half that. They were highly social, swam in coordinated groups, were keen on communication and apparently expressed emotion.

Now the thing about living in rivers is that they are often pretty murky places. But the baiji, like the other river dolphins, developed the most excellent sonar: they found their way around the murk of the Yangtze in the same way that a bat flies freely through the blackness of a deep cave (see Chapter 64 on bats). The baijis created sound pictures of the environment and that enabled them to live in 1100 miles (1780km) of river in the middle and lower Yangtze, and in lakes and rivers that were part of the catchment. The snag, however, was that 12 per cent of the entire human population of the world also lived and worked within the Yangtze catchments.

Last of their kind: baijis painted by Jeremy Saxton.

What caused the extinction of the baiji? Practically everything. The Great Leap Forward of 1958–62 was Mao Zedong's attempt to industrialize China at immense speed. It didn't go as well as hoped: half a million people were killed to expedite the plan, a three-year famine followed and then came the Cultural Revolution (1966–76), a period of great chaos in which those who caused the famine were supposedly punished.

Dolphins were killed to feed a desperate population. Fishermen began to use wider and larger fishing nets, which found dolphins as by-catch and drowned them. The illegal but widespread practice of electric fishing was indiscriminate in those it killed; dolphins were routine victims. There was the loss of habitat caused, among other things, by the massive Three Gorges hydro-electric scheme, construction for which began in 1994. The roads of the area were poor, but the Yangtze was a highway that could be relied on, and it carried more and more traffic. This led to regular incidents of ship strike; dolphins must surface to breathe every few minutes, and they were routinely stuck by the propellers of boats.

The river became less and less suitable for sustaining life as it was dredged to make the going easier for the boat traffic, and banks were reinforced with concrete. The industries that line the river flung their pollutants unchecked into the river.

That affected the other species that lived in the river. It's reckoned that the fish population of the Yangtze was reduced to one-thousandth of its level in the previous two or three millennia; dolphins eat fish. And we still haven't touched one of the main causes of the extinction: noise pollution.

How can you build a sound picture of your environment if your environment is howling and shrieking with sound? Douglas Adams, author of *The Hitchhiker's Guide to the Galaxy* (see also Chapter 38 on dolphins), also wrote *Last Chance to See*, with Mark Carwardine, in which they looked for species on the brink of extinction. It was published in 1990. They wanted to record the sound of the river, from the dolphin's point of view, but they hadn't brought an underwater microphone. So they improvised one with a condom (the book contains a long and splendid aside on the problems of getting hold of one) and eventually made their recording. Adams discovered to his surprise that it wasn't 'the heavy, pounding reverberations' of the boats but 'a sustained shrieking blast of pure white noise'.

The lives of the last baijis, then, were a good approximation of hell. It's been reckoned that there were around 5000 baijis alive a couple of thousand years ago. In 1970 it was made illegal to kill them; in 1978 the Freshwater Dolphin Research Centre was established on the Yangtze. There was a reserve set up for an insurance population, but it was hard to catch the dolphins on purpose; ironic since so many had been caught by accident, but the snag was that the process killed them. The few that were caught didn't survive more than a few months in the reserve; so in effect the policy made the situation worse. By the time an action plan was set up in 2001, it was already, as it turned out, too late. A six-week expedition of 2006 had no sightings, and in 2007 the animal was declared functionally extinct: that is to say, if there were any surviving individuals, there were not enough to breed and make a comeback.

The following year a video was made of what looked very like and almost certainly was a living baiji. This was exciting, but not consoling; it was more than likely that this animal was a member of what conservationists refer to as 'the living dead': the last members of a doomed species. In 2016 an amateur expedition of enthusiasts made another search for the baiji. Their leader, Song Qi, claimed a sighting: 'No other creature could jump... like that,' he said.

But one or two survivors do not add up to a sustainable population: and it is doubtful if the river could now sustain a viable population even if one could be established. Here, then, is a line-in-the-sand species: and the line in the sand has been crossed. We have grown used to stories of diminishing populations and species in danger: no doubt we shall soon get used to news of further megafauna extinctions. It will take considerable changes in the way the planet operates to prevent them.

SEVENTY-FIVE
CRANE

*'I will write peace on your wings and
you will fly all over the world.'*

Sadako Sasaki

They have been called the birds of heaven. They stand tall as humans, they dance like ballerinas, they call to each other with the sound of bugles, and they fly in the manner of angels, with long slow beats of their big broad wings. Wherever they turn up they seem too fine, too precious for the coarseness of this Earth: when they alight they do so with an altogether unexpected lightness of touch, as if the surface of the Earth was repulsing them, in the manner of the like poles of a magnet.

There are fifteen species of cranes in the world; they are found on every continent except Antarctica and – for no very obvious reason – South America. In every place where they share space with people they have become part of lore and legend, for they are birds that can't be overlooked, and they touch humans on a deep level. Once when writing about cranes, my spellchecker, by a classic happy accident, produced a reference to 'the most numerous species of crane': surely a hard thing to judge, since all fifteen species have an air of being touched by the light of God.

Cranes come in four genera. Some species migrate, other don't; some populations within some species migrate while others don't. Cranes include the world's tallest flying birds: the smallest of them, the demoiselle crane, is around 35in (90cm) long, but the Sarus crane can be 70in (175cm) long. All cranes have long legs, long necks and streamlined bodies; when they fly, they stretch out their necks in front of them, unlike the herons (this group includes egrets, see Chapter 18) which tuck their necks in. Outside the breeding season, cranes love to gather in flocks; I once saw a group of 40,000 Eurasian cranes in Spain.

Cranes are big birds that feed mostly on small stuff: invertebrates, seeds, leaves, occasionally small vertebrates. They are very keen on being half of a pair: once established, a pair will mostly stay monogamous for the rest of their lives. Some get together as early as two or three years of age, but only start breeding a few years later. Sometimes these early starters attempt to breed and fail; in such circumstances there are occasional divorces. These are much rarer once a pair has started successfully rearing young.

Birds to revere: Japanese six-fold screen (twentieth century).

Most crane species dance. This is not a fanciful interpretation: it is a plonking literalism. They spread their wings and jump, doing so with slow, gravity-defying grace. Their dancing is connected with courtship and breeding and also has elements of flock solidarity. You will sometimes see a wave of dancing pass through a flock, like the gust of wind rippling a field of corn: one bird will dance a few steps, another will take it up and then a neighbour will do the same thing. In courtship, dancing is more prolonged and elaborate: and dancing is part of the behaviour rituals that keep pairs together. It is both a quest for and a celebration of togetherness.

This can hardly fail to grab the human imagination. In Hokkaido in Japan, the Ainu women traditionally dress up as cranes and dance: the Japanese dancing cranes in the Hokkaido snow have become a celebrated image across the world. There is a story of how the great Sanskrit poet Valmiki chose to bathe in a clear stream where he saw two cranes mating. The male bird was killed by an arrow and the female died from shock. Valmiki upbraided the hunter, and the words he chose turned out to be a form of verse called *solka* (it's been compared to the English-language blank verse used by Shakespeare): and Valmiki went on to write the great epic *Ramayana* in that form (see also Chapter 30 on monkeys and Chapter 84 on deer).

The Greek word for crane is *geranos*: the scientific name for the plant cranesbill is *Geranium*. (This has been misappropriated for the ornamental plant, usually found in the UK in pots and window boxes, loosely referred to as geranium but botanically these are *Pelargonium*.) There is a story 2600 years old in which the poet Ibycus was attacked by thieves and left for dead. Ibycus called to a passing flock of cranes, and they followed his attacker to the theatre – open-air as you would expect – where they circled over him: overwhelmed, the thief confessed his wrongdoing. Pliny the Elder reported that when cranes sleep they always leave one bird awake to watch over them. This chosen bird holds a stone in his foot; should he fall asleep himself and drop it, the sound of the falling stone will wake him. Aristotle describes the migration of cranes.

Perhaps the West's second-favourite image from the art of the East (after *The Great Wave off Kanagawa* by Hokusai) is Shen Quan's *Pine, Plum and Cranes* of 1759. The picture's nostalgia for the wild world gives it a powerful contemporary feel.

Cranes were common in English wetlands, but went extinct in that country 500 years ago; they took part in too many medieval banquets. But they came back quite spontaneously in the 1970s: landing at Horsey Mere in the Norfolk Broads. Here they stayed and eventually they started to breed. They never migrated. They have, very slowly, begun to increase. In more recent years there has been a project to reintroduce cranes on the Somerset Levels on the other side of the country; the birds have stayed and started to breed.

There are two species of crowned crane, both found in Africa. The southern or grey crowned crane is found in Zambia, where their call is traditionally heard as 'Nimvela owa!', a Nyanja phrase which translates as 'I've got the fear'. When the British arrived and English became a lingua franca that crossed the tribal boundaries, the cranes' cry was adjusted to 'Olwan! Olwan!' All one: all creatures sharing the fear, sharing the same understanding of the fragility of life. Our ancestors who first walked the savannahs with lions (see Chapter 1) could certainly relate to that statement: the cranes remind what it was like to be human before civilization and farming began.

The whooping crane of North America has gone very close to extinction: a combination of habitat destruction and uncontrolled hunting brought the population down to fifteen adults in one migratory flock and thirteen more in a sedentary flock. The second population was more or less wiped out by a hurricane in 1940. In 1941 there were twenty-one wild birds left plus two more in captivity.

The plight of this bird – a bird so pleasing to human eyes – was one of the rallying points for conservation in the late 1950s and 1960s. The methods used to bring the wild populations back up to robust and sustainable levels were much discussed and there were a number of experiments: in those days, conservation was a novelty and people trying to conserve species were working blind. Sometimes captive-bred birds were unable to mate with their own kind, having spent too long around humans. Eggs were taken from whooping crane nests and given to sandhill cranes to rear: this didn't work. Most reintroduction projects were characterized by high rates of mortality and low rates of reproduction.

Hunting was and still is a problem, but at least this had become illegal in 1973, and the laws were at times enforced. But habitat protection was the main advance. With migrating populations, the trick is to protect the breeding grounds, the wintering grounds and, just as important, the stopover places on the migration route. With better protection given to such places, the whooping crane population has slowly risen: in 2017 more than 550 wintering birds were found at Aransas on the Gulf Coast of Texas. It seems that humans want to be in a world in which cranes still exist; that has made them important birds in the history of conservation.

Origami, or Japanese paper folding, requires calm and confident hands. One of the classics of origami is the crane: and if you make 1000 origami cranes you can make a wish that will be granted. Sadako Sasaki loved this idea, and so she folded her own wish-granting flock of paper cranes before she died, knowing as she did so that her death was imminent. She died of leukaemia at the age of twelve, as a result of the bomb that fell on Hiroshima in 1945. The origami crane is now used as a symbol of Hiroshima: a symbol not of death and destruction, but of hope.

Standing tall: European crane by Edward Lear, for John Gould's The Birds of Europe *(published 1832–37).*

E. Lear. del.

COMMON CRANE.
Grus cinerea (*Bechst*)

SEVENTY-SIX

MAMMOTH

'Thus life on this earth has often been disturbed by dreadful events.
Innumerable living creatures have been the victims of these catastrophes.'

Georges Cuvier

Mammoths are central to the history of humanity. First they kept us alive, and then, a few millennia later, they changed the way we understand the possibilities of life. They played an important part in the fact of our continued existence, they played an important part in the development of the way we think. The central concern of modern life was first realized from the teeth of the mammoth.

Mammoths are the best studied of all prehistoric life forms – with the possible exception of humans – because their remains have not only been found fossilized, but they have also been found frozen. We have been able to study their bones and their teeth, and we have also been able to study their stomach contents and their dung.

What's more, mammoths also taught us about art. They supplied both the materials and the subject for visual art: mammoths are the third most popular subject in prehistoric art, after horses and bison, and their tusks provided a surface for carving images. Images of mammoths have been found carved into mammoth tusks, too.

There are ten species of mammoths usually recognized; the woolly mammoth is by far the best known and the best studied, from individuals found frozen in Alaska and Siberia. The mainland population went extinct about 10,000 years ago but they survived on St Paul Island until about 5600 years ago, and on Wrangel Island until about 4000 years ago; both are in the Arctic. In geological terms that's the briefest possible moment. We twenty-first-century people only just missed them, in terms of Deep Time – the dinosaurs went 65 million years earlier. Modern humans lived alongside mammoths.

Mammoths were pretty much of a size with modern elephants, standing up to 11ft (3.4m) at the shoulder and weighing 6 tonnes. It's probable that they lived much as elephants do: highly social, in breeding herds led by a matriarch, the males largely separate. They ate vegetation: the woolly mammoth, a species that mostly lived in a wide ring around the North Pole, fed largely on grasses and sedges. The extravagantly curved tusks do not make the best tools: modern elephants use their tusks for digging and many other tasks. It's possible that, in mammoths, their most important use was for display and for weapons: mammoths

*Feared and admired: mammoth
from Pech Merle, France
(c.15,000 BC).*

Painting by Charles R. Knight of woolly mammoths and rhinoceros Cenozoic, Europe in Late Pleistocene time, 100,000 years ago.

have been found with broken shoulders, perhaps victims of a whanging sideswipe of an opponent's head.

Humans came up from Africa into Europe 30,000–40,000 years ago, and there they encountered mammoths. They also encountered Neanderthals, normally reckoned to be a different species in the genus *Homo*, who were accustomed to coexisting with mammoths. And for both *Homo* species mammoths represented a win-double of food and shelter. More than seventy dwellings constructed from mammoth bones have been found. There is often a wide age difference between the bones used, implying that a good many of them were taken from long-dead mammoths rather than freshly slain ones. It is likely that the hides of mammoths were also used for such shelters, but bones preserve better.

Mammoths were also eaten. Bones have been found with clear marks of butchery. We can only speculate on the importance of mammal meat to human

development – whether it was a bonus or a staple – but such a protein mountain was unquestionably a wondrous thing for our ancestors: the more so in Ice Age conditions, when the meat keeps for as long as you care to save it, provided you can keep competing animals – human and non-human – away from your store. Mammoth tusks have also been used for the manufacture of daggers and spears. It's more than likely that active hunting of mammoths went on. There is a cave drawing of a mammoth caught in what looks like a pitfall trap; a mammoth has been found with a spearhead in its shoulder; the ancient method of herding animals over a cliff would undoubtedly have been practised where the terrain allowed.

The change of climate in a warming world was likely to have been an important cause in the extinction of the various species of mammoth, and that's not a novel concept to us twenty-first-century humans. The contribution of humanity to those extinctions can only be guessed at; perhaps the important point is that large mammals are more vulnerable than small ones: for a start there are fewer of them and their speed of reproduction is drastically slower.

Europeans were aware of mammoths in the seventeenth century, and weren't at all sure what to make of them. Perhaps they were the remains of modern elephants; perhaps they were creatures of legend, like behemoths. In 1728 Hans Sloane (for whom London's Sloane Square is named; he was also effectively the founder of the British Museum) examined mammoth teeth from a specimen found in Siberia and said that they belonged to some kind of elephant. He speculated that the creatures had all drowned in Noah's flood and that Siberia was therefore once tropical. Others went halfway with this idea: the animals drowned in the flood all right, but they were *washed* to Siberia. (Perhaps it's a good idea at this point to respect the integrity and curiosity of the people from the past who put together the stuff we know now: I wonder what our long-term descendants, if any, will make of the way that twenty-first-century humans understood the world.)

Georges Cuvier was a naturalist, his dates 1769 to 1832, so a man of the Enlightenment. He has been called the founding father of palaeontology, the study of fossils. His greatest advances were in comparative anatomy: he compared fossils with the bones of extant species, and as he did so he expanded the work of Linnaeus (see Chapter 8 on blue whales, Chapter 11 on platypuses, Chapter 13 on *Tyrannosaurus rex*, Chapter 30 on monkeys and Chapter 70 on orang-utans).

In 1796 Cuvier wrote a paper on living and fossil elephants, and said boldly that mammoths were (a) different from living elephants and (b) extinct. This was dangerous stuff; it was widely accepted (see Chapter 19 on the dodo) that extinction was impossible: if God went to the trouble of making a species he would not allow it to vanish. Another eighteenth-century naturalist, the Comte de Buffon, opposed Cuvier's view, saying that mammoths must be living still in the tropics somewhere. Cuvier responded with basic common sense: okay then – where? If mammoths were still alive, living humans would surely have seen them, he said: they are impossible to miss.

Cuvier did not accept evolution, in the sense of transmutation. Early notions of this subject were being put about, most notably by Jean-Baptiste Lamarck (whose clever but wrong notions on the subject sometimes come in for unfair ridicule today). But he was convinced that extinction happens, and that a series of catastrophes shaped life on Earth.

The notion of extinction is one of the most important topics – if not *the* most important topic – on Earth today, for it encompasses the possibility of the extinction of our own species. It is an idea that began with the mammoth. In 2015 a project to sequence the genome of the woolly mammoth was completed; there is speculation as to whether the species can be recreated; the technology to do so is not here yet. And besides, when it comes to the question of extinction, it is possible that there are more urgent tasks awaiting humanity.

SEVENTY-SEVEN
GOAT

'The lust of the goat is the bounty of God.'

William Blake, *Proverbs of Hell*

Goats were one of the earliest creatures to be domesticated. For more than 10,000 years they have brought us meat, milk, leather and wool. We have made cheese from the milk; we have used their dung as fuel. They helped us become what we are. Yet we identify the ungodly with goats, while the righteous are sheep, and this essential truth is spelt out in the Bible. Since then we have taken this idea much farther: we present Satan – the devil himself – with the horns and the beard of a goat. In classical mythology the bottom half of Pan, the nature god, is all goat. The same is true of fauns and satyrs: collectively they represent forces of nature we seek to control, that we seek – and so often fail – to rise above. More than anything else, the goats of legend are about sex.

If you sit still in the mountains of Armenia and elsewhere in West Asia, you will often hear the sudden clatter of falling rocks. You may not see them, but you know they are bezoars: agile and nerveless climbers that subsist in a countryside that is often harsh and demanding and short of obvious nutrition. These are the ancestors of the domestic goat: there are more than 300 breeds recognized, and in 2011 the United Nations Food and Agriculture Organization estimated that the world population of goats was 924 million.

Perhaps the most significant thing about goats is that they are not sheep (see Chapter 68). The difference has for millennia been an overwhelming preoccupation of humankind. Sheep are primarily grazers: they eat grass for choice. Goats are primarily browsers: they eat taller vegetation, preferably the growing tips of bushes and trees. The two species look similar – some breeds of sheep are almost indistinguishable at a casual glance from some breeds of goat – but their different lifestyles make them very different creatures to manage. The at-a-glance difference is that goat tails go up, while sheep tails go down.

Goats have a reputation for voracity, creatures who will eat anything. That is an exaggeration, but they are bold and experimental when it comes to food, and will try just about anything that might be a plant, including paper, cardboard and the paper labels of tin cans. They can also consume without harm plants that are toxic to other animals. They are fussy about contaminated food and water,

Taking away our sins: The Scapegoat *by William Holman Hunt (1854).*

however, and that causes problems in husbandry, for when they are confined together such contamination is inevitable.

Goats do better when they are free-ranging, but that too causes problems. For them, food gathering requires a certain amount of agility and initiative: they are accomplished climbers, the only ruminant you are likely to find up a tree. Sheep on a pasture have only to lower their heads to find all the food they need, but goats need to be more engaged with their environment. Goats are less herdy than sheep, harder to drive and control. They are social animals all right, but they tend to respond as individuals rather than as obedient and submissive herd members. This soul-deep difference in temperament has caused humans to see sheep as good guys and goats as creatures born to make trouble.

Goats are hard to confine. They are adept at getting over, under and through any fencing, and their curious wandering nature makes escape natural and inevitable behaviour. Does that make them evil? In the St Matthew Gospel, Jesus says: 'And he shall set the sheep on his right hand, but the goats on the left... Then shall he say unto them on his left hand, Depart from me, ye cursed, unto everlasting fire prepared for the devil and his angels.'

Sheep and goats are both associated with food and therefore with sacrifice: surrendering something valuable to God, in the hope of future favours or in thanksgiving for favours past. The great example of such a sacrifice in goat form is the scapegoat, which is found in the biblical book of Leviticus, and is, of course, also part of the Jewish Torah; Leviticus is concerned with what is forbidden and what is not. There are two goats: one to be sacrificed and eaten, the other to be turned loose, bearing with it the sins of the community. It is easier to turn a goat loose than a sheep – sheep being far less inclined to leave the safey of the herd. 'But the goat, on which the lot fell to be scapegoat, shall be presented alive before the Lord, to make atonement with him, and to let him go for a scapegoat into the wilderness.' This has been interpreted as a prefiguration of Christ's death on the cross: dying to take away the sins of the world.

William Holman Hunt, the pre-Raphaelite painter, travelled to present-day Israel during a crisis of faith and painted a picture called *The Scapegoat*. It was a work that famously divided the crowd: for the artist it was a deeply personal work about salvation, but, to many of those who saw it, it was just a funky landscape with an inexplicable goat.

The half-goat god Pan from Greek mythology represents nature at its least inhibited, and he is half admired for that reason, and also half reviled. He gave us the word panic, and it was recognized that everything to do with Pan is both thrilling and dangerous. Pan is the only god not worshipped in temples but in rough-and-ready places out of doors: caves and grottoes and groves. Pan, like the goat-legged satyrs, was renowned for sexual incontinence. He chased the goddess Syrinx with lustful intentions, but she turned into a reed; from the sound of the wind in the reeds Pan invented the panpipes as a consolation prize. He more successfully seduced the moon goddess Selene. He is traditionally represented as an altogether unreliable figure, coupling with a goat or groping Aphrodite. According to Plutarch, Pan is the only god that dies: G. K. Chesterton commented: 'Pan died because Christ was born.' Christ's birth, then, is explicitly linked with the conquest of nature: civilization, and the civilizing forces of religion and morality, have conquered the goat within.

Goats are also associated in the classical world with tragedy; the word itself means goat song. It's been speculated that a goat was given as the prize for the best performance of choral singing and dancing; or perhaps the dance was performed before the goat was sacrificed.

Goats are uninhibited in their sexual behaviour. Both sexes bear horns, which are more prominent

Goatish appetite: Aphrodite, aided by Eros, fights off Pan with her sandal, marble sculpture from Delos (first century BC).

than those of sheep: this horniness can be seen as proof of a licentious nature. The equation of goats with sexuality led to the identification of goats with the devil. The famous engraving by Albrecht Dürer, *Knight, Death and the Devil*, shows both Death and the devil in different goatish forms, Death goat-horned and bearded, waving an hourglass, and the devil with a goatish skull apparently in the grips of self-destructive mania. An evil god by the name of Baphomet is represented in the form of anthropomorphic goat, very like the conventional Satan image we are all familiar with. The secret crusader order of the Knights Templar was – it is generally assumed falsely – accused of worshipping Baphomet, which made it acceptable to persecute it and break its power. The black mass involves the worship of Satan in the form of a goat; when the life-affirming pentagram is inverted, it is said to resemble the head of a goat.

But none of this association with evil has stopped goats being kept as domestic animals, for the many benefits they bring. Their skins made the most effective containers for liquids, which greatly expanded the distance humans were able to travel in dry places. It also meant that humans could carry intoxicating fluids a fair old distance as well.

Goat's cheese has become fashionable in recent years, but goats are primarily associated with poverty. Many charities, Oxfam for example, have encouraged donors to finance the purchase of goats for impoverished communities. A friend rang up for more information:

'But what do the goats eat?'
'That's the beauty of it, the goats subsist on the environment.'
'So what happens to the environment?'

No answer was offered. Perhaps there is no answer. But it's a fact that destruction of the environment by overgrazing is a major global issue. It is unfair to blame goats for all of this: all grazing livestock is involved. It's just that goats tend to come in when the process of desertification is relatively advanced, and not much else can make a living in these denuded pastures. Goats have been used to apply the *coup de grâce* – is that a pun? – to an already depleted landscape.

Desertification is a long-term and historic process. It has been suggested as the reason for the fall of the empires of Carthage, Greece and Rome; and the cause of major disruption to historic human populations of Mesopotamia, the Mediterranean and the Loess Plateau in China. Since 1900 the Sahara Desert has expanded southwards by 150 miles (250km) across a front of 3700 miles (6000km). This has been caused by drought, climate change and tillage for agriculture – but the main cause is overgrazing. Goats subsist on the environment: and when they have finished subsisting it's important for future generations – of goats, of humans, of everything else – to have something left.

SEVENTY-EIGHT

LOA LOA WORM

'All things sick and cancerous,
All evil great and small,
All things foul and dangerous
The Lord God made them all.'

Eric Idle, *Monty Python's Contractual Obligation Album*

We have two very solid, effective and helpful ways of looking at nature: we believe (1) that nature is benign, generous, kind, full of beauty and wonder and capable of sustaining us throughout eternity; and (2) that nature is hateful, hostile and brutal, and we are engaged in a perpetual war against it. We maintain these notions simultaneously and they support us very well. Their contradictory nature seldom troubles us. In a sense, after all, they are both right, even if they are also both wrong.

At one point Charles Darwin was working with hermaphrodite jellyfish, trying to establish the fact that they cross-fertilize, rather than propagate by means of a single individual fertilizing itself, which would lead to a loss of vigour in the species. He corresponded with his great friend and champion Thomas Huxley, wondering if the sperm wasn't washed by water into the creatures' mouths. Huxley responded: 'The indecency of the process is to a certain extent in favour of its probability, nature becoming very *low* in all senses among those creatures.' This rather knockabout reply pleased Darwin, and he passed it on to another great friend, the botanist Joseph Hooker (see Chapter 61 on barnacles). He then added a line to this letter that has become one of his most famous sentences: 'What a book a devil's chaplain might write on the clumsy, wasteful, blundering, low and horribly cruel work of nature!'

Darwin wrote later, in words quoted in the excellent biography *Darwin*, by Adrian Desmond and James Moore, that his own lovely garden at Down House was nothing less than a battlefield. 'One may well doubt this [when viewing] the contented face of a bright landscape or a tropical forest glowing with life... Nevertheless the doctrine that all nature is at war is most true. The struggle very often falls on the egg & seed, or on the seedling, larva & young; but fall it must sometime in the life of each individual.'

A life in blood: loa loa worm in a blood smear.

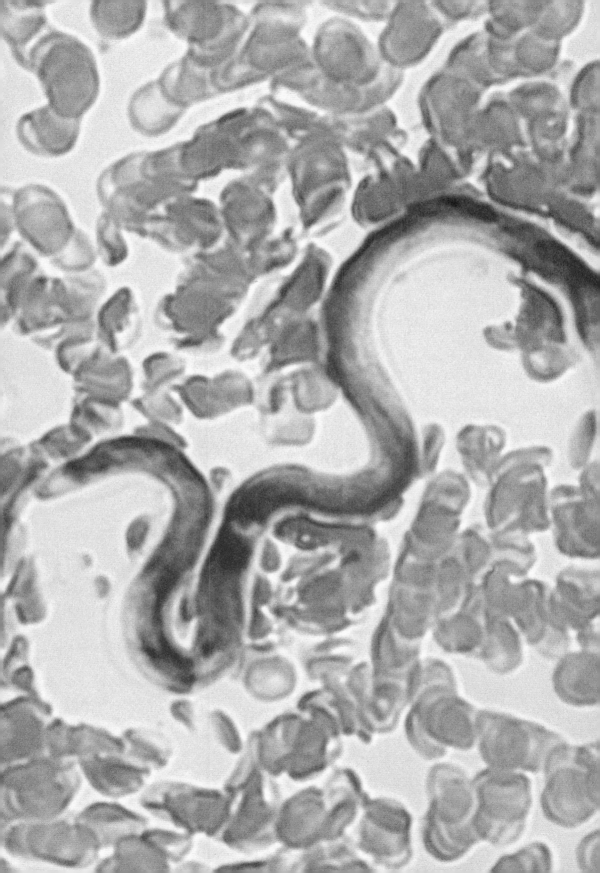

So let us take this notion of the hostility of nature a step farther with some lines of David Attenborough:

> My response is that when a creationist talks about God creating every individual species as a separate act, they always instance hummingbirds or orchids or sunflowers and beautiful things. But I tend to think instead of a parasitic worm that's boring through the eye of a boy sitting on the bank of a river in West Africa that's going to make him blind. And [I ask them]: are you telling me that the God you believe in, who you say is an all-merciful God, who acts for each and every one of us individually, are you saying that God created this worm that can live in no other way than in an innocent child's eyeball? Because that doesn't seem to me to coincide with a God who's full of mercy.

Which is a long and quote-filled way of bringing us to the loa loa worm (as a point of linguistic interest, the last three words all mean the same thing, so I have – correctly – named the creature worm worm worm). This is a nematode worm found in the rainforests of West Africa. There are 28,000 species already described in the phylum of Nematoda or roundworms, and 16,000 of them are parasites. It's been estimated that 80 per cent of all the individual animals alive on the planet right now are nematodes; you can find 1 million in a cubic yard (metre) of soil. And the loa loa is the worm that Attenborough was describing. It has evolved to exploit humans as its definitive host: it is here, on a human body, that it reaches maturity and produces young – technically microfilariae.

The process begins – insofar as a circular process can be said to have a beginning – with the biting of an infected human by a deer fly or a yellow fly. The fly becomes infected in its turn and then bites another human. The worms reach maturity inside this fresh human body, causing great suffering to the host. The larvae introduced to the human become adults, the female produces microfilariae, and she is capable of doing so for some years, for this is a long-lived species, one used to playing a long game. These larvae can establish themselves in blood, spinal fluid, urine and sputum. It takes a larva five months to become an adult, and it can do so only inside a human. It creates painful swelling over the joints – Calabar swellings – and, when it reaches the eye, it causes painful inflammation and sometimes blindness. It can also be found in the penis, testes, nipples, bridge of the nose, kidneys and heart. It is comfortably at home all over the human body.

A worm in the eye can be removed with forceps with the aid of local anaesthetic, but this is a tricky business: you need to be fast with the forceps or the worm will retreat out of reach. An infected person can be treated with antiparasitic medication, but this can cause neurological damage. The infection is called loiasis, and is categorized sometimes as one of the 'neglected diseases'. There is no control and no vaccine.

The loa loa is included here not to demonstrate the implacable hostility of nature, but to demonstrate the limitations of the opposing view. There is something about the deeply personal and intrusive nature of the loa loa worm's relationship with humanity that disturbs us. We'd like to think we were immune to such assaults from nature, but if we live in the forests of West Africa we can find out at first hand that we are not.

The eternal kindness and generosity of nature are best appreciated from the sitting room rather than the wild places of the world. Mrs C. F. Alexander published *Hymns for Little Children*; it was in its sixty-ninth edition by 1900. It contains words known across the English-speaking world:

> All things bright and beautiful
> All creatures great and small,
> All things wise and wonderful
> The Lord God made them all.

Hymn writer: Mrs C. F. Alexander (1818–95), author of 'All Things Bright and Beautiful'.

The loa loa worm is there to balance the view. The fact is that nature is not benign. But it is also a fact that nature is not malevolent. We maintain these two views, at the same time polarizing and complementary, because we humans see ourselves as both unique and outside nature. We do not care for the notion that we are one species among many, and that we, like every other living creature in the kingdom Animalia, are involved in the daily struggle for existence. Nature is not against us any more than it is against the dolphin and the crane: and it is no more intended for our special benefit than it is for that of the locust or the loa loa worm. The loa loa can live nowhere but inside of a human body: it has no choice, it can do no other. For a loa loa, humanity represents the overwhelming generosity of nature; for humanity, it is the other way round. And both views are wrong.

Nature is not two things at once: good things and bad things. Nature is simply everything. I have been bitten by a tsetse fly while revelling in the beauty of a leopard. C'est la vie. C'est la nature.

SEVENTY-NINE
PEAFOWL

'The pride of the peacock is the glory of God.'

William Blake, *Proverbs of Hell*

Peacocks are symbols of royalty; peacocks are symbols of divinity. Across the centuries they have represented an ideal of beauty: beauty beyond the scope and perhaps beyond the imagination of humanity. For some, the detonation of glory that is the peacock's train – beauty apparently for the sake of beauty, beauty without function, beauty without any meaning save beauty itself – was all the proof anyone could need for the existence of God. Charles Darwin, in a famous

Always magnificent: In the Conservatory *by Heywood Hardy (1842–1933).*

letter, wrote: 'The sight of a feather in a peacock's tail, whenever I gaze on it, makes me sick.'

Peacocks are more correctly termed peafowl, since the species has females as well as males, these females being peahens. These are drab and lack the great map of the stars that you can find in the peacock's train, but there is a case for saying that they are the more interesting half of the peafowl population. Certainly Darwin thought so.

There are three species of peafowl usually recognized: the Indian or blue peafowl; the Southeast Asian or green peafowl, sometimes known as Java peafowl; and the Congo peafowl. They are heavy omnivorous ground birds which make their living by finding food on the floor and getting up into the lower branches to roost at night. They feed mostly on plant matter, but will take large invertebrates and small vertebrates when they can get them. It's a familiar pattern, met before in the red jungle fowl, ancestor of the domestic chicken (see Chapter 29) and the pheasant (see Chapter 60). Both are spectacularly sexually dimorphic – that is to say, there is a massive difference between male and female – or, at least, the differences would have been considered spectacular if we were talking about any species but peacocks in this chapter.

Both male and female peafowl are sizeable birds, the females as big as the males if you forget the train. Not that anybody is ever likely to, whether human or peafowl. It is one of the most extraordinary things in nature. I remember circling a large bush in India, having heard the sound of what I thought was a thrush-size bird in the middle of it. And then the bush exploded, and a vast train-bearing peacock emerged in a great whirring of wings; he seemed to take about ten minutes to fly past. Flight is hard work for a peacock, but it's a near-unbeatable escape ploy when your enemy is wingless.

The train itself is an astonishing piece of physics in terms of its coloration. The colour comes not from pigment but from refracted light: the principle of the prism and the raindrop, though the physics are a good deal more complicated than that, based on what is called structural coloration. The iridescent colours shift with the angle of the light: the train can be black and then, in almost literally the blinking of an eye, it can be all the impossible shades from the peacock palette.

So extraordinary a bird has inevitably come into mythology many times over: the bird itself seems a good deal less likely than many of the fabulous monsters that we have invented across the millennia and the civilizations; dragons and griffons are much less exotic than peacocks. Four Hindu gods are associated with peacocks, most notably Kartikeya, the god of war, but also, in one of those wild paradoxes that you find in Hinduism, Lakshmi who is kind and compassionate. There are peacocks in Buddhism, and the bird is auspicious. In the seventeenth century, Shah Jahan, the Mogul emperor of India, commissioned a jewelled seat of power: the peacock throne. The last shah of Iran also sat on a peacock throne. To associate yourself with a peacock, then, was quite a statement.

In Greek mythology, Argus was a hero with a hundred eyes. He was appointed to watch over Io, Hera's priestess, who had been transformed into a cow, but in this office he was slain by Hermes. Hera then transferred the eyes of Argus to a peacock. The Greeks also liked the idea that the flesh of the peacock didn't decay, so the bird was a symbol of immortality. In Christianity both ideas were picked up: the unrotting bird stands for immortality again, and also as a symbol of the all-seeing God whose eyes are everywhere. For Ashkenazi Jews the peacock is an emblem of joy and creativity. Everywhere that peacocks crop up, they have inspired humans to respond to their extraordinary and gratuitous beauty, in stories and images. A peacock can be found in the painting of Eden by Peter Paul Rubens and by Jan Brueghel the Elder: the peacock is witness of Eve's offering of the apple to Adam: here too the peacock is all-seeing.

Peacocks are pretty hardy and can cope with the cold damp winters of Europe and at least parts of North America. Hardly surprising, then, that they became prestigious accessories for the owners of country houses. You can purchase a peacock easily enough: prices are quoted between US$35 and US$275 for adult birds. Naturally, peacocks escape and form small localized feral populations: in a previous house I used to have regular visits not just from peacocks but also from a peahen with peachicks. They thrived for a while and then died out; peacocks have never properly established themselves as a continuing self-supporting population in the UK.

In the nineteenth century the search for new understandings of life gathered pace. But Darwin's principle of natural selection could not explain the train of the peacock. A peacock's train didn't help it to survive: that was obvious. Quite the reverse. The big theory could explain the opposable thumbs that all of us primates share; it could also explain the wing of a bat and the fin of a whale and the hoof of a horse, and why they are all based on the same structure. But natural selection can't explain beauty. The answer could no longer be a creator simply enjoying himself: art for art's sake, as it were. There had to be a reason for the evolution of the peacock's train: but Darwin couldn't work it out.

But the rock-crusher mind kept grinding away at this insoluble problem until it was powder in his hands. He wrote about this in *The Descent of Man, and Selection in Relation to Sex*: 'The sexual struggle is of two kinds; in the one it is between individuals of the same sex, generally the males, in order to drive away or kill their rivals, the females remaining passive; while in the other, the struggle is likewise between individuals of the same sex, in order to excite or charm those

Glory of gods: Kartikeya, Hindu god of war, from Sri Mariamman Hindu temple, Singapore.

of the opposite sex, generally the females, which no longer remain passive, but select the most agreeable partners.'

Once again, in his polite, restrained way, Darwin had lobbed a hand grenade into the self-protective assumptions of the nineteenth-century world. The peacock has a beautiful train because that's what the peahens – the females – want. Female choice not only operates actively in nature, but it also actually dictates the way a male looks and lives. A peacock must accord with a female's idea of beauty – or at any rate desirability – or give up all hope of becoming an ancestor. Females have power over males. Females have the power to shape nature.

It is not surprising that the Victorian scientific establishment mostly rejected the idea. Darwin had gone far enough when he said that humans and monkeys shared a common ancestor. But the idea that females had power over males was altogether too much.

It was a long time before it was accepted that competition for mates within a species was a driving force of evolution, and that female choice was a crucial element of this. These days, when we listen to the best of what are called the repertoire singers – the species of birds in which the males sing complex songs with much variety between individuals – it is generally accepted that the most complex and (if you like) the most beautiful songs make the male more attractive to females (see Chapter 40 on nightingales). The males make the music, but it is the females who decide what music is the best.

There are various other theories with alternative explanations of the peacock's train. The Israeli evolutionary biologist Amotz Zahavi proposed the handicap theory: a long tail makes life much tougher for a male, so it is an honest signal of his own toughness and fitness for the task of surviving. The train means good genes. It doesn't entirely explain why the train needs to be beautiful as well as big, full of structural coloration. It's also been suggested that this raised tail is based on the need to intimidate predators: what is called an aposematic display; in which case, it seems unfair and perhaps inexplicable that the females don't have them. A researcher, Marion Petri, worked with a free-ranging population of peafowl in Whipsnade Zoo in Bedfordshire, in England, and found that the number of eyespots was crucial to mating success: when she snipped off eyes from a peacock's train the bird in question became less successful at winning females.

It seems that Darwin was, at the very least, more right than wrong. Beauty is part of evolution. Many people are troubled by the notion of the blind forces of evolution, fearing that such a view is brutish and soullessly functional. But it also involves an active quest for beauty: for visual display in forms of impossible loveliness and for music of glory and wonder. And such stuff comes down to decisions made by females. We must look at this, as Darwin was eventually able to look at peacock's tail, without nausea.

EIGHTY

GOLDFISH

'I wouldn't mind turning into a vermilion goldfish.'

Henri Matisse

Looking at fish is nice. It's something humans like doing. Fish are not just for eating: they are also there to delight the eyes: to delight the brain at a relatively deep level. Research at Plymouth University and the University of Exeter found that looking at fish in an aquarium reduces the heart rate and lowers the blood pressure. The soothing hypnotic nature of staring at fish has measurable physiological benefits. It is a meditative experience that we seek intuitively in many forms, some more efficient than others.

It's nice watching fish in open water, but it's quite hard to do so. Fish are not easy to see, mostly because it is in their best interests to be hidden. Fish that stand out from the crowd or from the background are an easy target for herons and pikes. That's why most fish you see in rivers and lakes are grey-silver: the colour of water in the daylight.

Fish have been important in human diets for as long as we have been able to catch them (see Chapter 17 on cod and Chapter 66 on salmon). Dead fish don't last very long as viable sources of nutrition, but if you can keep them alive for a while in some relatively accessible place, you make life a good deal easier for yourself. So after a successful freshwater fishing trip, you eat some but chuck the rest into a handy pond: perhaps one you have dug yourself. Here the fish will survive, at least for a while, and you can help yourself when you feel hungry: a system that has surely been invented by every civilization with access to fish. If you can keep them in conditions good enough to allow them to reproduce, you have a continuous supply of food.

The Chinese established such a system during the Tang dynasty (AD 618–907). Asian carp were kept and bred for food. These fish are grey-silver, but occasionally they will throw out a mutation, in red, orange or yellow. These were nice to see: standing out from the crowd. So – especially when you had more than one – you could try to breed a few more of these pleasing freaks. It became the practice to build ornamental pools, in which you could look at the brightly coloured fish for no reason other than the quiet meditative pleasure of it all. And from there it became the practice to take one or two from their pond and put them indoors, placing them in a nice porcelain container. This was never

Captive beauty: The Goldfish Bowl by Charles Edward Perugini, c. 1870.

intended as a permanent home; it was temporary accommodation that allowed you and your guests to take a more intimate pleasure in the spectacle of these pleasant and soothing animals.

By the Song dynasty (960–1279), selective breeding of goldfish was a well-established practice. You can find glorious and ancient scroll paintings of these fish. In 1162 the empress had a pool created to display her red and gold fish. It was forbidden for people outside the imperial family to keep yellow strains of goldfish, this being the imperial colour. It is still the case that yellow goldfish are scarcer than orange, even though the yellow types are easier to breed.

Things got more rarefied still in the Ming dynasty (1368–1644), as breeding got more elaborate. Fish were raised for their mutations, and some were so odd in shape they couldn't survive in ponds. Fish with great fancy tails were kept indoors, in bowls of porcelain. Not only were they lovely to look at, but they were also highly auspicious things to have about the place. The Chinese love of puns conflates the Mandarin word *yu* meaning fish, and *yu*, pronounced slightly differently (a different tone, like a different musical note) and meaning abundance: how could you afford not to have a goldfish in your house?

At the beginning of the seventeenth century, goldfish reached Japan, and from there they were taken up by Portuguese travellers who brought them home. From Portugal they spread across Europe. It became the practice to give a goldfish to your wife on the first anniversary of your wedding, as a promise of the prosperous years to come, but as the fish got more and more popular and widespread they lost any hint of prestige and the custom died out.

In the nineteenth century, keeping fish at home became infinitely simpler with the development of the aquarium. It was invented by Philip Gosse, who developed earlier work by Jean Villepreux-Power, a marine biologist who wanted to experiment with captive fish. Gosse worked out that plants oxygenate the water, and therefore fish could live for sustained periods in a well set up and well-managed aquarium. There were elaborate and extravagant aquaria on display at the Great Exhibition of 1851 in London, and that started a craze in Victorian households. In 1853 Gosse established the aquarium at the London Zoo, and in 1854 he wrote the first manual for fish keepers: *The Aquarium: An Unveiling of the Wonders of the Deep Sea*. In 1908 the mechanical air pump was developed, powered by running water; when electricity became available, the whole business was a great deal simpler.

Goldfish are, then, perhaps the easiest creatures in the world to study. They have tetrachromatic (four-colour) vision, which means that they can see ultraviolet light; they live in a more colour-intense world than humans, who see in only three colours. They can distinguish individual humans and will beg for food from those who normally feed them. They are gregarious, and love to form schools, being highly tolerant of each other's proximity, seldom causing each other deliberate harm. They have been trained to perform tricks. And they have a good memory: it has been shown by experiment that they can remember things for at least three months, so the notion of the minute attention span of a goldfish is just one more myth.

Another myth is that goldfish are a short-lived species. Under good conditions they will survive in captivity for up to fifteen years, but providing those conditions is not easy. You need to supply 17 gallons (77 litres) of water per fish; the fish require clean, well-oxygenated water and good nutrition if they are to reach sexual maturity and breed. The traditional goldfish bowl, with its narrow top allowing as little oxygen as possible to enter the water, is a classic piece of wrong-headed design: no wonder most pet goldfish died early.

Specialists have created many fancy breeds of goldfish: with names that include black telescope, bubble eye, celestial eye, fantail, lionhead, pearlscale, pompom and butterfly tail. When humans take on artificial breeding, there is a great tendency to go for extremes.

Funfairs used to give out prizes of a single goldfish swimming in a plastic bag full of water: large numbers of doomed fish were handed out because they are cheap, colourful and small. It happens less these days, since people have developed reservations about the practice, but it is legal in the UK so long as the recipients are over sixteen or accompanied by an adult. The goldfish in the plastic bag can be seen as a symbol of the way in which we humans so often deal with non-human lives: as something that simply doesn't matter.

Forms most preposterous: fancy goldfish from a Chinese silk scroll.

EIGHTY-ONE
CANARY

'A Robin Red Breast in a Cage
Puts all Heaven in a Rage.'

William Blake, 'Auguries of Innocence'

We started to build cities because there was really no help for it. We needed to bring a lot of humans all together to expand the possibilities of civilization: and so we built cities for human convenience. Other species didn't matter much. But as soon as we succeeded in creating a human hive, we began to miss the things we had lost. And what we missed, we missed above all with our ears. City life didn't feel right because it didn't sound right.

Even before the invention of the motorcar, cities were noisy places, filled with the shouts of men, the playing of children, the crying of babies, the movement of transport and the yells of commerce: the suffocating sounds of large numbers of a single species all living in close proximity in a claustrophobic monoculture. We craved relief from this: and we found it in birdsong. Not so much the sound of wild birds: there wasn't much space for them in the press of humanity. So people went out and caught wild birds; they chose the ones that could sing and the songs lightened the darkness of city living.

The canary gives the name to this chapter, because the canary is the archetypal caged songbird: the coloratura singer behind bars – even though we have taken it into our homes comparatively recently, within the last 500 years or so. But the chapter must also take in all the birds we have caged for their song. Pliny the Elder wrote: 'The knight Marcus Laenius Strabo was the first to introduce aviaries containing birds of all kinds, at Brundisium. Thanks to him we began imprisoning creatures to which Nature had assigned the sky.'

It is likely that keeping songbirds is as old as cities, and certainly it took place in many civilizations beyond the West. The Chinese tradition of aviculture is as ancient as that of keeping goldfish (see Chapter 80). When Hernán Cortés arrived in the Aztec capital of Tenochtitlan in 1519, he found that Emperor Montezuma had a House of Birds, in which the birds lived in great splendour, with much gold, most of which Cortés managed to steal.

Caged birds were taken on sea voyages by sailors of many cultures, including Babylonians, Hindus, Polynesians and Vikings. Not only was the presence and

the sound of the bird comforting for those out of sight of land, but the birds could also be released. If they flew off in a certain direction and didn't return, you could follow the course they had taken and, with luck, find land; if they came back down to the ship to be recaught, it was a safe assumption that no land was in sight, even from the greater elevation of a bird on the wing.

But it was for their role in improving city life that caged birds were most prized. Not that everyone approved of this. Giorgio Vasari, in his sixteenth-century work *Lives of the Most Excellent Painters, Sculptors and Architects*, writes of Leonardo da Vinci: 'He delighted much in horses and also in all other animals, and often when passing by the places where they sold birds he would take them out of their cages and, paying the price that was asked for them, would let them fly away into the air, restoring them to their lost liberty.'

That didn't mean that Leonardo didn't also watch them in flight with the utmost attention. He made notes on the way they flew away from him, for flight fascinated him, perhaps more than anything else in his wildly varied life. He made a memo to self: 'You will make an anatomy of the wings of a bird together with the muscles of the breast, which are the movers of these wings. And you will do the same for man...'

Mozart kept a caged starling, buying it after he had heard it singing a fragment of his own music – one that the bird could only have picked up from a visiting pupil, whistling the notes as he cast his eyes over the birds for sale. It's been proposed that the theme of Mozart's *A Musical Joke* was suggested, if not actually composed, by the starling. Mozart had a full-scale funeral laid on for the bird when he died, with what degree of irony it is impossible to know.

Originally, most of the birds that sang so beautifully in the cities of the world were wild-caught. It was comparatively easy to catch them: the trick was keeping them alive. In 1622, a book called *Uccelliera* was published, written by Giovanni Pietro Olina, and it is full of sound practical advice about how to look after birds in cages: what conditions they need and what food they like, including what food is best to prompt them into song. In the section on nightingales he recommends that: 'grains of pasta... shall be put in with them to one side, in *little boxes or drawers* and, on the other side, heart... spread out on a little square tablet of stone'. The nineteenth-century Russian author Ivan Turgenev (*Fathers and Sons; A Sportsman's Notebook*) wrote an informative essay on the best way to keep nightingales.

In the fourteenth century, Portuguese sailors reported that the islands of the Atlantic, along the coast of Africa, were remarkable for their birdsong. The birds were found on the Azores, Madeira and the Canaries, of which the main island is

People pleasers: The Yellow Canaries by Joseph Caraud (1821–1905).

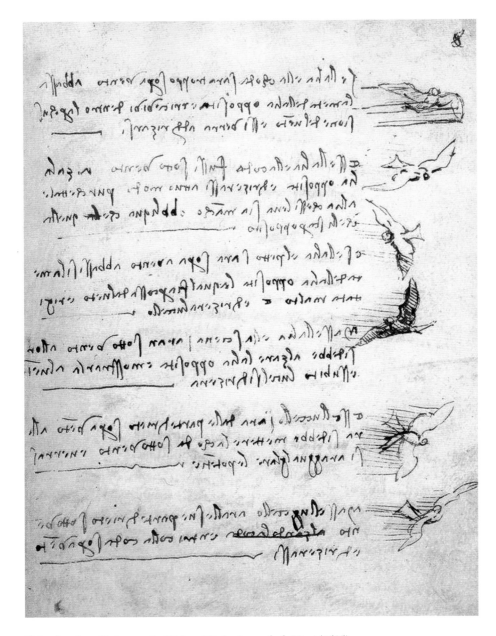

Flying free: from Codex on the Flight of Birds, *Leonardo da Vinci (1505).*

Gran Canaria, which means the island of dogs – so the pretty yellow birds with their pretty singing voices are named for a dog. These canaries were brought back to Europe for the cage-bird market, and rapidly became a success, for their looks, their song and for a certain resilience desirable in a cage bird.

Originally the merchants on Gran Canaria tried to sell only the male birds – only the cockbirds sing – to keep the monopoly, but the birds are hard to sex and the females got out. Canaries became immensely popular, and were bred not only for their voices, but also for their colour and for their shape. You can today find breeds of canaries bred for their song, such as Spanish Timbrado, German Roller, American Singer, Russian Singer and Persian Singer. In *Doctor Dolittle's Caravan*, the doctor stages *The Canary Opera*, starring the greatest singer of them all, Pippinella, and the work receives praise from Paganini himself.

It was in 1690 that Austrians first started the practice of taking canaries down coalmines, because their more delicate metabolism saw them suffer at much lower densities of carbon monoxide and methane than human miners could take: and so a distressed, dying or dead bird gave early warning to get out. The practice was prevalent for all but three centuries: the last canary to go down a mine in Britain did so in 1986. Since that time coal-mining canaries have become one of Life's Great Metaphors: the various species across the world that are heading towards extinction because of anthropogenic causes are compared to canaries down the mine: early warnings of the planet's condition that we humans disregard at our own peril.

Canaries, being fully domesticated, have been used for scientific experiments: on the way that neurons develop in the brain; how songbirds work out and produce their songs; and how the vertebrate brain learns.

In the sixteenth and seventeenth centuries, people came to East Anglia fleeing religious persecution in the Low Countries. They brought their looms, their weaving skills – and the birds that kept them company while they worked. Norwich became a centre of the weaving trade – and also of canary breeding. Two local breeds, the Crested Norwich and the Norwich Plainhead, were recognized. At a peak the city exported 2000 birds to New York every week. The local football team, Norwich City, plays in yellow to this day, has a canary on the badge and carries the nickname of the Canaries. I heard of a supporter of their East Anglian rivals, Ipswich Town, who refused to go on holiday to the Canary Islands out of loyalty to his own preferred football club.

Meanwhile, the practice of keeping cage birds has declined, at least in the West. Indonesia has a thriving trade in caged songbirds, and there are national competitions for the best singers; this has led to unsustainable exploitation of wild populations. The wild species, mainly in Asia, are therefore suffering because of the illegal trade in birds caught to sing in cages.

Cities are noisier than ever with the constant roar of more and more engines. But these days we cope with the mechanical noises of the city with a counter-attack of recorded music.

It seems that we don't need birdsong any more or, at least, we think we don't.

EIGHTY-TWO
REINDEER

'*More rapid than eagles his courses they came,*
And he whistled and shouted, and called them by name.'

Clement Clarke Moore, 'A Visit From St Nicholas'

W hat are the most recognizable images in the world – across all nations and all ages? The crucified Christ, the Buddha – and perhaps, after that, reindeer. Provided, of course, that they come with a gravity-defying sleigh and a white-bearded man dressed in red. The gift-bringing sleigh, bearing Father Christmas and towed by a team of reindeer, has become one of the great archetypal images, almost universally recognized, even though it has been around for only a couple of hundred years.

Before that time, reindeer were a great concern only for comparatively small numbers of people: the indigenous people who live around the North Pole. To appreciate the nature of reindeer and the humans concerned with them, it is helpful to look down on the world from a position directly beneath the Pole Star. We are used to understanding the world by means of the Mercator Projection, which seems to show an immense distance between east and west even in the Arctic Circle: but look down from above and you will see a comparatively small white circle of chilled land and sea that takes in parts of Siberia, Alaska, Canada, northern Russia and northern Scandinavia – and it's all pretty much the same place.

A single species of deer has made a speciality of this northern circle around the Pole. It is mostly called caribou in America and reindeer in Britain, but, although they can be divided up into different populations and subspecies, they are all the same species, *Rangifer tarandus*. They are impressively attired creatures: of all the deer species, these have the largest antlers in proportion to their body size. Females mostly carry them as well as the males.

They are superbly adapted for the greatest and most obvious problems of the far north: cold. They have a circulation system that keeps them warm with remarkably little expenditure of energy, and it's one of the many great wonders of evolution. It doesn't get as much attention as the cheetah's speed or the albatross's soaring ability, because the hidden engineering doesn't make the great pictures that appeal to the human mind. It's a counter-current heat exchange system and it keeps reindeer warm in the extreme conditions they evolved to live in.

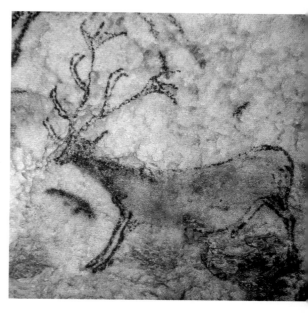

But not Rudolph: reindeer from Lascaux Cave, France (c.17,300 years old).

Reindeer are great travellers, moving north in the spring to exploit newly thawed pastures, and moving back down south when these are exhausted. Some populations travel as much as 3000 miles (4800km) in a year, which makes them the best travelled of all terrestrial mammals, beating the more famous wildebeest of the Serengeti. Their hooves are wide and crescent-shaped, to spread their weight as they cross snow, soft ground after the thaw and swamps in the brief Arctic summer. As they move they give out a series of sharp clicks as the tendons in their legs snap. It's been suggested that these have a social function: the better the click, the more formidable the reindeer. But it's possible that there is another use: clicking reindeer can hear their neighbours and know exactly where they are, even when they travel through a whiteout. The reindeer's eyes are also adapted for extreme weather: they can see deeper into the ultraviolet end of the spectrum than we can, which gives them much better perception in what to us is a pure white landscape.

Reindeer made human life possible in these boreal latitudes. They were the only readily available way of converting the scant plant resources of the land into protein accessible to humans. The indigenous people that still live in the far north, in all the nations that circle the Pole, have a profound relationship with reindeer. Some of the groups subsist on reindeer for twelve months of the year: reindeer is nothing less than life itself. There is an Inuit saying: 'the reindeer feeds the wolf, but the wolf keeps the reindeer strong'.

Throughout history, reindeer herds have been watched and managed and exploited: for food and also for clothing, shelter and milk. In some cultures, the reindeer are pretty well domesticated, the only species of deer to be so. They have been used as draught animals; these days their main job is pulling sleighs full of tourists, though some of the Siberian reindeer are big enough to ride. There is a complex economy for draught animals in the far north: a sled pulled by twenty reindeer will give you 12–15 miles (20–25km) a day, as compared to 4–6 miles (7–10km) for a person on foot, but a dog sleigh will give you 43–50 miles (70–80km) with cargo and an astonishing 93–110 miles (150–180km) unladen – but there is, as always, a payback. You need a team of five, six or seven dogs, and each one needs 4½lb (2kg) of fresh meat a day to keep going.

There are many herds of reindeer worked by humans to this day: modern reindeer herders use diesel-driven electricity generators and diesel-powered snowmobiles to make this way of life work. Reindeer meat is popular in Scandinavian countries; you can buy tins of reindeer meatballs, and reindeer sausage is available in supermarkets. The antlers are used in Asian medicine as an aphrodisiac.

It is the more picturesque notions about reindeer that captured the world's imagination. Unlike the subtle and cumulative development of most of such wide-reaching images and ideas, this one can be dated with pinpoint accuracy. The first recorded example of Father Christmas and a sleigh pulled by reindeer comes from 1823, and a poem entitled 'A Visit from St Nicholas', more often remembered from the first lines as 'Twas the Night before Christmas'. It was written by Clement Clarke Moore (though this has been disputed) and it was published anonymously in the *Troy Sentinel* (in upstate New York) on 23 December. The poem has been claimed as the best-known lines of verse written by an American, and it details the Christmas Eve visit of St Nicholas, before he was widely known as Father Christmas or Santa Claus:

> When, what to my wondering eyes should appear,
> But a miniature sleigh, and eight tiny rein-deer.

The poem names them, too: Dasher, Dancer, Prancer, Vixen, Comet, Cupid, Dunder and Blixem, though the last two sometimes have their spellings altered. The story is that Moore wrote the verses while make a shopping trip – by sleigh.

No Rudolph in the poem, though. He was the later invention of Robert Lewis May, who in 1939 was commissioned to write some verses for a colouring book. He lived in Chicago, and when a fog wafted in from Lake Michigan he had his eureka moment and set the poem on a foggy Christmas Eve. In 1949 May's brother-in-law Johnny Marks adapted the verses to make a song, and it was recorded by Gene Autry, also known as the Singing Cowboy. It told 2.5 million copies in the first year, 25 million in total:

> Rudolph with your nose so bright
> Won't you guide my sleigh tonight?

The floating sleigh towed by flying reindeer has become an emblem of childhood, innocence, the gullibility of children, the certainty of Christmas presents for all: everything we would like life to be.

Real life is tougher both for humans and for reindeer. In recent years there has been a continuing decline in global reindeer populations. Reindeer numbers tend to fluctuate in the natural course of things, but the current slump is alarmingly

Dash away all: illustration from 'A Visit from St Nicholas' by Clement Clarke Moore (first published in 1823).

prolonged. A 2018 report by the United States body the National Oceanic and Atmosphere Administration revealed the decline. The report's chief author, Don Russell, said: 'They're at such low levels you start to be concerned… if we return in ten years and numbers have gone down further, that would be unprecedented.'

Note the alarm within the scientific caution. Causes of the decline are various, but for once habitat destruction isn't at the top of the list. One problem is likely to be overhunting: we have already seen in this book (especially Chapter 44 on passenger pigeons and Chapter 17 on cod) that there is a tendency for humans to think that any resource is infinite until it's too late.

But the main reason is almost certainly climate change. You'd think that warmer summers would be a good thing to those that live around the North Pole, but the added heat brings drought, more flies and more parasites, while heat stress leaves these cold-adapted animals vulnerable to disease. There is also a problem when rain falls instead of snow: it forms sheets of ice, making movement more difficult (remember these are travelling animals) and food becomes harder to find.

An animal we associate with cold is suffering because of heat: anthropogenic heat. An animal that we associate, above all, with good times – with *the* good time of the year – is having a bad time. We can see reindeer as one more example of the canary in the coalmine: as one more last warning.

EIGHTY-THREE
TURKEY

'Therefore let us keep the feast, not with the old leaven,
neither with the leaven of malice and wickedness, but
with the unleavened bread of sincerity and good truth.'

1 Corinthians 5:8

The idea of the feast goes very deep in human life: it is perhaps the ultimate celebration of being human and alive. Feasting is part of every culture, though some cultures do it better than others. A feast is not just food as nutrition, not just food as pleasure. It is about the marking of a shared joy by means of excessive consumption. The tradition goes back to our hunter–gatherer days, when the gathering brought in the staples but the occasional successful hunt was an occasion for rejoicing. Since the problems of keeping meat edible in a hot climate are considerable, there was really no option: a feast it had to be, a treat for everybody, sharing and togetherness, a rare moment in a difficult life when, suddenly, everything is easy and everybody has more than enough.

The feast makes the point that not all days are alike. Some days are special. The church celebrates feast days with cattle (see Chapter 7). We have festivals, more or less the same word; we have festive occasions: and they are all about the glorious relief from hunger and uncertainty that our ancestors knew – but knew only every now and then. We have taken the memory of those occasions and dared to make them regular and predictable. Perhaps the triumph of civilization is that we became able to choose when to feast: to know that the feast will come.

A feast needs a centrepiece. And it has become the tradition in the culture of the West to centre a feast round a single dead animal: one that has been slain to make the joy possible. The fatted calf was slain to celebrate the return of the prodigal son: these days most of us delegate the slaying, but the tradition of centring the feast around one entire corpse continues.

The great feasts of Christmas and also, in North America, Thanksgiving both require a turkey. Not part of a cow or part of a sheep or a goat, but a single whole animal, to be cut up in front of us, rather than cooked in its parts. The dramatic entrance of the dead and cooked bird – sight, sound, smell, which predict texture and taste – is an essential part of the feast, as is the carving of it, not done in the privacy of the kitchen but out there on the table with the guests assembled and

Still wild: turkey engraved by William H. Lizars for James Audobon's The Birds of America *(1827–38).*

waiting. This beast has been sacrificed for us: to make our feast. No one is allowed to remain in a moment's doubt about that central fact.

The turkey is a native to Central and North America. There are two species: the ocellated turkey of the Yucatan peninsula, and the other one, normally called just turkey, or sometimes wild turkey when it happens to be wild. (I should add for clarification that the Australian brush turkey is not closely related to either of these.) Turkeys are, like many birds we have met in this book – chickens, pheasants, peacocks – ground-dwelling omnivores, powerful and well-muscled walkers (making the legs particularly desirable for their meat) that fly reluctantly, to escape from predators and to find a safe roost for the night.

They were an obvious target for domestication. The earliest known date for domestic turkeys is 800 BC, but it's likely that it happened a good deal earlier, and probably at least twice, by the Aztecs of Mexico and by Native Americans in what is now the USA. They have been bred for their feathers as well as for their meat: ceremonial robes made of turkey feathers have been found. The first Europeans to eat turkey were the Spanish conquistadors of the sixteenth century; chroniclers wrote of finding turkeys in the marketplace of the old Aztec capital Tenochtitlan. From there turkeys were brought to Europe and bred. The English navigator William Strickland brought turkeys to England in the 1520s; in 1550 he adopted a coat of arms bearing 'a turkey-cock in his pride proper', the last word meaning that the colours were lifelike.

There are two theories as to why the birds are called turkeys. The first is that they are similar to guinea fowl, two similar species of African birds which were imported into Europe via Turkey, and so were called turkey-fowl, or turkey-cocks, as we have seen. The second is that turkeys themselves first reached Europe by way of West Asia, again imported via Turkey. Either way, turkeys are not Turkish but North American in origin.

Horatio Walpole, the first earl of Orford, was very keen on turkeys and he introduced them to his Norfolk estate, into the woods around Wolterton. This established Norfolk as the centre of England's turkey farming; and that is still the case for premium stock, with the breeds of Norfolk Black and Norfolk Bronze. I visited such a farm, with free-ranging turkeys in hundreds, rounded up by a thrillingly enthusiastic collie; it was an impressive operation. At one stage I held one of the larger birds in my arms, hand clasped firmly – but not too firmly – around the legs to keep it from leaping off. The weight and the liveliness of the bird were equally remarkable.

It took a while for turkeys to become an essential part of the feast of Christmas. In 1573 a farmer, Thomas Tusser, wrote of the need to supply turkeys for Christmas, but turkey was for years very much seen as an exoticism: something not everyone could afford.

But from the seventeenth century, those who could afford turkeys for Christmas generally did so. The birds travelled from Norfolk to London on foot: they were shod and they walked all the way, 150 miles (240km) or more, along a series of green lanes. This was time-consuming and tiring for all concerned, but it got live birds to the market. It was worth it because live birds keep fresher than dead ones. For the less affluent, beef or goose was the traditional push-the-boat-out Christmas meal; in Charles Dickens's A Christmas Carol of 1843, Bob Cratchit and his family have a goose.

The situation changed in the mid-twentieth century with the introduction of intensive farming, cheaper and more accessible feed, the option of keeping birds in more crowded conditions, and with advances in refrigeration. Turkeys and Christmas became, in many places, inextricable. These days there are ethical concerns about overcrowding turkeys along with concerns about the pre-emptive use of antibiotics in all forms of meat production.

Keeping fresh: turkeys to market on a Norfolk road (c.1931).

The Thanksgiving Festival – or festivals – of North America have always slightly mystified Europeans. In England we associate harvest festivals with school and church, and with hymns like 'We Plough the Fields and Scatter'. But even to urban North Americans, Thanksgiving, another harvest festival, is a real feast laced with sincere feelings of rejoicing. The celebration is dated in Canada back to 1578, and in the United States to 1619, where it is associated with the Pilgrim Fathers, the first Europeans to settle in what is now the United States, and the gratitude they felt at getting the first harvest in successfully. It didn't just mean the end to quite a lot of very hard work: it also brought with it the assurance of survival through the winter. This celebration is a heartfelt thing in many American homes in which no one has ever seen a field of corn. It goes deep: and tradition demands that it can only be celebrated by the death and consumption of the most enormous bird you can put on the table. A turkey.

Turkey farmers sometimes wish that people ate turkeys at other times of the year, that their business was not so maniacally seasonal. Turkey meat has been praised for its low fat content as well as for its flavour, and it is possible to find the birds or parts of them (turkey breasts for example) out of season. But there is a widespread feeling that it would be inappropriate to consume a turkey at some ordinary, non-special, non-celebratory time of the year. The idea of eating a turkey at the wrong time of the year is almost blasphemous: trying to make Christmas less like Christmas, weakening the things that make the time special.

The death of a turkey is the birth of a celebration. In the twenty-first century we humans feel the urge to feast as strongly as our ancestors did. And when we feast, at least in the West, it is to the turkey we turn.

EIGHTY-FOUR

DEER

'Robin Hood, Robin Hood, riding through the glen,
Robin Hood, Robin Hood, with his band of men...'

Carl Sigman

Hunting and gathering sustained early humans across countless millennia, but the trend of advancing civilization moved us away from such activities. Farmed crops and domesticated animals provided the wealthy with all they needed in terms of nutrition. Once this point was reached, it was clear to everyone what had to be done: what was the rightful task of any man of wealth and power. That was, of course, to go hunting. And the king of all the beasts available to the hunter was the deer: the stag, magnificently crowned to show both his masculinity and his royalty.

The advance of civilization is the story of hunting: hunting not for necessity, but for fun: for social life: for the exercise and display and wealth and social position. What humans had once done in order to survive, they now did to show how truly civilized they were.

Deer were at the heart of it all, though other quarry was taken on. In the first and most glorious hunt in the great, fourteenth-century poem *Sir Gawain and the Green Knight* a deer is the great prize. The idea of pursuing and killing a deer has remained ever since, as recounted by the great Tom Lehrer (see also Chapter 22 on pigeons/doves and Chapter 98 on vaquitas):

I went out to hunt some deer
On a mornin' bright and clear.
I went and shot the maximum the game laws would allow
Two game wardens, seven hunters and a cow.

Deer have always been quarry for humans; there are ninety representations of stags in the cave paintings of Lascaux in France from about 17,000 years ago. In China *Homo erectus* fed on sika deer; early humans hunted red deer in what is now Germany; the reindeer was a staple of the Cro-Magnons. Deer have been valued: because of the palatability of their meat – and the very large amount of it that can be found on an adult red deer and similar-sized species like the elk (not moose) of North America and the sambar of Asia; because of the excellence of their skin (buckskin) as clothing; and because their antlers make nice handles for knives.

Noble beast: The Monarch of the Glen *by Edwin Landseer (c.1851).*

There are more than ninety species of deer worldwide; they are found in every continent except Antarctica and Australia (though they were later introduced into Australia and New Zealand). There is only one species in Africa, though: the Barbary stag, which lives in the Atlas Mountains and is a subspecies of red deer. Elsewhere, Africa is dominated by antelopes; the most obvious difference between these two groups is that deer shed their horns and regrow them every year, while antelopes keep them the year round.

All male deer wear horns apart from water deer; in this species the males grow exaggerated canine teeth for settling disputes. There is a huge range in size between species: moose (called elk in Europe) can stand up to 8½ft (2.6m) at the shoulder, to the northern pudú, as small as 13in (32cm) for an adult.

Deer are primarily browsers, taking leaves from trees and bushes, but they will also eat grass, lichens, moss and some fungi. They have adapted for a number of habitats, from the frozen tundra (we have met the reindeer in Chapter 82) to tropical rainforest. If you were to generalize, you would say that their most typical habitat is open forest with wide grazed clearings. More recently they have adapted for less obvious habitats; in the UK they hide up in woodland during the day and raid crops at night. In many places they live in comparatively large numbers, usually because they have few or no major predators. In most places near large human population centres, there are not many wolves.

There are deer in the *Rig Veda*, deer in the Bible. A golden deer plays an important part in the Hindu epic, the *Ramayana:* Sita is much taken by this beautiful creature, so Rama goes off to catch it for her; in his absence Sita is kidnapped and the deer is revealed as a demon. Deer appear again and again in the world's mythologies: Buddha at one point becomes a deer; deer are seen in some Celtic myths as fairy cattle. A deer was the sole companion of the contemplative St Giles. Artemis, the virgin huntress in Greek mythology (Diana for the Romans), was seen bathing by Actaeon, who was transformed into a stag as a result. One of the labours of Hercules (Heracles) was to catch the Ceryneian Hind, which was sacred to Artemis.

The invention of hunting as a leisure activity for the elite seems to have cropped up in many civilizations at round about the same time: at a point when civilization had become so far advanced that rich people had leisure: and, with that leisure, the opportunity to express nostalgia for an earlier stage of human development. Hunting was once part of humanity's survival package: you do not have to go so terribly deep to reactivate this.

It is worth pausing here for a moment to ask why deer were never fully domesticated. They have long been kept in parks as ornaments to a great house, where an idealized rural landscape was incomplete without a lake, noble trees with a horizontal browse line and a herd of deer to do the browsing. But reindeer apart, deer have never been fully domesticated in the manner of cattle. There are problems with keeping deer confined, because of their immense agility and leaping skills. Stags with their extravagant headgear and their headstrong behaviour of a male in rut (sexual excitement) can be dangerous. These are not insuperable problems: wild aurochs were ferocious and enormous and they were domesticated pretty satisfactorily as the homely cows we know today (one of which Tom Lehrer's hunter shot) (see Chapter 7 on cattle).

The answer is probably that we never bothered. Deer have plenty of meat on them, but cattle have more. Deer are quite big, but oxen and horses are bigger and make much better draught animals. There are, after all, only a certain number of animals worth the trouble of domesticating, and those that we have are the result of thousands of years of selective breeding. No doubt we could come up with many different breeds of domesticated deer over the next few millennia, if we wished to and needed to. But we don't and besides we have always rather liked the idea of the deer as a wild thing: a noble creature worthy of being hunted by noble humans.

In European culture, hunting as a noble pursuit was developed by the Merovingian and Carolingian rulers of the Frankish people (centred in what is now France) from the sixth through the eighth century. Charlemagne, king of the Franks from 768, was an enthusiastic hunter when not involved in more exciting pursuits.

Feast of venison: Robin Hood and His Merry Men Entertaining Richard the Lionheart in Sherwood Forest *by Daniel Maclise (also known as Alfred Croquis) (1839).*

Hunting became a central activity and was pursued to the point of obsession. The legend of St Hubert, now patron saint of hunters, tells of a hunter so enthralled by the ritual of slaughter that he goes hunting on Good Friday rather than going to church. He is confronted by a stag bearing between his antlers a crucifix. Hubert devotes the rest of his life to the church. The story of excessive hunting is memorably told by Gustave Flaubert's 'The Legend of St Julian the Hospitaller', in which, during Julian's mad orgy of slaughter, a stag curses him, so that he kills his own father and mother. (The story is in the same volume, *Three Stories*, as that of the parrot in 'A Simple Heart'; see Chapter 71 on parrots).

This Frankish tradition of the hunt was exported into England by the Norman Conquest of 1066, and a great deal of England became a hunting reserve as a result; it is estimated that, by the twelfth century, one-third of all southern England was a royal forest, subject to forest laws that forbade local people to kill any forest animal, to keep dogs or to own weapons. All of what is modern Essex and Huntingdonshire were royal forests.

These hunts were as much about ritual as they were about slaughter: doing things the right way was more important than doing them the efficient way. It was a deep communion with the social order of the time: a validation of yourself as hunter and a person of significance. It involved such warlike skills as horsemanship and use of the longbow, the crossbow, the spear and the lance. It was also quite dangerous: a number of royal people were killed hunting, including William II of England, who was victim of a stray arrow in 1100. (Tom Lehrer was singing about an ancient tradition: 'you just stand there looking cute, and when something moves you shoot'.)

All this led to the most significant legend of the English, a people who value concealment above all else. This is the story of Robin Hood, who lived in Sherwood Forest and dined nightly on the king's deer. The London–York Road runs through Sherwood Forest, and here Robin and his Merry Men were able to rob the rich to help the poor. The story always makes much of his outlaw status. An outlaw was beyond the protection of the law, and if he was killed it was not considered murder, merely the right thing to do. To help an outlaw was an offence; outlaw status existed in the UK until 1938.

The hunting of deer continues to this day. In 2006 the US Fish and Wildlife Service said that the amount received for licences to shoot deer totalled US$700 million for the year, while the deer-shooting industry was worth US$11.8 billion annually. In the UK in the nineteenth century it was customary to hunt 'carted' deer; deer that were taken to a likely place and then released so that mounted humans and dogs could chase them. Hunting with dogs was banned in England and Wales in 2005. Deer stalking – hunting of deer with a rifle – is pursued for recreation in the UK, particularly in Scotland. Here deer have been allowed to build up in large numbers, since there are no natural predators and there is money in producing suitable targets for stalkers – these being mature stags bearing large antlers with many tines. The result is that the hills of Scotland, largely stripped of trees for their timber for shipbuilding over the course of centuries, have been unable to regenerate in most places because the deer eat the shoots. Projects for restoring the great Caledonian Forest, such as Abernethy, depend on deer exclosures – fences to keep the deer out.

Hunting deer is still part of human life, but the practice is not universally admired. The 1942 Walt Disney film *Bambi* tells of the death by hunters of Bambi's mother, which is sometimes claimed to be the saddest sequence in the history of film. The plot tells of Bambi's eventual emergence as the new Great Prince of the Forest. Like so many of the stories about deer, *Bambi* values the deer above the hunter and portrays an idealized natural world, in which the only blemish is humanity. This can be seen as a forward-looking theme: looking to a time in which nature is there for itself and for all humanity, rather than the exclusive preserve of the rich, the powerful and the heavily armed.

EIGHTY-FIVE
RABBIT

' *"Now my dears," said old Mrs Rabbit one day, "you*
may go into the fields or down the lane, but don't go into
Mr McGregor's garden: your father had an accident
there; he was put in a pie by Mrs McGregor." '

Beatrix Potter, *The Tale of Peter Rabbit*

Vulnerability and innocence, protein and warm clothing: rabbits have filled human needs, both spiritual and physical, across the centuries. And all the time we have confused them with hares, who operate a very different strategy for staying alive: Easter Bunny is a hare; Bugs Bunny is a hare.

There are a number of species of both, all found in the order (we belong in the order of primates) of Lagomorpha and the family of Leporidae. But hares give birth in the open to good-to-go young, capable of seeing and running from the very start; while rabbits give birth to blind and naked little blobs deep in the darkness of a burrow. It follows that hares are very good at running for extended distances, with many rapid turns, to avoid capture, while rabbits bolt to the nearest burrow. That makes them easier to keep, control and domesticate.

European rabbits have been introduced to every continent except Antarctica, because humans like them and because they are a very convenient source of protein. Pliny the Elder writes of rabbit-keeping; technically cuniculture. The Romans found these creatures in Spain and, recognizing their handiness, started to keep them and then to transport them. Rabbits don't need much looking after.

Like cows and sheep, rabbits are highly capable of turning grass into protein accessible to humans. They are considered unclean in Judaism, on the grounds that they chew the cud and yet lack a cloven hoof: but that information is inaccurate. What rabbits chew are their own droppings; cud-chewers chew their own vomit, to express things in a rough-and-ready way. Rabbits produce two different kinds of droppings: they scatter hard droppings outside the burrow, but produce soft droppings in the burrows and chew them: in other words, they have a second go at partially digested food and so get double value from their food. Like chewing the cud, this is an elegant and effective strategy.

The easiest way to keep rabbits in useful numbers is to surround a pleasant area of grass with a nice wall, put the rabbits inside and let them get on with it.

This form of management persisted for centuries and explains why so many places and properties in England have the word 'warren' in the name. The catch is that rabbits, being great diggers, will sometimes succeed in burrowing beneath the wall and getting out: they have done so time and again in the history of their relationship with humans.

But never mind. The answer is simple: stop the holes and rely on the remaining rabbits to do one of the things that rabbits are supremely good at: making more rabbits. A female rabbit can reach sexual maturity as early as three months; she can breed at any time of year; and she has a short gestation period (the young being born as half-formed things) of 28–35 days. She will produce a litter of 4–12 kits, and can become pregnant the following day. She is quite capable of producing sixty young in a year: so if you can provide enough food in a walled space safe from ground predators, you are unlikely to run short of rabbits.

Rabbits were brought to Britain by the Normans after the conquest of 1066, though it's possible that the Romans also brought them over. They have been introduced to many other places, including Australia. Their adaptability and their capacity to increase with great speed have often made them a problem for humans, taking the grazing from domestic animals and damaging walls, buildings and roadways with their digging; rabbit holes are a problem for horses travelling at speed.

As a result, rabbits have frequently made the familiar transition from cherished and valued adjuncts to human life, to enemies to be exterminated. We have waged war on rabbits by gassing, shooting, fencing, snaring and ferreting. Sometimes their burrows are blown up. Rabbits damaged but not killed by these processes are traditionally despatched with a single blow: a rabbit-punch. These methods are frequently ineffective: if you have land suitable for rabbits, then getting rid of the rabbits only creates a vacancy: neighbouring rabbits will move in, breed and exploit the resource.

But rabbits were nearly wiped out in the UK by myxomatosis, a natural disease that was deliberately introduced in Australia, Chile and France in the 1950s. The disease spread and reached the UK, where diseased rabbits were deliberately introduced to the burrows of wild rabbits, a practice now illegal. The UK population crashed – it is reckoned that 99 per cent of the population was killed, and when I was young I seldom saw a rabbit. However, the population recovered; as we have already seen, rabbits are extremely good at that.

Rabbits or hares? The American jackrabbit is a species of hare. The notion of Easter Bunny, the jovial creature that gives us eggs so that we can celebrate the new life given by the resurrection of Christ, also comes from hares. In early spring when crops are low, hares are very visible. People who pursued them would often

Rash rabbit: Peter Rabbit, painted by Beatrix Potter in 1902; the book was published the previous year.

Oh my fur and whiskers! The White Rabbit *by Sir John Tenniel, watercolour by Gertrude Thomson,* *from* Alice's Adventures in Wonderland *(1889).*

find the nests of lapwings, ground-nesting birds that like the same big open fields that hares favour. These, obviously, were the rabbits' eggs: a miraculous gift from a benign creature. Pliny the Elder, among others, believed that hares were hermaphrodites and could fertilize themselves and so produce young without loss of virginity, so they were also symbols of purity and of the Virgin Mary.

Rabbits and hares have an association with churches, partly for that reason. The famous device of the three hares is found in many English churches, particularly in Devon, and the motif is repeated the length of the Silk Road from Europe to China (see Chapter 58 on silkworms). It shows a circle of three hares, each apparently bearing two ears, as is only right and proper, but the ingenious design shows only three ears in total, each one common to two different hares.

In Africa the scrub hare, referred to in local languages as *kalulu*, has become incorporated into a much-loved network of folk stories, as Kalulu. Kalulu is a trickster: small and vulnerable but ever capable of outwitting his enemies – especially the hyena – by his intelligence and deviousness. These stories crossed the Atlantic in the slave ships, and were told and retold, with the cast of animals changing to fit the new land of America. The tales were collected, elaborated and retold by Joel Chandler Harris in the *Uncle Remus* books, the first of which was published in 1881, and featured the adventures of Brer Rabbit, who constantly outwitted Brer Fox. The tales were told in dialect, in a manner subsequently regarded as patronizing. In perhaps the most famous example, Brer Rabbit, when caught by Brer Fox, says: ' "I don't keer w'at you do wid me, Brer Fox," sezee, "so you don't fling me in dat briar-patch. Roas' me, Brer Fox," sezee, "but don't fling me in dat briar-patch," sezee.'

So Brer Fox flings Brer Rabbit, only for Brer Rabbit to skip away with the derisive call:' "Bawn en bred in the briar-patch, Brer Fox…" '

Not hard to see that these stories of the underdog's survival by trickery had an added relevance to an enslaved people. The same notion of the trickster rabbit is found in Bugs Bunny, created by Leon Schlesinger Productions, later Warner Bros. Cartoons, in the late 1930s. Bugs Bunny, being an American (with a Bronx accent), is clearly a jackrabbit, which as we have seen is a species of hare.

Rabbits have also cropped up time and again in books for children: because of their easy maintenance rabbits were considered ideal pets and have been bred into many unexpected shapes and colours; there are more than 300 breeds recognized. In *Alice's Adventures in Wonderland* by Lewis Carroll (see also Chapter 19 on the dodo and Chapter 67 on the oryx) Alice pursues a white rabbit who is consulting a pocket watch; she later meets the March Hare at the Mad Tea Party. March hares are considered mad because of their extravagant leaping, chasing and boxing during the month of courtship.

Rabbits turn up several times in the stories of Beatrix Potter, most famously with Peter Rabbit, who disregards maternal instructions, raids Mr McGregor's garden and is pursued vigorously, losing his blue jacket and his shoes in the process: a cautionary tale of what happens if you don't do what your mother tells you; Peter is no trickster rabbit. Perhaps the most astonishing rabbits in fiction are in *Watership Down*, a work by Richard Adams comparable to the *Aeneid*, which tells the tale of a group of rabbits who escape the destruction of their colony and go on to found a new one elsewhere.

Rabbits are seen as prey animals who must live surrounded by enemies with only their wits to help them survive (or, in Peter's case, the wit of his mother). It is not hard to see why humans can relate to that: you need only to turn back to the opening chapter of this book, which is devoted to lions. We humans were once prey animals, forever surrounded by bigger, faster, stronger creatures. We too had to learn to hide and we had to use our wits. Our success is the most dangerous thing in the course of our evolutionary history: meanwhile, rabbits (and hares) continue to live all around us, finding ways to hang on and breed – and breed again – in the landscapes that humans have created.

EIGHTY-SIX
SPARROW

'Who sees with equal eye, as God of all,
A hero perish or a sparrow fall.'

Alexander Pope, *An Essay on Man*

Ubiquitous, plentiful, living alongside humans but causing little offence, inspiring if anything a kind of wry affection for their apparent cheerfulness: a sparrow is a small creature admirable for its humility, for its lack of pretension to be anything other than consummately ordinary. The sparrow became the bird best able to share its place and its life with humanity, and is – or perhaps was – the most recognizable bird in the world.

Sparrows have lived alongside humans for as long as there have been anything even close to permanent dwellings, for they are adept feeders on scraps and forgotten trifles. With the invention of agriculture they were plentifully supplied with grain. They could also make their nests in human dwellings, where their understated cheeping in the thatch was an asset rather than anything else. The eighteenth-century Japanese poet Basho wrote:

Sparrows in eaves
mice in ceiling –
celestial music.

The sparrows in question belong to the group known as Old World sparrows, the Passeridae. This includes forty-three species in eight genera, but we are really mostly talking about two of them: house sparrow and tree sparrow. Their natural range covers most of Europe and Asia; they have been introduced to the Americas and to Australia; they can be found in Africa: there is scarcely a centre of human population that doesn't share its space with sparrows.

They have become people specialists, and right from the start they were well equipped for taking on this rich and variable niche. Sparrows are very social birds; they will nest communally and form sizeable flocks outside the breeding season: they positively revel in crowds. They are naturally seed-eaters that also take invertebrates: they are anatomically distinct from finches because they have

Even a sparrow… : La Madonna del passero (*Madonna with sparrow*) by Guercino (c.1619).

Tree sparrows, not house sparrows: Sparrows in the Moonlight *by Yang Shanshen (1913–2004).*

an extra bone in their tongue that enables them to hold seeds in position for cracking and eating. They are versatile feeders, and this lack of fastidiousness, as we have seen in other species already in this book, is a helpful trait if you wish to exploit the resources of humanity.

In most of Europe we associate house sparrows with cities and tree sparrows with the countryside, but in many parts of Asia, and especially in China, tree sparrows fill the traditional house sparrow niche; when I lived in Hong Kong I had tree sparrows round my house.

One of the reasons for the sparrow's ability to make a living in cities was horses. Horses need food to power them, often in the form of grain, and its storage and spillage were a constant opportunity for sparrows. These resources become available all over again after they have passed through a horse: a horse's digestion is a rough-and-ready thing that requires a large throughput – they are not cud-chewers like cows and sheep – and so what's passed out is, if you are a sparrow, a concentrated food resource. No wonder cities in the long millennia of horse-drawn transport rang with the cheeps of sparrows.

So sparrows became a living example of the inescapable omnipresence of non-human life; common as sparrows, nigh on worthless to any but hemselves, but still for that reason afforded some grudging respect and admiration.

In St Matthew's Gospel, Jesus tells his disciples what they must do to spread the word, which is, on the face of it, a very difficult and dangerous activity. He consoles them: 'Are not two sparrows sold for a farthing? and one of them shall not fall on the ground without your Father… Fear ye not, ye are of more value than many sparrows.' God is so great and so clearly omnipresent that he even has time to care for each individual sparrow: a striking image of the hands-on nature of the Christian god.

Sparrows have also been seen as images of love and lust, and are associated with the Greek god of love, Aphrodite, while Chaucer describes the horrible Summoner: 'hoot he was and lecherous as a sparwe'.

The Venerable Bede takes the idea of the insignificant sparrow as an image of the fleeting and insignificant nature of the earthly life of humanity in his eighth-century work *Ecclesiastical History of the English People*: 'O King, the present life of man on earth is like the flight of a single sparrow through the hall where, in winter, you sit with your captains and ministers.' A seventeenth-century painting by the artist known as Guercino, *La Madonna del passero*, shows the Madonna and child, with a sparrow perched on the virgin's finger as the child watches spellbound: even a sparrow – even a sparrow – matters to one such as he.

Even a sparrow… the theme continues in *Hamlet*, as the prince discusses the nature of fate:

There's a special providence in the fall of a sparrow.
If it be now, 'tis not to come;
If it be not to come, it will be now;
If it be not now, yet it will come: the readiness is all.

Life is fleeting for us all, sparrows and humans: best be ready for anything.

The combination of sparrows and Shakespeare brings us to the extraordinary New Yorker Eugene Schieffelin, who loved birds and Shakespeare. He was a founder of the American Acclimatization Society in 1871, which was dedicated to bringing non-native birds into America and letting them go to see if they would settle in and start breeding. He also founded an organization called Friends of Shakespeare. These two passions have been conflated in subsequent mythologies, with the story that it was his ambition to introduce every species of bird mentioned by Shakespeare into the United States. The idea is so pleasingly bizarre that it is now an accepted truth.

In 1850 the trees of New York City were full of the caterpillars of linden moths, and in 1852 Schieffelin imported some house sparrows, in the hope that they would eat the caterpillars and so protect the trees at his family home in Madison Square. There were other releases at the same time and for the same reason, and, as is so often the way with attempts at biological control, control was the most significant thing the project lacked. The sparrows succeeded not wisely but too well (a phrase from Shakespeare's play *Othello*); at one time there were reckoned to be 540 million of them in the United States; by 2017 they were down to 77 million. (In 1890 Schieffelin released sixty European starlings in Central Park, New York City; there were once more than 200 million in the USA, now 85 million. He was also involved in less successful releases of goldfinch, chaffinch, skylark and nightingale.)

Sparrows have been treated as agricultural pests and killed as a result, stealers of grains and breakers of the stalks that bear the ears of corn. They have often been taken as food and in some places still are, but sparrows are widely tolerated as part of human life. This is not universal. In 1958, as part of the Great Leap Forward, Chairman Mao Zedong introduced to China the Four Pests Campaign, in which it became the duty of citizens to kill four species identified as the 'public animals of capitalism': these being rats, flies, mosquitoes and sparrows. People chased sparrows from their roosts by banging pots and pans until the birds dropped from the air from exhaustion; their nests were destroyed, eggs broken, chicks killed. The sparrow population dipped close to national extinction.

At the same time, and not by coincidence, locusts and other harmful insects proliferated and rice yields dropped sharply. It was all part of the ecological recklessness of the Great Leap Forward, which involved deforestation and profligate use of chemicals, and it led to the great famine that killed 15–45 million people (see also Chapter 74 on the baiji). The sparrow-killing policy was put into reverse in 1960, after advice from the ornithologist Cheng Tso-hsin.

But sparrow numbers in many places in the world have plummeted in recent years, and done so without any campaigns of deliberate hostility. In the UK, populations of house sparrows are down 71 per cent since 1977. The causes of their decline in rural areas are to do with changes in agricultural practice: stubble fields were once left over the winter and provided food and shelter for many species; these days the fields are mostly ploughed shortly after the harvest.

The reasons for their decline in urban and suburban areas are less obvious. Possible problems include less food, more pollution, loss of nesting sites, prevalence of disease and even increased predation, some of it by cats. Research has shown that the decline is less sharp in poorer areas of cities, where there is more waste ground, and fewer gardens treated heavily with chemicals; fewer home improvements means more traditional nesting sites available. A study carried out in 2003–2004 by the British Trust for Ornithology showed that gardens and allotments are important for the survival of urban sparrows: and that gardens with an element of wildness are an important resource. In the meantime, sparrows – the ultimate symbol of commonness – are now on the UK List of Conservation Concern and have made it to the IUCN Red List.

EIGHTY-SEVEN
BUTTERFLY

'Kill not the moth nor butterfly,
For the Last Judgement draweth nigh.'

William Blake, 'Auguries of Innocence'

The devil's chaplain has had his say in these pages, with relevance to such creatures as the loa loa worm, the mosquito and the tsetse fly (see Chapters 78, 23 and 45). But let us now turn to the butterfly: a creature of simple, undeniable, infinitely fragile beauty. Surely the human unable to see beauty in a butterfly has yet to be born (though perhaps will come soon enough). Butterflies have been seen as proof of the existence of a benign creator and a demonstration of the existence of the soul. There is a Roman sculpture of a butterfly leaving a man's mouth at the moment of death. The Ancient Greek word for butterfly is *psyche*, which also means mind – or soul.

A forest clearing in Colombia, and a lecture on the benefits of ecotourism… my attention shattered by the appearance, for a second, no longer, of a blue morpho butterfly, not so much a fluttering shape as a lightning-swift beam of impossibly intense blue light, vanishing the instant it appeared because of the nature of iridescence and the laws that attend refracted light… as beautiful a thing as I have seen in my life.

Throughout history we humans have loved butterflies, and desired them, too, wanted to preserve their ephemeral beauty for ourselves, to share it, to make it live with us for ever. In some ways they are the living, fluttering proof that God loves us: scraps of passing beauty that seem to have no function but to please the sight of humans, without, even for a second, threatening us with harm. (We naturally divorce the adults from their larvae: caterpillars are capable of damaging crops grown by humans, as the catch-all term 'cabbage white' indicates.)

The order Lepidoptera includes butterflies and moths; the first fossils date back 190 million years and to the Late Triassic, but the first butterflies, evolving from moths, came later; there are no clearly identifiable fossils of butterflies before 55 million years ago, 10 million years later than the end of the dinosaurs. In the main – though there are plenty of exceptions – butterflies belong to the day and moths to the night; moths are camouflaged in subtle and drab colours, butterflies rejoice in glamorous brightness; butterflies rest with their wings upright, moths

Collection mania: antique display case with random Lepidoptera.

with their wings folded. There are around 18,000 species of butterfly, about 170,000 species of moth.

Lepidoptera means scaly wing; butterflies and moths both travel by means of a fluttering flight on large, scale-covered wings, four of them in two pairs. Both live the classic insect four-stage life: the egg is laid on an appropriate food plant, the larva or caterpillar grows, often very rapidly, being basically an eating machine, and when it has finished it pupates and then hatches out as a fully formed butterfly or imago; the transition from pupa to this final stage is technically called imagination.

Some butterflies will go through several generations (egg to adult) in the course of a single year; others a single generation in that time. Some cold-climate species take several years to make the same transition. An adult can live for as little time as a week, and for a full year, depending on the species. Adults exist only on liquids, to maintain energy levels: their priority is making more butterflies, which involves sex and the laying of eggs in the right place. They mostly feed on nectar, but will also take nourishment from sap, rotting fruit, dung and decaying flesh: the sight of a lovely butterfly avidly feeding from a dog turd always jars with our understanding of what is fitting. Enthusiasts lure purple emperor butterflies – a species of singular beauty much sought-after in the UK – down from the canopy by means of foul-smelling treats, based on idiosyncratic or even secret recipes; fermented shrimp paste is usually involved.

It is generally accepted that butterflies are called butterflies after the male brimstone butterfly, which is a pale, eye-catching yellow, the colour of the best butter. (Brimstone means sulphur, which is yellow.) The reasons for the colours and patterns of butterflies have long been an evolutionary fascination. Part of this is to attract mates (though courtship also involves the use of pheromones, so butterflies can find and recognize each other through scent). Some species take on dangerous toxins as caterpillars and advertise their unpalatability as both larvae and adults with bright colours; other harmless species mimic them. (The famous migratory monarch is poisonous, the mimicking viceroy is not.) Patterns and colours often imitate those of dangerous predators: the eye pattern on the wings of the peacock butterfly resembles a startled little owl (Athena's species, see Chapter 51 on owls); the butterfly will also rub its wings together to mimic the sounds of the hiss the owl makes. Philip Howse in a series of books including *Seeing Butterflies* and *Butterflies: Messages from Psyche* has brilliantly extended this line of observation, by finding British butterflies that mimic goldfinches and a South American species that looks like a piranha.

The lives of butterflies can be bewilderingly complex. Monarch butterflies of North America are famous for their migration, and for their swarming behaviour. The painted lady butterflies of the Old World migrate twice as far, also in a series of generations, travelling from Africa up into the UK and, on big years, north as far as the Arctic Circle. It was once thought that this was a doomed generation, an evolutionary mistake or cul-de-sac: but in recent years advanced radar has tracked flocks of painted ladies flying back south, to lay eggs on the appropriate food plants at stopping points on the route, to create the next generation, which will continue the southward trail back towards Africa. Butterflies navigate by the sun, which requires an accurate understanding of the time of day; they can see polarized light so they are aware of the exact location of the sun even in cloudy weather.

Butterflies are unmissable, almost recklessly beautiful. They dance before humans in entrancing colours. As soon as a human community can look beyond the problems of survival, its members find delight in butterflies, proof of a benign world. There are butterfly images found in Egyptian art 3500 years old. The

Dreaming of being a butterfly: fourth-century BC sage Chuang Tzu, by Japanese artist Ike no Taiga (1723–76).

Aztecs believed that butterflies were the souls of dead warriors: butterflies often seem to lie halfway between life and death, seen as disembodied but clearly still-living spirits. In Japan there is a superstition that a butterfly entering your house and perching heralded the arrival of the person you love most. There is a traditional Chinese story of the Taoist sage Chuang Tzu, who died in 289 BC. He dreamt he was a butterfly; when he woke up he wondered if he was a butterfly dreaming he was a man.

In the eighteenth century a widespread fascination with butterflies seized many parts of the world. The Aurelian Society of London was in full swing by

1745, having been founded probably a few years earlier. Its members were dedicated to the collection of butterflies. Butterflies, and their capture and collection and display, became a widespread passion. The beauty of butterflies was celebrated by their death: here was a beauty that a human could not only savour but also pursue and possess. It is a process that we find a little sinister these days, but it is unhelpful to judge people by standards two hundred years later.

Back then the idea of extinction, or even shortage, was unthinkable: everyone knew that nature was inexhaustible, and that human civilization was a triumph against nature and against the odds. It was also accepted that nature was utterly at our disposal: everyone had a right to a butterfly and taking one from the wild could do no possible harm. And besides, to collect butterflies was to labour on the side of beauty. The classic example – what naturalists might call the type specimen – is Gilbert White, who worked as a clergyman in Hampshire for most of his life during the eighteenth century, producing the classic of observation *The Natural History of Selborne,* which has never been out of print. Charles Darwin thought of becoming just such a person before his voyage on the *Beagle* changed him.

The craze for the collection of butterflies combined both art and science, just at the time when the two things were becoming decisively separated. Butterfly collecting was a way of understanding the world: of doing so through nature, at a time when nature was becoming simultaneously depleted and easier to access.

Collection mania was a part of this, and it's an easy thing to despise. The idea of collecting, and of attempting to complete a collection, is a way of seeking knowledge: if you like, seeking to encompass all the world's knowledge. In the UK there are birdwatchers who seek out lost, gale-blown rarities that turn up for a few hours or days: such people are called twitchers and they compete with each other for the longest possible list. They can be mocked and despised, but Darwin was a twitcher of a sort: when he was young he collected not the sightings of rare birds, but the bodies of beetles – 'Whenever I hear of the capture of rare beetles, I feel like an old war-horse at the sound of a trumpet,' Darwin said in his middle age.

Butterflies were mounted and displayed in wooden cabinets; many such collections succumbed to the invasion of mites and beetles. They were countered with mothballs, which, being made from poisonous naphthalene, are now illegal in most countries. Walter Rothschild, the British banker, politician and zoologist, had a collection of 2.25 million butterflies, along with a great deal else. Part of his many collections was sold to the American Museum of Natural History, to pay off a blackmailing mistress, while on his death the bulk of it went to the Natural History Museum in London (at that time part of the British Museum), the largest ever such bequest.

Butterfly collecting was a serious business; Neville Chamberlain and Winston Churchill, former prime ministers of the UK, both collected butterflies. Butterfly farms were set up to breed and supply specimens to collectors. Butterfly collecting still continues and is mostly legal. In the UK it is legal to capture, kill or trade in some species, but twenty-three of the fifty-nine species that regularly breed there are protected.

But these days most collecting is done with a camera. This change of attitude to butterfly collecting illustrates changing attitudes to non-human life. This was expressed in *The Collector*, a John Fowles novel of 1963 (the year after the publication of *Silent Spring*, see Chapter 23 on mosquitoes). In its pages the butterfly-hunter Frederick (who prefers to be known as Ferdinand) captures and kidnaps – collects – the beautiful Miranda as a human butterfly. It is a creepy tale about what is widely considered a creepy practice. These days there are still many people who pursue butterflies, but few use a net, preferring close-focusing, high-spec binoculars and digital cameras.

Butterflies give easy delight even to the least proficient observers of nature. As we modify our countryside they are becoming ever scarcer. In the UK 76 per cent of all species are in decline; in the United States the western monarch has undergone a decline of 99 per cent. Butterflies are a frail yet powerful link between human and non-human life. They are incontestably bright and beautiful. They were a symbol of the soul: now they are perhaps still more eloquent as a symbol of loss.

EIGHTY-EIGHT

FRUIT FLY

'Every cell in my body has it all writ down.'

Mike Heron of The Incredible String Band

Charles Darwin established the principle on which life operates; Gregor Mendel, with his pioneering study of genetics, established the mechanism by which it does so. Everything else we owe to the fruit flies. The combined work of these two humans is sometimes referred to as the neo-Darwinian synthesis, or just neo-Darwinism. It has been developed in ways not only beyond the nineteenth-century imagination but in ways that boggle twenty-first-century minds: and, again, all by means of fruit flies.

There are a good few insect species called fruit flies; it's a loose term for any flying, two-winged insect with a taste for fruit. Here we are talking about the genus *Drosophila* and, in particular, the species within this group called *D. melanogaster*, which is sometimes called the common fruit fly, sometimes the vinegar fly. We are most of us familiar with it as a wild creature, on and above fruit that's gone past its best; a musty fruit bowl will release a rather cloying cloud of them, creatures that seem to have a special attraction for human eyes and nostrils. They are minor pests in homes and restaurants, anywhere there is readily available food. For convenience we will use the term fruit fly for this species for the rest of the chapter.

Fruit flies have become the classic 'model organism': a species you can experiment on, in the knowledge that such research will be applicable to many other species as well. They are easily found wild on all continents, including islands, and they have become one of the most widely used of all laboratory animals. It has been suggested that they are the best known of all eukaryotic organisms: that is to say, creatures built from cells that contain a nucleus; the domain Eukaryota contains animals, plants and fungi. Eight Nobel Prizes have been awarded for work with fruit flies.

It's a process that began in 1910, when Thomas Hunt Morgan instigated experiments on heredity at Columbia University in New York, setting up in what became known, accurately if unimaginatively, as the Fly Room; Morgan was the first to link inheritance to a single chromosome. The flies were bred in milk bottles and fed on banana paste: already the reasons for choosing these flies begin

to become clear. They are dead easy to keep: in fact, it's probably harder to get rid of them than to keep them and breed them.

But they have many other advantages. They take up very little space. It is easy and safe to anaesthetize them. They have short generations: about fifty days from egg to death. They breed enthusiastically. They are sexually dimorphic: the male is easily distinguished from the female because he has a distinct black patch on his abdomen: if it has a black bum it's a male. They have only four pairs of chromosomes, which simplifies matters; humans usually have twenty-three. These chromosomes are large enough to be independently examined.

Fruit flies embody the principle of simple complexity: and that is not entirely a contradiction. It can be illustrated by comparing the fruit fly's brain to that of a human: a human brain weighs 3lb (1.3kg) and contains 100 billion neurons; a fruit fly's brain takes up a cubic millimetre and contains 150,000 neurons. In other words, by studying fruit flies you can find important principles about the complexities of the brain without getting bogged down.

It should be added here that fruit flies don't excite too many ethical problems. Few people will get too worried about cruelty to fruit flies. So far as most people are concerned, you can do what you like with them and, to a very large extent, that's what scientists have done. As a result a very considerable amount of what we know about life we owe to fruit flies.

They have been used for the study of genetics, physiology, disease and evolution. They have taught us about sex-linked inheritance, about gene-mapping, about the basic principles of heredity. Their genome was sequenced in 2000; they have 60 per cent of genetic material in common with us humans, and 75 per cent of known human diseases also affect fruit flies, making them a handy species to work on when researching human diseases. Research on immune systems has been conducted with fruit flies. Scientists have isolated the fruit fly genes for vision, smell, hearing, learning and memory, courtship, pain and longevity. Fruit flies have been used for research on cancer, and on bacterial and fungal infections. They have taken part in experiments on the effects of radiation. They have also featured in work on sleep patterns and the circadian rhythm: one of the fundamental aspects of daily life.

Where do you stop? That is perhaps the most relevant ethical question. We may be relaxed about the discomfort and unnatural life suffered by laboratory fruit flies, but we are less easy in our minds about using them to help with the creation of completely new forms of life. Experiments on the introduction of transgenic material into fruit flies have been performed: that is to say, the addition of genetic material from quite different species. This is part of the growing

Unlocking the secrets of inheritance: Thomas Hunt Morgan (1866–1945) studying fruit flies.

research into genetically modified (GM) organisms: crops and, by extension, genetically modified farm animals.

GM crops routinely harvested include soy plants that are tolerant of herbicides and maize that is resistant to insects. GM crops are now a fact of life in the United States, Canada, Australia, Brazil, Colombia and Argentina; one form of GM cattle feed is accepted in the EU.

In Canada a GM salmon is now farmed. It matures in eighteen months, twice the rate of a naturally developed specimen. There is continuing research on the development of disease-resistant pigs, bird flu-resistant chickens, hornless dairy cattle and more productive sheep. This is not an overnight job; the salmon took twenty-five years to develop.

GM crops have been a fact of life for a couple of decades now, but we are no nearer working out the ethics of it all. Proponents claim that GM produce of all kinds is the only way forwards for humanity, particularly with the ever-rising human population. If we are to feed the arriving billions, we must take the GM route – the route shown to us by the fruit flies – or accept starvation and planet-wide social unrest.

The opposition comes in two forms. The first is religious, and states that humans have no right to meddle in the sanctity of species. Artificial selection, by means of selective breeding, as humans have been doing for 12,000 years, is acceptable to such fundamentalists; but direct modification is not. The second form of opposition relies on the principle of the unintended consequence. Genetic science is in its infancy and to allow GM species to dominate the planet will give rise to consequences that we cannot foresee, because we lack a full understanding of the subject.

One of the main problems facing those who wish to advance GM crops is public unease. Some proponents – you might call these fundamentalists of science – see such unease as foolish and retrograde, a simple terror of progress; others see the push for GM farming as reckless and irresponsible. Either way, the central issue is that, by means of fruit flies, humans have taken a major step towards the usurpation of God. The Book of Genesis told us that God created all the species on Earth: that will be – perhaps already is – no longer the case, and perhaps it will become less and less so. The Bible postulates a God of infinite wisdom. Perhaps infinite wisdom is required of anyone who seeks to create a species: here, alas, the fruit fly cannot help.

EIGHTY-NINE

SAOLA

*'Astonishing though it may seem to many wizards, Muggles have
not always been ignorant of the magical and monstrous creatures
that we have worked so long and hard to hide.'*

Newt Scamander, *Fantastic Beasts & Where to Find Them*

Conservation is one of the arts of peace, and perhaps the greatest of them. When a country is at war, people long only for the cessation of war: when a peace, however uneasy, has been declared, they set about living rather than surviving, with the hope of making their country better than it was before. And that is what happened in Vietnam after the Americans pulled out in 1975 after nearly twenty years of war.

Rebuilding the country after nearly half a century of warfare – for the conflict predated American involvement – is a long, hard process that's still continuing. In 1986 the Vũ Quang area of more than 200sq. miles (500sq. km) was declared a forest reserve; it became a National Park in 2002. Once it had been gazetted there was a natural curiosity to find out what was actually in it. Forests had not been safe areas, for scientists or anybody else, for many years: but an expedition set up by the Vietnamese Ministry of Forestry and WWF explored the biodiversity of the place in 1992. From a local hunter they acquired the skull of what looked like an antelope, but, as it turned out, was nothing of the kind.

The following day they managed to get hold of the remains – or part of them – of a second animal of the same species. But what species? This was a creature of a serious size, but no one apart from the local people had had any idea of its existence. Five years later it was photographed in the wild for the first time. It was the first new species of a large terrestrial animal that had been discovered for more than fifty years.

This was the saola, sometimes known as the Vu Quang ox; it's a bovid that looks like an antelope. It has also been called the Vu Quang bovid and the spindlehorn; it has also been referred to as the Asian unicorn, presumably to avoid confusion with all the other species of unicorn. Just about the first thing that was discovered about this creature was that it was Critically Endangered.

The saola is the only member of its genus, which is named *Pseudoryx*, because of its superficial similarity to the oryx (see Chapter 67). It is significantly different

Forever elusive: illustration of a saola by Barry Road Carlsen.

to all other bovids in its antelope-like qualities. And it can reach a hefty size: 33in (84cm) at the shoulder, 5ft (150cm) in length, and with horns that can be 8in (20cm) long, unbranched and curving gently backwards. The coat is a deep brown and the face is picked out with white patches: perhaps to break up the outline in shadowy forest; perhaps for recognizing each other in dim light; perhaps a warning, in the manner of a badger's three white face patches, informing potential attackers that this end is dangerous: as a badger bites, so, perhaps, a saola can use its horns, which both sexes possess.

I can speculate in this way as much as I like, because very little is known about the saola and so I have as much chance of being right as anyone. Saolas are found in wet forest, both evergreen and deciduous; they like rivers and valleys. They are found only in the Annamite Mountains of Vietnam and Laos: their range, so far as this is known, is about 1950sq. miles (5000sq. km), an area that takes in four nature reserves. They seem to be mostly solitary, but they have been seen once or twice in groups of three. It seems clear that they dislike the proximity of humans: they are hunted by local people and they are a prized quarry; in Laos the saola is 'the polite animal'.

A few saolas have been caught and kept in captivity, though never for long, and they have been studied in these unsatisfactory circumstances. A female, with the somewhat fateful name of Martha (see Chapter 44 on passenger pigeons), was caught in 1998 and kept in a menagerie in Laos. All that is known about the saola comes from remains, the odd captive, camera traps and local knowledge.

The saola remains unique as the only large land animal that has never been seen in the wild by a field biologist. It is an animal of thick forest, active at night and either ends of the day, avoiding the hotter middle hours of daylight. There are very few of them, though no one is sure what that means in terms of numbers. A few hundred? Less than a hundred? Make your own guess, but it's easy to see that this is an animal on the edge.

One of the many reasons for its elusiveness has been war. War is sometimes a good thing for wildlife conservation, creating no-go areas for humans and, more importantly, for large commercial enterprises. In modern times no one is going to try to clear-fell a forest for profit in the unstable times of warfare. But in the Vietnam War the forest was the friend of the insurgents of North Vietnam, a hiding place for their guerrilla soldiery.

That made the forest an enemy to the Americans, and so they waged war on it. Their main weapon was the defoliant known as Agent Orange. It was sprayed from above, to kill off the forest the enemy soldiers hid in. The stuff was also used on crops, in an attempt to destroy the countryside that supported the North Vietnamese. The results of this form of biological warfare are obvious to this day: large areas of what the locals sardonically call 'American grass'. Where the great forests once stood, you now find vigorous tussocks of grassland, a great lumpy mat that allows nothing else to grow. Rainforest cannot regenerate from nothing: a rainforest is a closed system that nourishes itself; astonishingly the soil it grows on is often poor in itself, rich only in the forest it supports.

Now, instead of miles of saola habitat and great forests that air-condition the planet, we have American grass. There are schemes to regenerate parts of the forest, run by the local NGO Viet Nature: I have visited courageous projects that aim to undo the damage of half a century ago. 'It's great,' I said. 'And it should be really nice in about 500 years.' The CEO, Tuan Anh Nguyen, smiled at me: 'Not so much,' she said. 'Maybe only 200 years.' It is right and wise to think on such a scale.

Trophy: a hunter from the Katu tribe in Vietnam, with a saola skull and horns.

One of the problems with the saola is that no one knows very much about it, and therefore no one knows what is needed to conserve it. The animal is scarce therefore data is scare; data is scarce therefore the animal is getting scarcer. But the greatest problem that the animal faces is obvious enough: habitat destruction. Its future, if it has one, lies in the protection and regeneration of forest, particularly in the areas that lie between the four reserves in which the saola has its being.

The saola was the first large terrestrial animal found for half a century. It is likely that it is also the last new land animal of any great size that we will ever find. The saola was the land's last secret. The age of finding is over; we are already deep in the age of losing. Those who are disturbed by this truth are invited to turn to the epilogue of this book.

NINETY

GIANT SQUID

*'Bond stared down, half hypnotized, into the wavering pools
of eye far below. So this was the giant squid, the mythical
kraken that could pull ships beneath the waves...'*

Ian Fleming, *Dr No*

The saola stayed hidden for a long time, but when it at last turned up in 1992 no one doubted that the creature was real. No one suggested that the party from WWF and the Vietnamese Ministry of Forestry were off their heads. But when Captain Peter McQuhae, commanding HMS *Daedalus*, described the creature he saw between the Cape of Good Hope and the island of St Helena in 1848, the highest authority on non-human life, in the UK if not the world, implied that he was not fit to hold the queen's commission.

Richard Owen was a great biologist, best remembered these days as the person who coined the word dinosaur, which means terrible lizard. He was also an outspoken critic of Charles Darwin; he accepted that evolution took place but not by the method that Darwin described. Owen loved a good controversy, was never shy of using the weight of his reputation and, eleven years before the publication of *The Origin*, he took against McQuhae. What McQuhae had seen was the creature we now know as the giant squid.

But Owen decided that McQuhae had seen nothing more exciting than a large seal. He said it was likely 'that men should have been deceived by a cursory view of a partly submerged and rapidly moving animal which might be strange only to themselves'. The sneering and patronizing tone was calculated to exasperate. Any seaman who had seen a giant squid was well advised to keep quiet about it.

But the giant squid has haunted human imaginations across the centuries, sometimes accepted as a fact, at other times as one of the classic tall tales of the unfathomable ocean. Aristotle mentioned the giant squid in the fourth century BC. Pliny the Elder wrote of the squid in the first century AD, and noted that its 'head' was 'as big as a cask'.

The giant squid also turns up in mythologies: the fact that something appears in a myth doesn't mean that it's not true. Scylla is a squid-like monster who appears opposite the whirlpool Charybdis in the *Odyssey* (see also Chapter 37 on octopuses). The kraken turns up in Scandinavian mythologies: a many-tentacled

Unspeakable monster: giant squid illustration from a twentieth-century edition of Twenty Thousand Leagues Under the Sea *by Jules Verne (first published in 1870).*

monster of the deep who is found off the coasts of Norway and Greenland. Fear of the dark and the depths, fear of being out of sight of land, and our innate fascination with monsters – and most especially the monsters that emerge from the ocean – added fanciful details to a real creature, multiplying its size, so that it routinely pulled down ships with its tentacles and devoured the hapless sailors. A good few of the stories about the great sea serpent can be put down to badly seen giant squids.

For centuries the giant squid existed in the no-man's land between fact and myth. The exaggerations of earlier times meant that later and less fanciful observers of the wild world were inclined to disbelieve the creature's reality. Georges Cuvier (see Chapter 76 on mammoths) said that the woolly mammoth must have gone extinct, because if it existed humans would surely have managed to clap eyes on it. That argument was used for the giant squid, but the ocean hides things as nowhere on land – not even the rainforests of a nation with a long history of war – could ever hope to do. A squid scientist, Clyde Roper, once said: 'We know more about the moon's backside than we do about the ocean's bottom.'

Linnaeus was in two minds about the giant squid. He put it into the first edition of his *Systema Naturae* in 1735 and named it *Microcosumus*. But it was omitted from subsequent editions; perhaps Linnaeus changed his mind about the relationship of truth and mythology. It has to be admitted that McQuhae's much later portrayal of his squid lacks the cold clarity of a disinterested field description. He said the creature was nearly 100ft (30m) long, as reported in *The Times*, and that it had a mouth that was 'full of large teeth… sufficiently capacious to admit a man standing upright'. A Danish scientist, Japetus Steenstrup, plumped in favour of the squid in 1857, basing his evidence on parts washed up on shore, including a large beak. He called it *Architeuthis dux*. Then in 1861 a French warship, *Alecton*, came across a giant squid, and even managed to bring a piece back – and so another sea captain was laughed to scorn; the experts said it was plant material.

But Jules Verne liked these stories. His science-fiction works often involve a reckless mingling of fact, mythology and speculation: in *Twenty Thousand Leagues Under the Sea* (published in 1870) the crew of the submarine *Nautilus* find the lost city of Atlantis, and they also encounter a giant squid, which attacks the *Nautilus* and devours one of the crew before it is driven off.

Then in 1873 a fisherman in Newfoundland named Theophilus Picot said that he hadn't exactly found one, but one had damn near caught him. He said that it had seized hold of his boat and tried to drag it down. This was a classic tale of the deep, but there was a twist: he and his men managed to hack off both feeding tentacles – that is to say, the two longer ones; there are eight shorter ones. They brought these to shore, where the local rector, Moss Harvey, realized that they had found something pretty momentous: the tentacles were 19ft (6m) in length. Try sneering your way out of that.

Last rites: giant squid washed up on a Californian beach in 2005.

After that it gradually became accepted that the giant squid was real: that the kraken really did inhabit the bottom of the sea. It was seldom seen because it lives at immense depths, feeding on fish. All squids and octopuses are molluscs, like the slugs in your garden: they are invertebrates and their soft parts don't last long after the creature has died. Should you ever get hold of a piece of giant squid, you will have a difficult job getting it to people who can examine it, especially in days without refrigeration.

These days the giant squid is part of zoology as well as mythology. A specimen has been found measuring 43ft (13m) long, with a mantle – the body and head – of 6½ft (2m) in length; as big as a cask, in fact. In 2004 the first photograph was taken of a giant squid in its deep-water natural habitat. In 2012 an adult was filmed, again in the depths, and another in 2019; you can find clips on YouTube, which show the creature's grace without giving a clear impression of size.

As the giant squid was accepted as biological orthodoxy, so talk and evidence gathered of a still more enormous creature. In 1925 tentacles too long for a giant squid were found in the stomach of a sperm whale; sperm whales, the largest of the toothed cetaceans, are the natural predators of giant squids. In 1981 a Russian trawler caught a large squid: again, too big to be a mere giant. This was accepted as a new species: the colossal squid, with a total length of 60ft (18m), and eyes 12in (30cm) across. This seems to be more formidable than the giant, perhaps an apex predator, too big to be preyed upon.

Humans have always understood non-human life by a mixture of properly observed facts and wild myths that usually tell us more about our own needs than the creature in question. But in the crowded and complicated part of the Venn diagram, where facts and myths come together, the myths sometimes tell us truths that science has problems in reaching.

NINETY-ONE
BEAVER

'So it was only common politeness when Susan said, "What a lovely dam!" And Mr Beaver didn't say "Hush" this time, but "Merely a trifle! Merely a trifle! And it isn't really finished." '

C. S. Lewis, *The Lion, the Witch and the Wardrobe*

We humans started modifying the landscape to suit ourselves around 12,000 years ago, as we have noted again and again in this book. Beavers have been doing the same thing, on a less extravagant scale, for a great deal longer: they have been creating wetlands, altering rivers, making ponds and changing the hydrology of the places where they live for millions of years. Many non-human animals inadvertently alter the habitat they live in – elephants being the most obvious example – but when it comes to the deliberate manipulation of landscape to your own advantage, then beavers are history's second best, enthusiastic and hard-gnawing runners-up to humankind.

Master builders: illustration of European beavers from a Soviet encyclopaedia (1927).

This has consistently put them at odds with the king modifiers. Beavers' ideas of landscape manipulation have frequently and increasingly clashed with our own. But in recent years there has been a kind of beaver revisionism. People are beginning to value the changes that beavers can make to a landscape and to a river system, realizing that there are many ways in which this benefits us. There has also been an increasing human delight in beavers for their own sake: in many places (including New York City) they have been welcomed back into areas from which they have been long absent.

There are two species of beaver: the North American beaver and the Eurasian beaver. They look pretty similar: the Eurasians are a little bigger, and there are a number of more subtle differences. When brought together artificially they are reluctant to breed and do not hybridize. But their behaviour and ecology are pretty well interchangeable.

They are rodents: the second largest two species after the capybara of South America. They live their semiaquatic lives and eat the vegetation of the wet places where they live. They don't touch fish; in fact they make conditions better for fish, though plenty of anglers, fiercely protective of their pleasures, are reluctant to accept this.

Beavers are great gnawers. As we have seen already with rats (see Chapter 25), gnawing is the great talent of rodents. Beavers have taken this as far as any species possibly could: their perpetually growing incisors have been turned on the environment in which they live, allowing them to eat, to store food and to create the conditions that suit them best. It starts with gnawing: in the felling of saplings and fully grown trees. It moves on to construction: famously of dams, which create deep pools. These pools protect the beavers from predators: a beaver in a lodge surrounded by deep water is safe from bears and wolves.

Beavers also construct canal systems, which they use for floating lengths of gnawed-off trees to where they want them, to be used for food and as construction material for lodges and dams. They don't hibernate; their food stores keep them going through the winter. They build their dams across rivers and streams, starting with the verticals, adding horizontal branches and then plugging the gaps with plants and mud.

Humans have always hunted beavers, less for their meat – though they were conveniently declared by the Catholic Church to be fish (like barnacle geese, see Chapter 61), so that they might be eaten on Fridays – than for their fur. Their semiaquatic lifestyle, especially in the more northerly parts of their range, demands efficient insulation, and beavers are insulated by their coats. Their underfur also makes the best felt, so beaver pelts were used for the finest top hats. (A bowler in cricket who takes three wickets – three outs – in three balls has taken a hat-trick; the archaic term for four wickets in four balls is a beaver-trick: a beaver being superior to a normal hat.)

But there is a further reason for the beavers' value to humans. They carry a substance called castoreum, which has been used in medicine for a couple of millennia and more; not to be confused with castor oil, which is made from castor beans. Castoreum can be extracted from glands that beavers of both sexes carry and use to scent-mark their territories. It comes from organs called castor sacs and has been employed as an analgesic and an anti-inflammatory. It was used in the twentieth century to treat complaints of the womb and to increase blood pressure and cardiac output. Some people recommend it to this day, for the relief of anxiety, restless sleep and period pains.

Aesop tells a story that a beaver, when pursued by a hunter, will bite off his testicles and offer them; Pliny the Elder liked that one and passed it on uncritically, but unlike the myth of the giant squid this tale is purely fanciful. Castoreum was also used to enhance food flavours, particularly vanilla, strawberry and raspberry. It's still a significant ingredient in the manufacture of perfumes: it gives a leathery smell, a favourite in male toiletries.

Both species of beavers have not only been hunted, but have also been persecuted for the disturbance they bring to environments that humans seek to control. Eurasian beavers went extinct over much of their range. In North America a beaver population estimated at 90 million at the time the Europeans arrived was reduced to 6–12 million by the late twentieth century. The demand for beavers' pelts was a powerful factor in the opening up of the American West: it was the hunters and trappers who led the way. The history (and the mythology) of the conquest of the West began with beavers.

And so it was thought, in Europe and in North America, that beavers were part of the past, representing a historical state from which humans have emerged. Outside national parks beavers were seen as unwelcome reminders of times we had risen above. But with the revisionism that came with humanity's too-complete triumph over nature, there was a rethinking about beavers. There were rerelease projects in parts of the former range of the European beaver, on the Elbe, the Rhine and in Scandinavia, while there was greater tolerance of existing populations in Poland, Slovakia and the Czech Republic. In North America there has been increasing public delight taken in beavers; with cleaner rivers, beavers reached urban areas including Toronto, Calgary, Winnipeg, and they have been seen in Chicago, San Francisco and on the Bronx River in New York City.

Beavers went extinct in Britain in the sixteenth century, but reintroductions and escapes have established a few small populations. These have been closely monitored, because there are still many who object strongly, fearing that beavers will create havoc on the manmade landscape. But studies on beaver impacts have shown many surprising and unexpected things. Beaver engineering makes for great water purity and by creating wetlands they increase biodiversity. Anglers

Beaver hunt: the beaver at the top is biting off his testicles (c.1230).

fear that their dams will stop fish migrating: but fish have been migrating for millions of years, long before humans started controlling beavers and, for that matter, rivers. Fish can get through and over dams, while the cleaner water and greater diversity is a positive asset to them.

Beaver engineering also helps humans to manage water. Traditional thinking in water management is about getting rid of it as fast as possible: rivers have been dredged and canalized to hasten the progress of rain to the sea. This leads to floods in vulnerable downstream areas and water shortages in times of drought. A site containing confined beavers in Devon has been closely measured: in times of drought it releases more water than it receives and in times of flood it holds more water than it releases. Reforestation of the headwaters holds water and releases it slowly, reducing the danger of flooding downstream. It's a system that works even better when there are beavers within the water catchment.

As a result of all this, beavers have become one of the polarizing animals: those who delight in a wilder country and those who believe in more, not less, control over non-human life. Releases in Scotland have given great pleasure to many without disastrous consequences; but in Wales objectors have maintained an over-my-dead-body stance. In the United States the beavers' return in many places has been regarded as a good thing; in other places they are still shot on sight.

The great attraction of beavers comes in that seductive concept of rewilding, an ill-defined term that in the eyes of some – both for and against the process – involves the release of wolves and other major predators into the farmed environment. The debate will continue, and will continue to polarize, but to many people beavers are gentle pioneers of a gentler way of managing the countryside, creating a landscape that is less brutally functional than it was before, bringing with it a mental landscape that is less rigorously human-centred.

NINETY-TWO
GUANAY CORMORANT

*'So twice five miles of fertile ground
With walls and towers were girdled round…'*

Samuel Taylor Coleridge, *Kubla Khan*

The guanay cormorant repeatedly shits in the same place. This apparently unremarkable fact led to war and famine, launched the imperial ambitions of the USA and prompted the invention of intensive agriculture, making it possible for more humans to live on the planet than ever before. It is a startling set of achievements for a bird with few ambitions beyond catching fish, raising young and attempting to become an ancestor.

Producers of a miraculous substance: (from left to right) cape cormorant, pygmy cormorant and guanay cormorant (artist unknown).

The guanay cormorants live on small islands and remote headlands off the Pacific coast of Peru and north Chile; there was a population off the Atlantic coast of Patagonia that no longer survives. Over the twentieth and twenty-first centuries there has been a steady decline in their numbers, from loss of nesting grounds and overexploitation of their fishing waters by humans. They breed all year round, but breeding peaks at times of maximum food availability.

Like all species of seabirds they must return to land to nest. They do so in large gatherings, alternating periods of solitude and high sociality. They are deeply faithful to the islands and other places where they nest: generation after generation is hatched and raised on the same island, forever returning to the same spot, their offspring and their distant descendants returning to the same island: to live, breed, sleep – and shit.

That adds up. Their islands at one time lay under uncountable millennia of droppings. That is all perfectly inevitable, but in the nineteenth century the people of Europe and North America discovered that these droppings were useful to humanity. More: they were revolutionary: a source of immense wealth and power. The nineteenth century has been called the Age of Guano.

The word comes from the Spanish, which in turn comes from one of the Andean languages. The secret of guano was known for 1500 years in South America before the colonial powers realized that it was something they could also exploit. Guano makes food grow. It allows the same area of land to produce more food. Instead of knackering your land by taking too much from it, you can use guano to put something back. That way the land produces more – rather than less – than it did before.

The droppings of the guanay cormorant contain nitrogen, phosphate and potassium. When it's been dried and baked on a tropical island it comes in an ideal form to apply directly to the fields. The Inca rulers placed a high value on it: anyone who tried to take guano from a colony or even disturbed the birds was put to death. Right from the beginning of human involvement, guano was a serious business. The guanay cormorant was not the only bird capable of supplying the treasured commodity: the droppings of the Peruvian pelican and the Peruvian booby were also used. They were exploited slowly, so that the business, if not exactly sustainable, was there for the foreseeable future. All that changed when Alexander von Humboldt sneezed during a walk round the docks.

Humboldt was the great German geographer, naturalist, explorer and polymath of the early nineteenth century, and a great deal of the way we see and understand the physical nature of the planet is down to him. It's been claimed that more species are named for him than for any other human; there is a Humboldt penguin and, for that matter, the Humboldt Current. And Humboldt, insatiably curious, wondered about the substance that was being unloaded from a ship in the docks in Peru, and what it was that made him sneeze so violently. It was the dry dust of

Bringing guano to Europe: Alexander von Humboldt, painted by Friedrich Georg Weitsch (1758–1828).

guano. He was told it was the droppings of birds and at first he was sceptical: there was surely far too much of the stuff. It took some leap of the imagination to realize that they were the droppings of the ages. He discovered what it was used for and made a point of acquiring some and bringing it back home, where chemists analysed it. It was tried out on depleted soils and the results were, it seemed, miraculous. More food was available to humans than ever before. Suddenly everyone wanted guano, because guano changed everything. It was, perhaps, the most important stuff on the planet.

It was boom time for those who managed to get in on it. In fifteen years the UK imported 2 million tonnes. It was a time of overnight millionaires. It followed that there was horrific exploitation of workers: digging dried bird droppings in tropical heat while inhaling the dust was a lethally dangerous task; it was often carried out by Chinese and Polynesian labour living on terms of de facto slavery. There were crooked attempts to cut the guano with sawdust and other materials; dockside analysts were employed to forestall the cheats.

Those who could find and exploit new sources of guano could make an instant fortune. Bat colonies could also provide rich pickings and were quarried. The island of Ichaboe off the coast of Namibia was found to be covered in guano in 1843; the following year it was visited by 450 ships. When the island was exhausted it was 25ft (8m) lower than it had been before the excavation began.

The British formed an alliance with Peru and, as a result, found it advantageous to import a new variety of potato from them as well as other goods. This may have had the unintended consequence of introducing potato blight to Ireland: the Great Famine took place between 1845 and 1849; 1 million died and more than 1 million more emigrated. The British alliance with Peru provoked the United States to pass the Guano Islands Act 1856 in which citizens were encouraged to find guano islands and claim them for the United States: a hundred islands were annexed and seventy remain United States possessions to this day. It was the beginning of the country's ambitions as an imperial power.

Australia, France, Germany, Japan and Mexico were all equally eager to annex guano islands. The trade prompted wars between South American nations, in which Chilean forces invaded Bolivia and made a pre-emptive strike against Peru; hostilities were ended by a treaty of 1884.

Intensive farming was now possible. After the invention of agriculture 12,000 years earlier, the exploitation of guano was the crucial second phase. But two things were happening at the same time. The first was that the world started to run out of guano: what birds had done over the course of hundreds of thousands of years was used up inside a century. The second was that humans learnt how to make synthetic guano. Artificial fertilizer. As this process continued through the nineteenth century the birds were phased out. They had shown humans what they needed to make intensive agriculture possible. Now humans were able to expand their numbers as never before.

NINETY-THREE

HOUSE MOUSE

'I hates those meeces to pieces!'

Mr Jinks, from the Hanna-Barbera cartoon *Pixie and Dixie and Mr Jinks*

The house mouse is perhaps the second most successful species of mammal on Earth. Like humans they have spread to every continent including a fringe population in Antarctica, and like humans they are capable of making a living in almost every environment. What's more – and also like humans – they are capable of adapting their behaviour and their social structure to suit changing circumstances.

In recent millennia the secret of success for non-human species has been to make an ally of humanity, whether humans like it or not. Mice have taken both routes: they exploit the dwellings and the food of humans, largely against human wishes, and they also thrive as pets and in the laboratory. Thriving is what mice are good at.

The house mouse is one of around 1800 species in the superfamily Muroidea, which includes all mice, rats, voles, hamsters and gerbils. The difference between mice and rats is informal: if it's big, we call it a rat. The smaller species we call mice, and some of them will enter human habitations and exploit human resources. But the species that has done it in greater numbers and with greater success than any others is *Mus musculus*, the house mouse. Attentive readers will have noted that it shares a specific name – the second part of the binomial – with the blue whale *Balaenoptera musculus*. Linnaeus was quite a wag in his quiet way.

The house mouse existed long before houses: their genius was to take up the opportunity presented when permanent human dwellings began to be a fact of life. They had, like all the other species that share our lives, backed a winner. They were originally plant-eaters, probably based in the north of India. From here they spread west, reaching the eastern Mediterranean about 15,000 years ago, and the rest of Europe about 3000 years ago. In other words they spread a little ahead of the invention of agriculture, but once agriculture and permanent human dwellings were established they began to increase and spread still farther: but not at express pace. It seems that human populations needed to reach a certain size before mice could exploit them successfully.

Eek, a mouse: Held Captive by Angelo Martinetti (1874).

It is the adaptable creatures that do best in radically changing circumstances, and humanity's spread on Earth was (and is) the most devastating change to the planet since the meteor strike that ended the dinosaurs 65 million years ago. The meteor made possible the rise of mammals and birds; humanity made possible the spread of domestic animals and their opportunistic fellow travellers. Of these, the house mouse is the most effective mammal.

They may have evolved as herbivores but they are by no means picky eaters. They will eat just about anything that has any nutritional value. The tender trap of specialization was not for them: present a mouse with anything that is edible and it will start gnawing; like the beavers, like all rodents, mice are highly talented gnawers. They avoid daylight: they are mostly nocturnal, though they also operate well at dawn and dusk. They have excellent hearing, as you would expect from the (comparatively) large round ears, and they also have an excellent sense of smell. A large mouse weighs 1½oz (45g); they can move with astonishing rapidity over short distances and jump well: 18in (45cm) from a standing start. The long tail gives them excellent balance; it is largely hairless, which helps with thermoregulation: cooling the blood that comes close to the surface of the bare tail. Mice are not long-lived: an individual that lasts a full year has done well, though domesticated mice can live two or three years when well looked after.

The mouse populations that exploit food resources provided inadvertently by humans are called commensal mice: mice that share our table. Such environments are characterized by a great deal of food, more than the mouse populations need to survive. In such circumstances there is usually a large population of mice living at high density; they are much less aggressive towards each other than other types of mouse populations. They go in for cooperative breeding: a male will often have two or more females, and these are generally related. The females will share the task of raising young, and continue doing so should one of the females die.

Non-commensal populations of mice live in much more open areas, with much larger territories that are enthusiastically defended. There is more aggression between females and between males. Males will tend to have a single female in these circumstances. These open-area mice build complex tunnel systems complete with escape routes.

The ability to adapt behaviour and diet made for the great success of mice. They have achieved this with comparatively small impact on human lives. Of course, they have caused considerable inconvenience, but without inspiring the deep hatred and fear that rats provoke. Mice eat human food before we have thrown it away, if they get the chance. They damage crops, and damage human structures by gnawing. They can gnaw through electrical and computer cables, causing all kinds of disruption; disasters that always have something ludicrous about them. They can pollute food supplies with parasites and faeces. They have

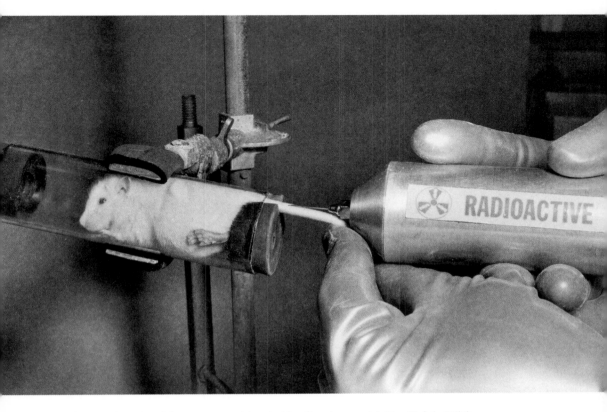

Cutting-edge rodent: laboratory mouse injected with radioactive material, New York (c.1953).

also caused problems when accidentally introduced to islands without a population of terrestrial mammals. They helped to hurry a number of New Zealand's bird species to extinction; on Gough Island they have been observed eating live albatross chicks, which may stand 3ft (1m) tall but are helpless.

Some people are afraid of mice; my mother shuddered even at the mention of them. It is a phobia particularly associated with women; perhaps it's about the fear that this invasion of home will be followed by an invasion of still more private parts. The terrifying idea that the mouse will run up inside a skirt has given rise to the classic – and, to all but the woman involved, comic – image of the screaming woman standing on a chair while mice get on with their lives below.

There has always been a slightly jocular response to the problem of mice; for all that, we have come up with many ways of getting rid of them. One of these methods is of course the domestic cat (see Chapter 2), which has the virtue of acting preventatively. Had there been no mice in the world, it is possible that humans would never have taken the trouble to domesticate cats: many lives would have been less happy as a result. The relationship between these two species is caught in language: we talk of a 'cat and mouse' situation when there is a protracted battle of wits, or when one of the protagonists seems to be toying with the other. A well-fed house cat doesn't need to kill mice for food, but a full belly doesn't blunt a cat's delight in the hunt (any more than it blunts the same urge in many humans). Sometimes a cat will catch a mouse and amuse itself by letting the mouse go and catching it again: behaviour you can see when lions and other big cats are teaching their cubs to hunt. The idea of a cat playing with a mouse has become an image of idle sadism.

But there is still a degree of jocularity in all this, even if the mouse (like the woman) doesn't see the joke in quite the same way. The eternal wars of Tom and Jerry illustrate this, in the cartoon series by William Hanna and Joseph Barbera that began in 1940. In episode after episode – 114, with the last one of the original series made in 1958 (there were later revivals and 164 films made in total) – the cartoon cat Tom is outwitted while the mouse Jerry punishes his opponent with sadistic glee. Mickey Mouse has become one of the most recognizable representations of a non-human animal in the world, even though the cartoons are seldom watched these days. He was a creation of Walt Disney, first seen in 1928, appearing over 130 short and feature-length films. Mickey's greatest role was as the sorcerer's apprentice in the 1940 film *Fantasia*, in which he learns about youthful ambition and the dangers of overreaching yourself to the music of Paul Dukas.

Mice have been domesticated as low-maintenance pets, and those who love them say they are playful and affectionate. They have also been used extensively as model animals in laboratories, in research on biology and psychology. They are the easiest and cheapest mammals for such work: their sometimes harsh lives cause us little concern because they are so much unlike us, while the results are valuable because they are so much like us: another of the classic contradictions to do with humans and our relationship with non-human life.

Mice have also been used in the study of genetics, with what is called the knockout mouse. This is a mouse who has had genetic material knocked out; that is to say, inactivated or replaced. This is a way of discovering the function of unknown genes: you observe the mouse and take note of the differences after the knockout.

Fictional mice are often brave and strong: it is a beloved contradiction. Perhaps the finest mouse in fiction is Reepicheep, the valiant martial mouse in *The Chronicles of Narnia*, who wins undying glory at the Second Battle of Beruna and then sails to the world's end on the *Dawn Treader*. Again and again we have taken a mouse as an emblem of courage against the odds: precisely the way we have always seen ourselves in our battle against nature. Like poor Tom in the cartoon, nature comes off worst in every episode.

NINETY-FOUR
STORK

'His mother should have thrown him away and kept the stork.'

Mae West

We have tried for centuries, with varying degrees of success, to keep mice away from our dwellings. At the same time many people have attempted to attract storks to their homes. A stork's nest on your building was and is considered an honour, a blessing, the best kind of luck – as well as an object lesson in all kinds of domestic virtues. Storks also brought humans – one of the great revelations about the way that nature operates: and, as ever, this revelation showed that nature is more marvellous than anything we can make up. As if that isn't enough, storks also showed humans how to fly.

There are twenty species of storks in six genera: all of them big, tall, long-legged and armed with an impressive beak. They mostly forage in open spaces and shallow water, and mostly take small animals of many kinds: beetles, grasshoppers, crickets, fish, frogs, small birds and mammals. The biggest is the marabou stork of Africa, which has a wingspan of up to 12¼ft (3.7m), and can weigh 20lb (9kg).

But one species in particular has inspired the attention, affection, imagination and myth-making capacity of humanity, and that is the European white stork. They were once widespread summer visitors to the Iberian peninsula, Germany, east and central Europe and West Asia. Their numbers are depleted but there are still plenty of them. They don't usually breed in Britain, but there is a record of a pair breeding on St Giles' Cathedral in Edinburgh in 1416 and a pair formed at Knepp Castle in Sussex in 2019, though there were no eggs. They were associated with all kinds of life-affirming things: how could they not be? They arrive in spring, actively seeking out human settlements on which to build their extraordinary nests, more than 6ft (1.8m) wide, on church towers, chimneys and roofs. The birds themselves are big enough to capture the attention: 3¾ft (115cm) in total length, an imposing presence.

Once established they get on with the task of making more storks, and do so by means of an ostentatiously equal partnership. They build, rebuild and repair the nest together and, in the first few weeks after their arrival, they engage in a series of spectacular rooftop copulations. Both sexes feed the chicks, and the birds

communicate with loud bill clattering; their voices are weak and hissy, but this percussion of their beaks fills their loose nesting colonies with sound. The clattering is amplified by their throat pouches; storks like a lot of noise and bustle around them on a nest site. On the nest the birds will preen each other: reciprocal altruism, pair bonding – and generally, to the humans below, everything they do looks like a domestic idyll.

It was assumed that the birds are models of fidelity; studies and prolonged observation have shown that this is not entirely the case. But the combination of their apparent domestic bliss and their appearance in spring – the arrival of the storks traditionally tells you when it is time to plant the crops – makes them important symbols of life and virtue. In Greek legend storks were associated not just with parental devotion to their young, but also with filial duties to parents: for that reason, storks were turned into humans as a reward for their good behaviour. Hera, wife of Zeus and queen of heaven, was linked with storks because of her own domestic virtues. The ancient Egyptians associated storks with the soul; in Slavic mythology storks carry souls down to Earth.

All of which brings us to the notion that storks bring babies into human families. The image of the stork flying in with a cloth-wrapped baby dangling from its beak has become part of the way the people of the West see the world. When Carl Jung's sister was born, Jung was told in all seriousness that the stork brought her. The story was no doubt helped along by the fact that in many places in Europe the storks arrive close to the spring equinox; which is nine months later than the summer solstice, a traditional festival of ritual licence. It is also the case that the warm weather of summer provides opportunities for naked privacy denied to people sharing crowded accommodation in winter. Babies were born in spring: storks arrive in spring: storks are examples of domestic bliss.

These truths became a classic lesson for training statisticians: figures showing the numbers of storks' nests and the rising human population can be put together and compared. The meaning of the lesson is that correlation is not the same thing as causation.

But the baby-bringing stork became an acceptable way of explaining the appearance of babies to children who were considered too young to be informed about sex and childbirth. There is sometimes a facetious or semi-facetious note to this: a baby delivered by forceps and bearing their mark after birth is said to have 'a stork-mark': a discoloration allegedly caused by the bird's beak as it delivered its burden.

There is a sinister interpretation of this legend in the Hans Christian Andersen story 'The Storks', in which naughty boys tease the storks, and in revenge they

Stork as life-bringer: Heartiest Congratulations, *colour lithograph (twentieth century).*

deliver a dead baby to the boys' family. There was, and perhaps in some places still is, a custom of young couples placing sweets on the windowsill to encourage the stork to pay a visit.

It is the dramatic arrival of the storks in spring that makes them so special to humans. What happens to them? Where do they go? How do they spend their winters? All kinds of ideas were bandied about. Perhaps they hibernated at the bottom of the sea. But the answer was delivered by the storks themselves in 1822 at the village of Klütz in Mecklenburg, in what is now Germany. A stork arrived bearing an arrow in its neck. The poor bird was then shot as a curiosity, and the arrow, when examined, was quite obviously from sub-Saharan Africa. The truth was unavoidable: storks go to Africa for the winter. The bird, complete with its arrow, was stuffed and mounted and can be viewed at the museum of the University of Rostock. Other arrow-bearing storks later turned up, to be shot in their turn. Such a specimen is known as a *pfeilstorch* and there are more than twenty of them. Thus the study of bird migration began; these days it is more relevant than ever because long-distance migrants are more at risk than any other group of birds in these changing times.

Storks are big heavy birds and powered flight over long distances is difficult for them, and very expensive in energy. But they don't migrate by flapping: they travel with still wings, a method that uses far less energy, gliding and soaring. They gain height without flapping, soaring on thermals created by warmer air rising over the land. Their wings, long and very broad, are perfect for the task, and the stork was a model for early gliders: photographs of soaring storks taken by Ottomar Anschütz were used by Otto Lilienthal to create the first gliders, which he operated from 1891 to 1896, when he was killed after stalling from 50ft (15m).

Johannes Thienemann pioneered the ringing (banding) of storks, and between 1908 and 1954 around 100,000 were ringed with 2000 recoveries. It's been learnt that storks migrate as little as possible over the sea, because thermals only form over land. So they funnel out, crossing the Mediterranean either at Gibraltar or over the Levant. In 1942 Heinrich Himmler wanted to use migrating storks to carry propaganda from Germany down to the Boers of South Africa. The scheme was called *storchen propaganda*: a brilliant idea, perhaps, but impractical and never actually tried.

Storks are creatures that humans unequivocally like. Their numbers have declined, through the intensification of agriculture and the spread of human dwellings, and consequent destruction of their feeding grounds, but they remain birds that we welcome into our lives and have an assured place in our mythologies.

Social whirl: stork nests in the city of Strasbourg, France (nineteenth century, artist unknown).

NINETY-FIVE

OYSTER

*'The kingdom of heaven is like unto a merchant man
seeking goodly pearls: Who, when he had found one pearl
of great price, went and sold all he had, and bought it.'*

Matthew 13:45–46

Oysters create an environment. They are capable of cleaning vast quantities of the oceans in which they live. For hunter–gatherers who lived on the coast they provided a high-protein meal that didn't run away: middens have been found with oyster shells 10,000 years old. When humans developed ways of gathering oysters in immense quantities they became cheap food for the working classes; when oysters were exploited to scarcity they became one of the ultimate luxury foods; perhaps it's something to do with the thrill of eating something while it is still alive. And for the past 2000 years or more, items that oysters produce naturally have become objects worth a fortune, for which people will gladly risk their lives.

Oyster is a term used informally for many different species of mollusc with twin shells, technically bivalves. Traditionally we have valued two kinds of oyster: those you can eat, and those that make pearls. But in more recent times we have come to value all species of colonial filter-feeding shellfish, because they make and maintain worlds. A single oyster can pass an astonishing amount of water through its system in the course of a single day: estimates are up to 50 gallons (225 litres). Oysters feed by taking in the water and passing it out again, extracting tiny items of nutrient as they do so. This action cleans the water and makes it better for every other species. Also, by building colonies or reefs, oysters provide places for many other species to live – sea anemones, barnacles, mussels – and therefore opportunities for those that prey on these items. Oysters create worlds. They also affect the wave action in coastal waters, and in times of changing climate and rising sea levels they are increasingly viewed as a good thing, helpful to humans and the coastlines we manage and exploit.

Chesapeake Bay, which straddles Maryland and Virginia in the United States, is famous for its oysters. It's been estimated that, when the colony was at its peak, the entire waters of the estuary were cleaned every three or four days; the modern figure, with fewer oysters, is closer to a year. Across the world, fertilizers used by

Food for workers: sketch for Oysters, Young Sir? *by Henry Perlee Parker (1795–1873).*

humans run off the land into the river systems and so into the sea, where this over-richness kills fish and causes algal blooms, dead zones in which nothing can live. Oysters remove these excess nutrients, so they are more necessary than ever.

Oysters begin as eggs, and then as free-swimming larvae. When these are big enough, they settle. Oysters are predated by crabs, starfish and seabirds as well as, of course, by humans. The natural tendency of many species is to gather together in immense numbers: and those that humans can eat were inevitably overexploited. They have been purposefully fished – though some prefer the idea of harvesting, as if they were a plant – since Roman times. Whitstable, the centre of the oyster business in the UK, has been fished for 2000 years.

Oysters fed New York in the nineteenth century: they probably created the city's great restaurant tradition. Oysters were regarded as cheap nutrition: in his poem 'The Love Song of J. Alfred Prufrock', T. S. Eliot writes:

> Let us go, through certain half-deserted streets,
> The muttering retreats
> Of restless nights in one-night cheap hotels
> And sawdust restaurants with oyster-shells…

Here is a cityscape that is sordid and depressing, and the crucial detail is the shells of oysters. But, after overexploitation, oysters became a delicacy: food to show off with, food to accompany champagne. It is traditional to eat oysters when there is an R in the month: that is to say, to avoid doing so during the warmer northern-hemisphere months of May, June, July and August. That makes good sense: in warmer weather it is harder to keep oysters fresh and the pathogens they can carry are more active. Oysters can be dangerous to eat: it is necessary in all seasons to clean them thoroughly to get rid of the human faecal matter that many of them contain, in a process called depuration.

Just about every species of shelled mollusc is capable of producing a pearl. A pearl is the way a mollusc deals with foreign particles that get stuck in its body. The oysters that we eat – most often the Pacific oyster, sometimes known as the Japanese or Miyagi oyster, but always *Crassostrea gigas* – make pearls all right, but they're lumps of stuff to which we attach no value. That's because they're not shiny. And we like shiny things.

But other species produce pearls that shine. The so-called true oysters come from the superfamily Ostreoidea; pearl oysters come from a quite different family, the Aviculidae, so they are not that closely related. And they do make pearls with lustre. The pearls shine because of the nature of the layers of nacre (sometimes called mother-of-pearl) with which the oysters surround the intruding irritant: it reflects, refracts and diffracts light. And that entrancing visual effect has made the oyster an object of desire across the centuries.

Pliny the Elder calls pearls 'the most prized of all jewels', and he also tells the best of all oyster stories. The 'headstrong woman' Cleopatra, queen of Egypt, boasted to Mark Anthony that she could put on a banquet more lavish than anything he ever could. She won her bet by taking a large pearl of immense value and dissolving it in vinegar before drinking it. (A pearl is made from calcium carbonate, which dissolves in weak acid.) Pliny also notes that the best pearls come from the Persian Gulf: from the species we now know was *Meleagrina vulgaris*.

Other species and, indeed, other groups of molluscs create pearls of value. One of the reasons for Julius Caesar's invasion of Britain in AD 55 was the British pearl beds: then-thriving populations of freshwater pearl mussels. There is a project to reintroduce them, for their ecological water-cleaning behaviour.

Pearls have been regarded as things of wonder in many cultures. The Quran explains that people who pass into paradise will be bedecked with pearls. Pearls have been seen as the tears of the gods. They have also been seen as the tears that Adam and Eve shed on their expulsion from Eden: Adam cried the black pearls, which are much rarer, he being more capable of controlling his emotions than a mere woman. There is a superstition that wearing pearls on your wedding day will stop you crying.

Royal pearls: Queen Elizabeth II and future Prince of Wales, Prince Charles, 1949.

There is not a pearl in every oyster, nor in every hundred. Estimates vary as to how many oysters you need to open to find a single pearl: the one certain thing is that an awful lot of deaths are required to make a pearl necklace; perhaps that again is part of the charm. The overexploitation of wild oysters created a gap in the market and so oyster farming began.

Cultivated pearls are made by introducing a bead into the mantle of an oyster and recovering the resulting pearl a few years later – eighteen months minimum. The only reliable way of telling artificial from more expensive natural pearls is by X-ray: a natural pearl shows more prominent growth rings. People still fish for natural pearls in the Persian Gulf and off the coast of Australia: natural pearls are premium products. A necklace of matched (that is to say, symmetrically graded) natural pearls will cost money beyond the imaginings of most of us. But to give an idea, in 1917 the jeweller Pierre Cartier bought a mansion on Fifth Avenue; it's the place where you can find the New York Cartier shop to this day. He swapped it for a double-strand of matched pearls, which was valued back then at US$1 million.

These days we need oysters for more than their pearls. We need them for the mitigating effect they have on the ocean and on the harmful things we routinely pour into it. In New York there is a scheme to protect the city from storms by using oysters. It began after the depredations caused by Hurricane Sandy of 2012, which caused an estimated US$62 billion of damage. The process, nicknamed oyster-tecture, aims to replace those lost oyster reefs and so make the surrounding waters less volatile and the city safer. We destroyed a vast and valuable asset without knowing we were doing so: and, as so often in this book, I am reminded of Othello, who, looking back on his murder of his wife Desdemona for the crime of infidelity that she did not commit, compares himself to 'the base Indian' who 'threw a pearl away richer than all his tribe'.

NINETY-SIX

JAGUAR

'Oh, hear the call! – Good hunting all
That keep the Jungle Law!'

Rudyard Kipling, 'Night-Song in the Jungle'

We like things that stand for other things. Metaphor, simile, symbol, emblem: it's part of the human condition, though it occurs, at least to an extent, in non-human species as well. When my horse is irritated with me, she swishes her tail: she considers me annoying in the manner of a fly. But we humans have rather specialized in this business and, as we have seen earlier in this book, we have used non-human animals to stand for God, for the Holy Spirit, for kingship, for peace, for many different nations, for love, for any number of virtues and vices. Sometimes these meanings change: the gorilla, once a monster, now stands for the peaceful idyll of the wild (see Chapter 3 on gorillas). And the jaguar, once a symbol of royalty and martial might, now stands for the vanishing rainforest: for the beauty that is slipping through our fingers: for human folly.

Jaguars are found in the Americas, from the southwest United States, where a few remain, down as far as the forests in the north Argentina. They are the third largest cat, after lion (see Chapter 1) and tiger (see Chapter 24). So many cats in this book: they are there because we humans love them. Jaguars, for the Aztecs, were embodiments of all kinds of fierce virtues. They have become a vivid image of one of the most precious resources we have on the planet: the pearl, richer than all our tribe, that has been thrown away again and again and is still being thrown away.

Jaguars can operate in many different habitats, but tropical and subtropical broadleaf forest is what they like best. Like most cats – lions and feral domestic cats are exceptional – they are mostly solitary, males holding territories that take in the territories of two or three females; the females will share their own territories with their growing cubs. They are stalking ambush predators, adept at crawling, climbing and swimming.

They look a lot like leopards, but are bigger and more powerful, though shorter and stockier. Their spots look different too: the rosettes are larger, there are fewer

Forest stalker: though the picture was taken in 2014 at a zoo north of Bucharest at sunset.

of them, and often there are black marks within the rosettes. Jaguars, unlike leopards, are apex predators across their range: top predators that nothing can dominate. There are records of jaguars attacking humans, but very few: their principal prey is capybara, the world's biggest rodent. Charles Darwin, when travelling on the *Beagle*, reported a local saying that there is no need to fear the jaguar as long as there are plenty of capybaras. But with spreading human populations jaguars have inevitably moved in on domestic stock, and routinely kill cattle; I was staying at a jungle station in Paraguay where a jaguar had recently killed one of the horses.

Jaguars evolved from the cats of the Old World, and their ancestors made their way into the Americas by way of the Bering land bridge, now the Bering Strait, and proceeded south. The jaguar's historic range went as far north as the Grand Canyon in the United States, and from there west as far as Monterey in northern California. The few individuals that remain in the southwestern states are threatened by plans to create barriers along the boundary between Mexico and the United States: as a completely isolated population, the US jaguars would almost certainly be unsustainable.

The largest populations are in the Amazon Basin, the Gran Chaco and the Pantanal, and it is the Amazonian population that has caught the world's imagination. The idea that rainforest matters – even to people who live miles away without the slightest interest in non-human life – is a comparatively recent understanding. It is a relatively complex concept, involving biodiversity, carbon sequestration and the transpiration of oxygen... so it's much easier to see an image of a beautiful spotted cat walking through the eternal shadows of the Amazon jungle, moving as if every joint has been bathed in a gallon of oil and glowing as if lit from within – and to say no, such a thing must not be destroyed.

In the crashing of the trees and the desperate graveyards of tree stumps that remain, the folly of rainforest destruction is obvious to us all. It is a woeful sight, so far as most humans are concerned: not least because it makes the jaguar homeless. The lifeless vistas of ruined forest seemed to be the prevision of a lifeless world.

The awareness of rainforests and their destruction is an example of humanity's shifting understanding of non-human life. Before the second half of the twentieth century, the term rainforest was seldom found: they were jungles: hostile to all save the most intrepid of human explorers and the most exotic of human tribes. The idea that rainforests are a valuable resource, not just for those who live in traditional ways but also for every living thing on Earth, first became widespread in 1984, with David Attenborough's *The Living Planet* (see also Chapter 70 on orang-utans). Since then, awareness has become part of popular culture: people use the notion of saving the rainforest as a facetious criticism of unnecessary verbiage of free newspapers, books, annual reports and junk mail, while the sentimental advocacy of their value can be written off as 'tree hugging'.

But even these negative examples show the proliferation of the rainforest meme in twenty-first-century life. The growing certainty that rainforests matter has powered a good deal of the growing unease about the way we manage the planet, and has helped to force this issue higher up the political agenda, with global protests becoming a matter of routine. The vividness and the mystery of rainforests and their inhabitants – perhaps jaguars most especially – and the equal vividness of their destruction have become potent ideas and images in modern life.

We know now that forests, and most especially the great rainforests that remain, act as the planet's air-conditioning system, and also as its water system that keeps us and our crops and our domestic animals alive. Rainforests absorb carbon, taking carbon dioxide out of the system and preventing it from building up as one of the greenhouse gases, those that force the Earth to get continuously warmer. The forests also pour oxygen back into the system. They transpire the water vapour from the liquid they hold in immense quantities and so create clouds, and the water produced in rainforests goes across the world. They store water and release it slowly, reducing vastly the risk of flood downstream. The release of water from rainforests feeds and nourishes local people and their animals: when the trees are taken away the clean water goes with them, the people cannot survive and must migrate into the cities. A quarter of all modern medicines came originally from rainforest plants; so far we have exploited 1 per cent of them. What else lies there? Perhaps we will never find out.

There is also the question of biodiversity. It has been suggested by some scientists that life on the planet is sustained by its variety: that the web of different forms of life is what gives life on Earth strength and resilience. A web is a good image: as we have seen in the chapter on spiders (see Chapter 57), break one strand and the whole structure is weakened. We are breaking strands again and again, replacing complexity with simplicity; replacing diversity with monoculture. Perhaps we will find a way of making such a system work, but to repeat the words of Edward O. Wilson: 'one planet, one experiment'.

And 70 per cent of all species are to be found in rainforest.

We are beginning to get used to the idea that rainforests matter. In many countries that still have extensive rainforests, the traditional response has been: the developed world destroyed all its forests in order to become developed: it is our right and our ambition to do the same. But rainforests are needed by the entire world: therefore it is in the interests of the entire world to safeguard them.

The jaguar has come to stand for that difficult and delicate issue: as the one unambiguous and irrefutable emotional argument in favour of preserving and renewing the forests.

NINETY-SEVEN
PINK PIGEON

*'Many people think conservation is just about saving fluffy
animals… what they don't realize is that we're trying to prevent
the human race from committing suicide.'*

Gerald Durrell

'What a piece of work is a man! how noble in reason! how infinite in faculty! in form and moving how express and admirable! in action how like an angel! in apprehension how like a god! the beauty of the world! the paragon of animals!' But in chapter after chapter of this book, humans have failed to live up to this summary in *Hamlet*. Angelic behaviour is rare; self-destructive folly is frequent. So, as we move on to the final stretch, let us make a reasonably big deal of demonstrating that humans are perfectly capable of acting with wisdom, compassion and effectiveness, and of doing so towards species of animal other than our own. The pink pigeon makes it immensely clear that if we apply our hearts and our minds, we can not only halt decline and destruction, but we can also actually put it into reverse. It is about human will.

Still here: pink pigeon, from The
IUCN Red List of Threatened
Species 2018.

The pink pigeon, like the dodo (see Chapter 19), is a species found only on the island of Mauritius. As seen before, islands tend to throw up unique species; it is in the nature of isolation. The dodo was one such: and was unable to cope with the destructive nature of colonizing humans. The pink pigeon is another: but since it is neither large nor flightless, like the dodo, it stood a better chance of holding on. It is a fine-looking bird, without the monstrosity of the dodo, but the colour gives it that slightly surreal touch that islands often go in for. It's a subtle pink, the sort of colour that interior designers like to go for in hotel lounges, smoky greyish pink rather than shocking; the colour is neatly offset by the red eye ring and the dark brown eyes. It is ever so slightly over-the-top: a first-time viewer would certainly ask whether or not that was really real – but then the expression, beady, inquisitive and just a little bit mad, makes it plain that this is a real pigeon all right.

Its nearest relative, the Réunion pink pigeon, from the eponymous island 140 miles (225km) away, failed to survive the arrival of humans and was extinct by 1700. But the Mauritius species, though set about with difficulties, managed to hold on. It was still there when humans started to come to terms with the idea of human-led extinctions, and were willing to do something about it – and the bird responded to human attempts to stop the extinction in its tracks. It was a damn close-run thing: in 1990 the world population of the pink pigeon was down to nine.

The problems faced by the pink pigeon will not come as any violent shock to those who have read the previous chapters. They were hunted by humans for food. There are continuing problems with introduced predators: those most damaging to the pigeons have been the black rat, mongoose and crab-eating macaque, a species of monkey; all these predate the eggs and nestlings. There are also problems with diseases that have come in with introduced species of pigeon. But, as usual, the greatest problem has been the destruction of their habitat: these are forest birds and most of the forests have been cut down for fuel and agriculture; around 1.5 per cent of the original Mascarene forest is left. There were further problems with the forest itself: it was invaded by non-native plants, notably privet and Chinese guava, and these inhibit the growth of the native species that the pigeons – and other species – need to survive.

It was clear that the pigeons were fast approaching the end in the 1970s. The central organization in the work to prevent this extinction was the Mauritian Wildlife Foundation, but a crucial player was the organization now known as the Durrell Wildlife Conservation Trust, which was founded by Gerald Durrell (as the Jersey Wildlife Preservation Trust). Durrell wrote a series of semi-autobiographical books about wildlife and conservation, and he did so with humour, panache, a great eye for caricature and marvellous knack for telling a story. His masterpiece is *My Family and Other Animals*, which tells of his idyllic childhood on the Greek island of Corfu; it remains one of the best wildlife books

Mauritian birds: Mauritius pink pigeon (centre left); Mauritius fody (centre right), Mauritius olive white-eye (below), Mauritius kestrel (above left), Mascarene swiftlet (above right), upland forest day-gecko (far right).

ever written. He established a zoo on the island of Jersey, which lies in the English Channel, and, in a revolutionary step for the time, its prime aim was conservation rather than entertainment. So it remains to this day, and it has made rather a success of taking on those island species that have evolved for one specific place and then, like the pink pigeon, found trouble. The Durrell organization operates a number of projects across the world and is still very much involved in Mauritius.

In 1986 there were twelve pink pigeons left, and they made five nesting attempts. Every nest was predated by rats. The following year there was a release of captive-bred birds, but at first there was no obvious improvement: as said earlier, nine birds were left in 1990. But with continuing efforts the population began to build up again, and new populations were established in different parts of the island. There are now nine populations, six of them in the Black River Gorges National Park. There is also one at the Ile aux Aigrettes, which lies offshore, and is now predator-free. In 2000 the species was downgraded from Critically Endangered to Endangered; by 2018 there was a global population of only 470 wild pink pigeons so the species was shifted once again from Endangered to Vulnerable.

Let us not be so rash as to think that is an end to the troubles of the pink pigeon. Because the forest has been so drastically degraded, there is not enough food in there for the pigeons, and supplementary feeding takes place – wheat and cracked maize, at feeding stations designed to keep other species out. These naturally attract the birds in good numbers, and it's feared that this may allow them to pass on disease. As ever, in a tiny population, and in particular from a population that has risen from just a few individuals, there is a pronounced lack of genetic diversity, and in diversity is strength and resilience and promise of the future. (Perhaps more humans should be made aware of that fact: there are people attempting to build racial monocultures all over the world, unaware that out-breeding is essential for the continuation of our own kind.) In 2019 three pink pigeons that had been bred on Jersey were released on Mauritius in an effort to create greater genetic diversity.

But perhaps the crucial problem is not any one thing. The fact is that, with a small population, there is no such thing as a small disaster. Fire or disease could bring about the end. In a world with ever more extreme weather events, a single cyclone could destroy the relict forests of Mauritius. A very small and isolated population lacks what an English football manager once called bouncebackability.

That is a lesson that spreads beyond Mauritius and across the world. Most wild species of non-human life are now found in small protected and isolated areas. Such places are threatened by inbreeding and the vulnerability of all small populations. The story of the pink pigeon, then, is a warning: but at the same time it is a classic example of the good things that humans are capable of. It is increasingly clear that human intervention is now crucial to the lives of many species: and that almost all species depend on human goodwill for survival.

We have a choice about this, but the time for making such choices is running out. We were just in time with the pink pigeon, and a number of dedicated individuals made it possible for the bird to survive. The moral is that saving species can be done: and also that the most important way of saving species is saving habitats. We did it with the pink pigeon: therefore it is perfectly possible for us to do it with every other species that has its being on this planet. If we want to do it, we can do it.

The challenges of replicating what happened on a small island of 788sq. miles (2040sq. km) over the entire Earth, with a surface area of 197 million sq. miles (510 million sq. km) (of which 71 per cent is ocean) are somewhat more demanding. But let's look at it as a mere matter of upscaling. We have proved that we can destroy anything: it is therefore possible to set about the more thrilling and rewarding challenge of saving everything that we have left.

NINETY-EIGHT

VAQUITA

'In conservation, the motto should always be "never say die".'

Gerald Durrell, *The Aye-aye and I*

I should offer a warning about the material in the next chapters: some of it is a bit glum. Sorry about that. I would very much prefer to offer nothing but good news. And please note that the previous chapter dealt with the pink pigeon, brought back from the brink of extinction, and showed what humans are capable of. If you feel oppressed by the glumness, then take an advance look at the Epilogue: for it is not my wish or my purpose to revel in gloom.

It's become a routine question when I give a talk: are you an optimist? I am tempted to respond: do I *look* stupid? Though there was a time when I believed that optimism about the future of the natural world was essential to engagement with it. The problem is that optimism in any sphere, let alone conservation, becomes less and less sustainable with increasing experience and, for that matter, evidence. The pattern of movement with many of the species in this book is relentlessly downwards. But that doesn't make me a pessimist. In my view neither optimism nor pessimism is helpful or, for that matter, relevant. It is not necessary to recapitulate Voltaire's Professor Pangloss, who believed that all's for the best in the best of all possible worlds, any more than we need to follow Private Frazer, from the British sitcom *Dad's Army*, who repeatedly told his fellow soldiers: 'We're all *dooooooomed!*'

Let us instead try realism. But not in a passive way. When I am asked the optimism question, I routinely say that I have rejected it, along with pessimism. I am very much aware of the realities that surround us. But it's not my job to predict which party is going to win: what matters for us all is to make sure that we are fighting on the right side.

And that brings us to the next extinction. Which admittedly doesn't sound so cheerful, but we saved the pink pigeon, at least for now, so perhaps we can win the still more difficult battle of the future of the vaquita. This is the hot favourite for the planet's next megafauna extinction.

Vaquita means little cow in Spanish but, a trifle unexpectedly, it's a porpoise: the world's smallest cetacean (whales and dolphins) with the females slightly larger than males, reaching as much as 4½ft (1.4m). They are found only at the

Not dead yet: vaquita, poster for the 2019 film Sea of Shadows.

northern end of the Gulf of California in Mexico, the area between the long peninsula of Baja California and the rest of Mexico, sometimes called the Sea of Cortés. Their home range has always been small; they are as close as a marine animal can get to being an island species.

They like the warm shallow waters of the gulf and its surrounding lagoons, and they pursue fish there, sometimes in waters so shallow their backs are on view for sustained periods. Porpoises are a sober version of dolphins, seldom leaping or indulging in other kinds of extravagant behaviour. That makes them hard to observe: they roll just clear of the water to breathe and then slide back below the surface. Vaquitas are even more discreet than the harbour porpoises occasionally glimpsed in northern waters, so they have never been animals to impose themselves on human minds. They weren't even described for science until 1958, on the basis of two skulls that had been found earlier in the decade.

Vaquitas can be distinguished from other porpoise species by the dark ring round their eyes. They seek fish in the sometimes murky waters of the gulf by means of their sonar, and they communicate with each other in high-pitched sounds. They feed and swim at an easy pace, and have learnt to keep clear of boats, no doubt because the engine sounds interfere with their sonar. They rarely go deeper than 100ft (30m). They are much less social than most species of dolphins and porpoises: usually seen alone, or as mother and calf. Harbour porpoises are mostly found in small groups and though they lack the relentless social activity of dolphins they are keen on each other's company. The vaquitas have taken a different option. They have no very close relatives, and have evolved as an offshoot to the main line, no doubt because of their isolation.

The geography of the Gulf of California makes obvious difficulties for them: for sea creatures they have an awful lot to do with land. They are virtually living in a lake. The gulf is 700 miles (1125km) long, making Baja California the second longest peninsula in the world after the Arabian Peninsula. It is mostly very narrow, from 30 miles (48km) to a maximum 150 miles (240km) wide. This cramped, elongated bay has 2500 miles (4000km) of coastline: it follows that everything in it is at the mercy of large and often poor populations of coast-dwelling humanity.

No one has got it in for the vaquitas. They have never been hunted. Given a choice, most people would sooner they were there in large numbers and thriving: engaging and endearing little creatures with a cheerful and friendly expression. No one has any malice towards vaquitas: it's just that they get in the way of people's livelihoods.

Vaquitas get caught in gillnets. Gillnets are nets stretched out across the water in vertical panels, kept in position by floats. They are indiscriminate: anything larger than the holes in the nets gets caught. If they are air breathers like vaquitas, they drown. And it happens that the main target at the vaquitas' end of the gulf is more or less the same size as a vaquita: a chunky, meaty fish called the totoaba, which fetches premium prices. One of the reasons for this is that Chinese people value the fish for their perceived health-giving properties. The swim bladder – the device that allows all ray-finned fish to remain stationary in the water while still breathing (see also Chapter 31 on Archaeopteryx) – is considered a delicacy with high medicinal value. And if the medicinal value is spurious, the financial value is not. The totoaba, like the vaquita, is Critically Endangered. Fishing for totoaba has been illegal since 1975.

The vaquita has been Critically Endangered since 1996. In 1997 there were 600 of them left; by 2014 this was below 100; in 2017 they were down to around 60. A survey in 2018 reckoned that the global population of the vaquita was 12–15 individuals. They are now restricted to an area of 12 × 25 miles (19 × 40km).

The Mexican government has spent serious money trying to conserve the vaquita. Gillnet fishing was suspended and then made permanently illegal. There has been talk about establishing a captive breeding population, and a feasibility study has been done. This was suspended when a female caught for this purpose died while being captured. A nature reserve has been established at the top end of the gulf, taking in the mouth of the Colorado River. Financial compensation has been paid to fishing people; effectively paying them not to fish.

But there are continuing problems: illegal fishing with gillnets continues, because the totoaba is a great prize for people living in comparative poverty. Boats removing illegal nets have been attacked by fishermen. The Mexican navy has been deployed, but ineffectively, for they are constrained by the need to act pacifically against their own people, while people involved in illegal activities are not. Ecotourism is increasingly important in the gulf: in season it holds many other species of cetacean, as well as spectacular populations of seabirds. When there is a powerful economic argument in favour of conservation, conservation is more likely to happen.

I was talking about optimism. The surviving members of this species are fit and healthy and ready and even eager to breed when opportunity arises. A female known to researchers as Ana has been seen twice, each time with a different calf. If the vaquitas can be allowed to get on with it, they will make some sort of a comeback. The problem is not loss of habitat: the waters of the gulf are still filled with nutrition. If there are echoes of the sad story of the baiji here (see Chapter 74) the parallel is far from exact. There are healthy individuals out there still living in a healthy environment.

In the early 1960s the American satirist Tom Lehrer, already much quoted in these pages, wrote a song called 'Who's Next', which was about the acquisition of nuclear weapons by one nation after another. The phrase in the title, as well as his depiction of the inevitability of human folly, has stayed with me across the years, and when I ponder the list of Critically Endangered animals – in 2019 there were 3086, so I shan't sing through them all – I can't help but ask it again. Who's next? If we restrict our interest to megafauna, the vaquita is in pole position. Optimists and pessimists must both take that fact in their stride: but both groups must remember Ana, who was still hard at it at the time of writing.

NINETY-NINE
ANT/TERMITE

*'It was a bright cold day in April, and the clocks were
striking thirteen.'*

George Orwell, *1984*

At the heart of the *Star Wars* galaxy is Coruscant, which is both city and planet, an unending capital of sky-reaching towers and buildings with cellarage that extends at least as deep below the surface: a place from which nature has been outlawed, for there is no longer any room. It was conceived as fantasy but with every passing year it looks more and more like the human masterplan for the one planet that we inhabit.

But it's not a new idea. The concept of intense sociality has been around for 120 million years and was developed by the insects. The technical term for such a way of life is eusociality, defined by the great scientist and science writer Edward O. Wilson as a system with 'group members containing multiple generations and prone to perform altruistic acts as part of their division of labour'. This definition, he states, in his 2012 work *The Social Conquest of Earth*, includes humans: and this, he says, is the core reason for the success of humanity: 'In this respect [they] are technically comparable to ants, termites and other eusocial insects.' Some would draw the definition of eusociality more tightly: with eusocial insects the division of labour includes the division of reproduction – only some individuals breed. (It is important to note here that though ants and termites live in very similar ways, they are not closely related. Ants are hymenoptera, and so related to bees and wasps, while termites are sociable cockroaches.)

Walk across the savannahs of Africa and you will encounter a termite mound before terribly long; it has been calculated that 1 per cent of all Zambia is a termite mound. When you find an active mound, you can help yourself to a small moment of wonder: just place the palm of your hand over one of the openings at the top. On a still day, you will feel a gentle warm draught. The day may well be hot – rising close to 50°C (122°F) at midday – but the air you can feel is still warmer: well, of course it is, or it wouldn't rise. This phenomenon is the result of the mound's air-conditioning system: inlets lower down create a convection current, which keeps the air moving and prevents the inhabitants from baking alive in their own castle. There might be a million living animals in a single

Human anthill: poster for the film Metropolis *by Fritz Lang, 1926.*

mound: and when you feel the hot breath of civilization rising from the vent you grasp something of the nature of this hyper-intense sociality: a trait that has sustained ants and termites and other eusocial insects across the ages.

Ants are better at it than we are, having been doing it for so much longer. They have also evolved ways of living that remind us of ourselves in a fashion that is genuinely alarming. In *The Social Conquest of Earth*, Wilson described the leaf-cutter ants of the Americas as 'the most complex social creatures other than humans'. They live by carrying carefully cut-out pieces of leaf, across (to them) colossal distances, all for the benefit of the hive: the image of the procession of ants, each one carrying a burden greater than its own body weight, is something that we find simultaneously admirable and disturbing.

City raiders: nineteenth-century engraving of termite mounds and a termite gatherer.

They don't use these cut-up leaves for food or for construction. They use them for agriculture. Once the leaves are in the safety of the nest – a nest defended by an aggressive, well-armed caste of soldiers, each one of them perfectly willing to make suicidal assaults on intruders – they are chewed into pulp, to make a mulch. They then fertilize this mulch with their own droppings. From this culture they grow fungus: a kind of fungus that is found nowhere else in nature, and that is what feeds the colony. In other words, though the colony seems entirely self-sufficient, it must get the energy it needs to power itself from outside. It is never explained in *Star Wars* how Coruscant feeds itself, but presumably it is able to exploit a nearby planet or two. That would be the best solution for humanity in the next century or so; the trouble is that finding a spare planet is such a hard job.

Some eusocial insects go in for husbandry rather than farming. Many species of ants exploit aphids, which suck sap from plants and then excrete a nutritional substance called honeydew, a process that makes the plant's nutrition available to ants. The ants feed on the honeydew, so much so that some species of aphid will not excrete until solicited by an ant. The ants care for and protect the aphids; they milk them; they carry them to new food sources when a plant is exhausted; they build shelters for them; they defend them against predators; they defend them pre-emptively by destroying the eggs of predators like ladybirds; and they carry aphid eggs to the ants' nest where they can overwinter safely. Some aphid species can only survive when looked after by a particular species of ant. The extent to which this can be called domestication depends on the way you define your terms: but, as with humans and many of the species already met in this book, both species profit from these associations.

Ants and termites are hugely successful. It has been calculated that, in 2½ acres (1 hectare) of Amazon rainforest, the total weight of all the termites and ants makes up two-thirds of the weight of all the insects in the same area. The ants alone weigh four times as much as all the terrestrial vertebrates. They provide an additional function of making inaccessible food resources available to other inhabitants of the same ecosystems; termites break down the cellulose in plants and become energy providers by way of their own esculent selves.

The numbers involved with the eusocial insects quickly assume the dizzying nature of those in astronomy. Wilson estimated the number of individual ants living on Earth at 10^{16}, or ten thousand trillion. But an examination of human numbers can also boggle the mind satisfactorily. The population of cities is difficult to estimate because it is always shifting, usually growing, and also poses questions of precisely where a city stops and becomes somewhere else. Do you, for example, separate Yokohama and Tokyo, even though they are contiguous? Figures for urban areas – for conurbations – are frightening: Tokyo 38 million, Delhi 35 million, Shanghai 34 million. Lower in the table you find New York with 23 million, Los Angeles 19 million and London 14 million.

In the course of the second decade of the twenty-first century a tipping point was reached: most people on Earth now lived in cities. That figure became 55 per cent in 2019; it is projected to reach 68 per cent by 2050. Coruscant here we come: a notion that has required the coining of a new word: such an enormous, all-encompassing city is an ecumenopolis: from the Greek and meaning world city.

The ants and termites are able to cope with their highly advanced level of sociality because they have evolved this way of life over millions of years. They have the march on humanity in this respect. The early signs of the newly extreme nature of twenty-first-century city living are that the process is causing humans problems of adaptation.

The most common cause of physical and mental illness in the twenty-first century, insofar as they can be separated, is stress. Dementia, diabetes, cardiovascular disorders and depression are all linked to the problems of stress. The easiest and cheapest method of reducing stress is by giving people access to nature. Tony Juniper, in his book *What Has Nature Ever Done For Us?*, estimates the annual cost of treating mental health problems in England at £105 billion, and the annual cost of maintaining 27,000 parks and green spaces in the entire UK at £630 million.

Experiment after experiment has proved the health benefits, both psychological and physiological, that come from contact with nature. People recover faster from operations if they can see out of a window, faster still if they can see a tree. People doing demanding proofreading tasks do better after a break; better still after a break and a walk; best of all after a break and walk in a green space. Children with learning difficulties and attention problems frequently do better with outdoor lessons; and better still when in contact with nature.

All of which appears to demonstrate that humans are not ready for the intense degree of eusociality that we observe in our role model species of ants and termites, and which we are in the process of imposing on ourselves. We have evolved technically with immense speed, such is the remarkable nature of our own species – so much so that we have outpaced the nature of our own minds, which retain a great deal of our past. Part of us remains with our hunter–gatherer ancestors on the savannahs of Africa (with all those termite mounds).

We are capable of creating Coruscant, creating an ecumenopolis, and that is certainly the direction in which we are moving: the human population of the Earth is, at the time of writing, 7.7 billion and growing at the rate of about 82 million a year. People have pointed out, and rightly, that the rate of increase has slowed down and continues to do so: but a slower rate of increase is still an increase. This can be understood as the greatest problem of them all: the ecological crisis that dare not speak its name. The solution, if there is a solution, lies in the education and empowerment of women: the example of Kerala in India is cited as a classic model of that principle in action.

But right now the world's human population is increasing, and the cities are growing and joining up. Our skills in eusociality brought us where we are: future generations will see where they take us.

ONE HUNDRED

POLAR BEAR

'And I looked, and behold a pale horse: and his name that sat on him was Death, and Hell followed with him. And power was given unto them over the fourth part of the earth, to kill with sword, and with hunger, and with death, and with the beasts of the earth.'

Revelation 6:8

Two images tell the story of the polar bear: the greatest revisionist animal of them all. The first is a painting by Edwin Landseer of 1864, one that represents the ultimate victory of nature over humanity. The second is the generic image of a polar bear swimming across an empty sea, looking, looking for the ice-sheet that is melting and retreating before his eyes: symbol of the ultimate victory of humanity over nature. Less than two centuries separate them. We have changed our minds utterly about the relationship of our own species with every other species on Earth: even the fiercest and strongest of them all.

Landseer was a great Victorian painter and sculptor of animals: his *The Monarch of the Glen*, showing a red deer stag, is almost as famous as his four lions, majestically maned and lying at the foot of Nelson's Column in Trafalgar Square in London. He also painted polar bears.

His subject in 1864 was the expedition of Sir John Franklin of 1845. It was an attempt to find the non-existent Northwest Passage: a shortcut through Canada to all the commercial opportunities of the East, and so to instant fame and fortune. But the expedition's ships got icebound in the northern seas, and the men never got away. They made camp, hoping to survive the winter and sail on when the ice had melted, but failed. Their remains were eventually found in 1854. They died from starvation, hypothermia, TB, scurvy and lead poisoning (which came from early examples of canned food). It was reported that they had also tried cannibalism, at which there was revulsion and furious refutation. Subsequent investigations showed convincing evidence that this was in fact the case. The story of the doomed expedition was, then, something that people felt strongly about.

Landseer painted an imaginative reconstruction of the end of the expedition at the hands – or rather the claws and the teeth – of polar bears, two of them enthusiastically ransacking the camp. Just in case that was an insufficient clue to the moral meaning of this classic picture that tells a story, Landseer gave it a title:

Nature's power: Man Proposes, God Disposes *by Edwin Landseer (1864).*

Man Proposes, God Disposes. God, in the form of nature, is still in charge. Men can do their best, but in the end God and nature will always be mightier. We need to be humbler, because for every human ready to change the world a polar bear awaits. You can't argue with facts like that.

Polar bears are fine animals all right. They are high up on many a bucket list, for these days we wish to see them and celebrate them, rather than run from them. They are the largest surviving land predators on Earth, and can be three times the size of a lion; a big male polar bear weighs up to 1550lb (700kg). The largest polar bear on record, reportedly weighing 2209lb (1002kg), was a male shot at Kotzebue Sound in northwestern Alaska in 1960. Polar bears live more or less within the Arctic Circle, and are so well adapted to the cold that they start to overheat at 10°C (50°F). They have a thick layer of fat and great thick pelt of insulating fur.

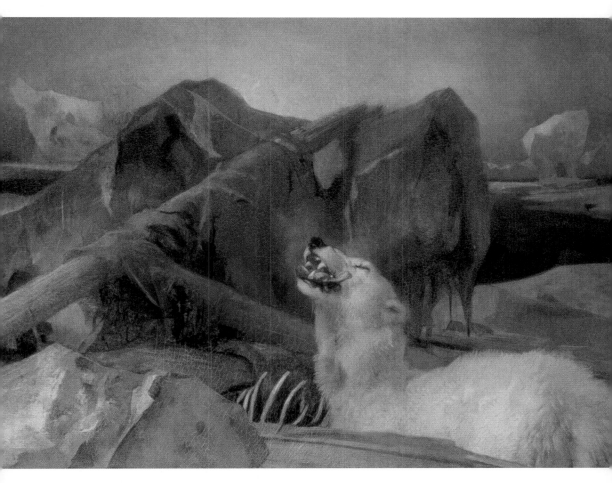

They retain more of their historic range than any other land-based carnivore, for they have chosen a habitat that humans have comparatively little enthusiasm for penetrating. Their range takes in five nations: Russia, Canada, United States by means of Alaska, Denmark by way of Greenland, and Norway by way of Svalbard. They are about snow, ice and open water, and their prey is seals.

They can be found on the ice around the continental shelf of the five nations of their range; they are somewhat scarce north of 88 degrees. Their layer of fat makes them extremely buoyant; they can almost be considered marine mammals, they are so much at ease in the water. Not that they can match seals for speed and agility when it comes to swimming; polar bears travel in a massive version of the doggy paddle. They don't attempt to catch seals by outswimming them: instead they use surprise; coming out of the water on the blind side of a seal basking on the ice is a favourite ploy. They are also adept at finding the dens in which seal pups are hidden: they can smell seal 1 mile (1.5km) away.

They can swim for miles, almost indefatigably, thanks to their buoyancy. A female was recorded as swimming in the Bering Sea for nine days without a break, covering 400 miles (650km). She survived the trip; her cub didn't. One of the problems of this environment is that there is no drinking water: polar bears get round this by their ability to metabolize the fats in the blubber of the seals they consume.

Polar bears have a reputation for unthinking ferocity, though people who spend time with them report that most of them tend to be cautious about confrontation. That, after all, is the nature of their hunting technique: they are sneaker-uppers rather than head-on chargers. Most polar bears see very few humans; when they do so, their response is unpredictable.

They have been hunted, initially for their fur, though this, for all its thickness, was considered less desirable than that of Arctic foxes and even reindeer. In more recent times they have been hunted for sport, from snowmobiles and aeroplanes, a method that doesn't put the hunter at any great risk. This practice mostly came to an end when the five polar bear nations signed the International Agreement on the Conservation of the Polar Bear in 1973. This is a relatively early date for such legislation, and it shows the immense popularity of the species, along with an awareness of the likely public response to any spectacular loss.

Polar bears are great seizers of the public imagination, for their vastness, their whiteness, the legend of their ferocity and the fact of their remoteness. They became hugely popular zoo animals: Brumas, born in London Zoo in 1949 and the first of her species successfully reared in the UK, became a national celebrity and in 1950 some 3 million visitors came to see her in the course of the year; she died in 1958. Another polar bear, Pipaluk, was born in the same zoo in 1967, but by this time there were growing doubts about the ethics of keeping the species in captivity. Pipaluk went to a zoo in Poland.

As our thinking about our fellow animals and the way we run the planet began to change, so, most dramatically, did the role of the polar bear. They ceased to be Landseer's wrath of God, and began to be victims of the wrath of humankind. Or if not the wrath, the unstoppable ambition.

They are now seen as an indicator of the health of the Arctic ice fields and, by extension, of the health of the planet. As the global temperature rises and continues to rise, so the ice of the Arctic melts. There is less habitat for polar bears; they are forced to swim farther in search of ice floes to rest on, and of seals basking on them, which depletes their energy. The ice is thinner, and in some places less capable of bearing the weight of polar bears.

All this tells us, in a sharp vivid fashion furnished with a sharp and unforgettable image, of the reality of climate change. This has come about because the heat of the sun is trapped in the Earth's atmosphere by increased amounts of carbon

Power of humanity: melting ice floes, and a polar bear swimming in the Beaufort Sea off Alaska (photograph by Gerrit Vyn).

dioxide and other gases (including methane, see Chapter 7 on cattle). This has mostly been the result of our burning of fossil fuels, a process that continues every time you get in the car. The rise in temperature has been measured at 0.8°C (1.4°F) since 1880 and, if that doesn't sound a lot when you are deciding whether or not to wear a coat, it matters vastly in the way the Earth is run. The oceans are warmer, and the sea level across the world is higher as more water is added by way of melting ice. In the meantime, we continue to destroy the forests that absorb carbon dioxide and cool the planet. The process of putting more energy into a system makes it (obviously) more energetic: we put more energy in the Earth's climate by retaining more heat and it responds by creating more extreme weather events. It is calculated that, if we carry on as we are, global temperatures will rise 1.5°C (3°F) by 2050 and 2.6–4.8°C (5–9°F) by the end of the century.

Al Gore, former vice president of the United States, popularized the image of the polar bear, forlornly swimming eternally towards a world that has now melted, as an image of climate change. This has subsequently been questioned; polar bear numbers are, at the moment, relatively stable; the species is categorized as Vulnerable, so not in any immediate danger of extinction. That doesn't invalidate the image or the arguments behind it. Climate change is not just an emergency for polar bears: it's a problem for everything that lives on the planet.

Polar bears, because of their imposing nature and their symbolic power, have helped us to understand that a state of emergency exists. In 2018 carbon emissions reached a record high, not a great stat for a planet that's supposed to be throttling back. Also in 2018 David Attenborough was invited to take a 'people's seat' at the UN Climate Conference in Katowice. He said climate change is 'our greatest threat in thousands of years'. The Black Death? World war? Ethiopian famine? Well, climate change is going to be a great deal worse: a greater threat than the lions that shaped the lives of our ancestors, as seen in Chapter 1 of this book. At the World Economic Forum in Davos in 2019, Attenborough told business leaders and politicians: 'We need to move beyond guilt or blame and get on with the practical tasks at hand.'

It's nothing new. Drastic climate change has been experienced before on the Earth. The best example can be found in the years that followed the Permian era. That's when, 251 million years ago, a series of volcanic eruptions in Siberia – or perhaps a meteor impact – caused a release of carbon dioxide and methane. Greenhouse gases. As a result, all the rainforests were destroyed, their beneficial effects were lost and the soil they once stood in was eroded away. The oceans lost their oxygen; 96 per cent of all species were wiped out.

The good news is that the Earth recovered. The snag is that it took 20 million years.

So it's worth paying attention to the twenty-first-century polar bear, swimming for ever to the island of nowhere. It's a destination we are all heading for.

Unless we do something. And unlike the creatures who lived on Earth during the Permian extinction event, we can make choices. We are in control of the current extinction event. We started it; we can stop it; we can put it into reverse.

Greta Thunberg, the teenage apostle for the planet Earth, said with perfect simplicity: 'We still have time to turn everything around... but that short period of time isn't going to last for ever.'

We have gone through a series of drastic changes in the way we understand other species, our planet and ourselves in the course of the short time we have lived on Earth, in the course of the shorter time since we invented civilization, and in the course of the very short time indeed since we invented industry, intensive farming and city living. We have now realized that we have reached a classic tipping point: either we act or we go extinct. The future, of our own species and of just about every other species, is in our hands.

EPILOGUE

We have all experienced those rare moments when the Earth stops revolving. Time is suspended. You scarcely need to breathe, your awareness is no longer restricted to humdrum things like seeing and hearing, and in the perfect stillness you seem to grasp things beyond your normal capacity for understanding.

I was in rainforest of course. Everyone else in the camp was asleep, but I sat, still fully dressed, in the folding chair in my hut, which was on stilts 6ft (1.8m) above the forest floor. It's been calculated that three-quarters of all species are rainforest species: in the dark I could sense them all, every single one of them, teeming all around: breathing, transpiring, dying, living, eating, being eaten, copulating, germinating, hatching, being born.

I was in Borneo, on the banks of the Kinabatangan River. It had been a good day. I had shared the river with twenty enchanting pygmy elephants, watched a huge male orang-utan build his shelter for the night, and counted six species of hornbill including the impossible rhinoceros hornbill. I had visited a forest restoration project; palm-oil plantations have destroyed a great deal of forest, but there is still some fine forest left and more is being replanted. I had visited areas of forest that the local conservation organization Hutan had recently acquired, with funding from the World Land Trust, a conservation organization based in the UK; I was there as a council member of that organization. We had talked fancifully of what Hutan would do if we could raise £1 million, and we had looked at an area of glorious riverfront forest that would be the dream acquisition. As I sat, still and silent in my hut, I knew that we would find this million and the tract of forest would soon be safe; a year later it was so.

Hutan began as a project to observe the orangs, but it was soon obvious that observation was not enough. So Hutan got active in conservation. The people from Kampung Sukau are now deeply involved: observing apes, planting trees, patrolling the forest to deter poachers and illegal loggers, guiding ecotourists and alleviating clashes between local people and wildlife, which mostly means keeping the elephants out of the fields. Money from these projects goes into the local community. Here is a template conservation project, doing great work in what is perhaps the most important habitat on Earth.

Despair is easy in a city. But when you are out there among the living trees and the life is teeming all about, you feel only the life. That's when you know, with a very deep certainty, that every habitat on Earth is saveable, and that every species on Earth is saveable. The rest is only a matter of will.

While the forest still stands: rhinoceros hornbill in flight (photographed by John Holmes).

INDEX

PICTURE CREDITS

Pg 1 Bridgeman Images; Pg 2-3 Bridgeman Images / Photo © Christie's Images; Pg 5 Wikimedia Commons https://commons.wikimedia.org/wiki/File:Jan_Brueghel_de_Oude_en_Peter_Paul_Rubens_-_Het_aards_paradijs_met_de_zondeval_van_Adam_en_Eva.jpg / Jan Breughel de Oude (1568-1625) and Peter Paul Rubens (1577-1640), *The Garden of Eden with the Fall of Man*; Pg 6 Digital image courtesy of the Getty's Open Content Program, 92.PB.82; Pg 10 Alamy Stock Photo / Ian Dagnall; Pg 12 Nat Geo Image Collection / Stephen Alvarez; /Courtesy of Ministère de la culture et de la communication, DRAC Rhône-Alpes, Service régional d'archéologie; Pg 16 Bridgeman Images / The Stapleton Collection; Pg 18 © The Trustees of the British Museum; Pg 20 Nature Picture Library / John Sparks; Pg 24 Getty Images / Royal Geographical Society; Pg 26 Getty Images / DEA Picture Library; Pg 30-31 Alamy Stock Photo / Everett Collection Historical; Pg 32 Wikipedia / https://en.wikipedia.org/wiki/File:Bison_skull_pile-restored.jpg; Pg 34 Alamy Stock Photo / Asar Studios; Pg 36 akg-images; Pg 39 Bridgeman Images; Pg 41 Alamy Stock Photo / World History Archive; Pg 44 Shutterstock; Pg 46 Alamy Stock Photo / Photo12; Pg 49 Alamy Stock Photo / National Geographic Image Collection; Pg 50 BluePlanetArchive.com / © Doug Perrine; Pg 52 Getty Images / Universal History Archive; Pg 54 Getty Images / Buyenlarge; Pg 57 Alamy Stock Photo / Dave Watts; Pg 59 Alamy Stock Photo / Florilegius; Pg 61 AKG / Hervé Champollion; Pg 62 SPL / American Philosophical Society; Pg 65-67 Getty Images / Field Museum Library; Pg 70 Getty Images / LMPC; Pg 75 Getty Images / De Agostini Picture Library; Pg 79 Alamy Stock Photo / Nature Picture Library; Pg 80 Getty Images / University of New Hampshire / Gado; Pg 83 Bridgeman Images / Photo © Photo Josse; Pg 85 The Metropolitan Museum of Art / Purchase, Lila Acheson Wallace Gift and Bequest of George Blumenthal, by exchange, 2015; Pg 87 Bridgeman Images / © Archives Charmet; Pg 89 Alamy Stock Photo / National Geographic Image Collection; Pg 92 Bridgeman Images; Pg 94 Getty Images / Photo 12, Pg 95 Alamy Stock Photo / Peter Barritt; Pg 97 Bridgeman Images / Mondadori Portfolio/Electa/Sergio Anelli; Pg 99 From Rudyard Kipling *The Jungle Book*, (The Reprint Society Ltd, 1955). Illustration by Stuart Tresilian (1891-1974); Pg 103 Bridgeman Images; Pg 105 Bridgeman Images / Purix Verlag Volker Christen; Pg 107 Wikimedia Commons https://commons.wikimedia.org/wiki/File:Verrocchio,_Leonardo_da_Vinci_-_Battesimo_di_Cristo.jpg; Pg 109 Getty Images / Universal Images Group / Windmill Books; Pg 112 Getty Images / Alfred Eisenstaedt; Pg 113 Alamy Stock Photo / Peter Horree; Pg 117 The Metropolitan Museum of Art / Rogers Fund, 1917; Pg 118 Alamy Stock Photo / Chris Howes / Wild Places Photography; Pg 121 Bridgeman Images / © British Library Board. All Rights Reserved; Pg 123 Bridgeman Images / Private Collection; Pg 124 Alamy Stock Photos / CPA Media Pte Ltd; Pg 127 iStockphoto.com / whitemay; Pg 131 Wikimedia Commons https://commons.wikimedia.org/wiki/File:Jan_Brueghel_de_Oude_en_Peter_Paul_Rubens_-_Het_aards_paradijs_met_de_zondeval_van_Adam_en_Eva.jpg; Pg 134 123RF.com / Bidouze Stephane; Pg 136 Bridgeman Images / National Trust Photographic Library; Pg 138 Minneapolis Institute of Art / The John R. Van Derlip Fund; purchase from the collection of Elizabeth and Willard Clark; Pg 140 Wikimedia Commons / https://commons.wikimedia.org/wiki/File:Pieter_Bruegel_de_Oude_-_Twee_geketend_apen.jpg; Pg 143 123RF.com / neyro2008; Pg 145 Alamy Stock Photo / MasPix; Pg 147 Alamy Stock Photo / Tom Bean; Pg 149 Getty Images / Corbis / Kevin Fleming 77902; Pg 150 Alamy Stock Photo / PjrStudio; Pg 153 akg-images / MPortfolio / Electa; Pg 155 Alamy Stock Photo / National Geographic Image Collection; Pg 159 Getty Images / Deb Garside / Design Pics; Pg 160 Getty Images / Stock Montage; Pg 164 Alamy Stock Photo / Peter Horree; Pg 169 Getty Images / ullstein bild Dtl; Pg 172 Alamy Stock Photo / Heritage Image Partnership Ltd; Pg 174 Bridgeman Images / © British Library Board. All Rights Reserved; Pg 177 Getty Images / DEA / G. Gagli Orti; Pg 178 *Juvenile female sponger*. Photo: Ewa Krzyszczyk; monkeymiadolphins.org; Pg 181 Courtesy National Gallery of Art, Washington, Rosenwald Collection; Pg 183 John Frost Newspapers / *Daily Mirror* October 9, 1961 front cover; Pg 186 Alamy Stock Photo / Old Images; Pg 190 Bridgeman Images; Pg 192 akg-images / Rabatti & Domingie; Pg 195 Alamy Stock Photo / A.F. Archive; Pg 197 Alamy Stock Photo / Nature Picture Library; Pg 200 Alamy Stock Photo / The Granger Collection; Pg 203 USFWS National Digital Library / https://digitalmedia.fws.gov/digital/collection/natdiglib/id/25278/rec/1; Pg 204-205 Walton Ford (b. 1960), *Falling Bough*, 2002. Courtesy of Kasmin Gallery, New York; Pg 208 Shutterstock / Jaco Visser; Pg 211 Alamy Stock Photo / Daisy Photography; Pg 213 Bridgeman Images / Photo © Christie's Images; Pg 216 Alamy Stock Photo / incamerastock; Pg 219 Alamy Stock Photo / Pictures Now; Pg 221 Alamy Stock Photo / Historic Collection; Pg 224 iStockphoto / USO; Pg 226 Minneapolis Institute of Art / Gift of Rhodes Robertson; Pg 229 Wikimedia Commons / https://commons.wikimedia.org/wiki/File:Jacques_Louis_David_-_Bonaparte_franchissant_le_Grand_Saint-Bernard,_20_mai_1800_-_Google_Art_Project.jpg; Pg 232 Getty Images / Fine Art; Pg 235 Bridgeman Images; Pg 236 Getty Images / Fine Art; Pg 239 Bridgeman Images / © Giancarlo Costa; Pg 243 Rawpixel Ltd / Elizabeth Gould (1804-1841) https://www.flickr.com/photos/vintage_illustration/41972947871; Pg 247 Alamy Stock Photo / Pictures Now; Pg 251 Bridgeman Images; Pg 253 Bridgeman Images / Florilegius; Pg 255 Library of Congress LC-DIG-ds-09858; Pg 257 Getty Images / Fine Art; Pg 258 Alamy Stock Photo / Dorling Kindersley Ltd; Pg 261 Alamy Stock Photo / Premiergraphics; Pg 263 Bridgeman Images / Look and Learn; Pg 266 Bridgeman Images; Pg 269 Alamy Stock Photos / Glasshouse Images; Pg 271 Wikimedia Commons / https://commons.wikimedia.org/wiki/File:366_Iceland_or_Jer_Falcon.jpg; Pg 273 Getty Images / DEA / M. Seemuller; Pg 275 Bridgeman Images / Photo Jose; Pg 277 Bridgeman

Images / Photo © Christie's Images; Pg 279 Alamy Stock Photo / Universal Images Group North America LLC; Pg 282 Charles Darwin (1809-1882), *A Monograph on the Sub-Class Cirripedia with Figures of all the Species*, (London: The Ray Society, 1854). Plate XXV; Pg 284 Getty Images / Photo Josse / Leemage; Pg 286 Getty Images / DEA / ICAS94; Pg 287 Flickr / Internet Archive Book Images; Pg 289 Alamy Stock Photo / Auscape International Pty Ltd; Pg 291 Bridgeman Images / © British Library Board. All Rights Reserved; Pg 294 Alamy Stock Photo / Chronicle; Pg 295 Getty Images / De Agostini Picture Library; Pg 297 Alamy Stock Photo / Historic Images; Pg 300 Getty Images / GraphicaArtis; Pg 302 Bridgeman Images / Photo © Christie's Images; Pg 303 Bridgeman Images / © Mark Adlington; Pg 305 Alamy Stock Photo / Niday Picture Library; Pg 307 Alamy Stock Photo / Godong; Pg 309 Bridgeman Imges / Lukas - Art in Flanders VZW; Pg 311 WWT; Pg 313 © The Sir Peter Scott Art Collection courtesy Dafila Scott. Photo: Alamy / John Keates; Pg 316 Alamy Stock Photo / The Picture Art Collection; Pg 318 Alamy Stock Photo / Nature Picture Library; Pg 320 Alamy Stock Photo / V&A Images; Pg 324 Alamy Stock Photo / Zoonar GmbH; Pg 325 Getty Images / De Agostini Picture Library; Pg 327 Getty Images / SSPL / The National Archives; Pg 330 Getty Images / Heritage Images; Pg 332 Alamy Stock Photo / Heritage Image Partnership Ltd; Pg 334 Alamy Stock Photo / Heritage Image Partnership Ltd; Pg 335 Jeremy Saxton, *Baiji*, watercolor, Courtesy Saatchi Art; Pg 338-339 Bridgeman Images / Photo © Christie's Images; Pg 341 Alamy Stock Photo / INTERFOTO; Pg 343 Alamy Stock Photo / Granger Historical Picture Archive; Pg 344-345 Getty Images / Field Museum Library; Pg 348-349 Bridgeman Images; Pg 350 Bridgeman Images / Luisa Ricciarini; Pg 353 Alamy Stock Photo / BSIP SA; Pg 355 Alamy Stock Photo / Alpha Historica; Pg 356 Getty Images / Fine Art Photographic; Pg 359 Alamy Stock Photo / Godong; Pg 362 Alamy Stock Photo / Painters; Pg 364 Bridgeman Images / Photo © Christie's Images; Pg 366 Bridgeman Images / Photo © Christie's Images; Pg 368 Alamy Stock Photo / The Picture Art Collection; Pg 371 Alamy Stock Photo / CPA Media Pte Ltd; Pg 373 Bridgeman Images; Pg 375 Alamy Stock Photo / Antiquarian Images; Pg 377 Getty Images / Topical Press Agency; Pg 379 Alamy Stock Photo / Niday Picture Library; Pg 381 Bridgeman Images; Pg 384 Alamy Stock Photo / CuriousCatPhotos; Pg 386 Alamy Stock Photo / Photo12; Pg 388 Alamy Stock Photo / The Picture Art Collection; Pg 390 Yang Shanshen. Photo Bridgeman Images / © Christie's Images; Pg 394 Alamy Stock Photo / Marc Tielemans; Pg 396 Bridgeman Images / Photo © Christie's Images; Pg 400 Bridgeman Images / National Geographic Image Collection; Pg 404 Barry Roal Carlsen, *Saola*; Pg 406 WWF / Jeremy Holden; Pg 408-409 Bridgeman Images / © Look and Learn; Pg 411 Getty Images / David McNew; Pg 412 Alamy Photo Stock / Ivan Vdovin; Pg 415 Bridgeman Images / © British Library Board. All Rights Reserved; Pg 416 Getty Images / De Agostini Picture Library; Pg 418 Wikimedia Commons / https://commons.wikimedia.org/wiki/File:Alexander_von_Humboldt,_by_Friedrich_Georg_Weitsch.jpg; Pg 421 Bridgeman Images / Photo © Christie's Images; Pg 423 Bridgeman Images / National Geographic Image Collection / Volkmar K. Wentzel; Pg 426 Bridgeman Images / Look and Learn / Valerie Jackson Harris Collection; Pg 429 Alamy Stock Photo / Falkensteinfoto; Pg 431 Bridgeman Images / Roy Miles Fine Paintings; Pg 433 Alamy Stock Photo / Everett Collection Inc; Pg 435 Getty Images / Dan Antoche-Albisor; Pg 438 Alamy Stock Photo / Life on white; Pg 440 Nature Picture Library / Julian Hume; Pg 443 National Geographic Creative / The Refinery; Pg 447 Alamy Stock Photo / Heritage Image Partnership Ltd; Pg 448 Alamy Stock Photo / World History Archive; Pg 452-453 Bridgeman Images; Pg 455 Nature Picture Library / Gerrit Vyn; Pg 458-459 Alamy Stock Photo / FLPA.

Every effort has been made to find and credit the copyright holders of images in this book. We will be pleased to rectify any errors or omissions in future editions.

ACKNOWLEDGEMENTS

Many thanks to everyone who made this book possible. At Simon & Schuster, Ian Marshall, for making it happen and Laura Nickoll for holding it all together. Thanks to Joanna Chisholm for a superb job of editing, to Keith Williams, the designer, to Liz Moore for researching the pictures, and to Lorraine Jerram for proofreading. I'm deeply grateful to Matt Shardlow, CEO of the invertebrate charity Buglife, for checking the manuscript. At Georgina Capel Associates, thanks to George as always, also to Irene. And back home in Norfolk, well, I couldn't have done it without Cindy, Joseph and Eddie.

First published in Great Britain by Simon & Schuster UK Ltd, 2020

1 3 5 7 9 10 8 6 4 2

Simon & Schuster UK Ltd
1st Floor
222 Gray's Inn Road
London WC1X 8HB

www.simonandschuster.co.uk
www.simonandschuster.com.au
www.simonandschuster.co.in

Simon & Schuster Australia, Sydney

Simon & Schuster India, New Delhi

Editorial Director: Ian Marshall
Design: Keith Williams, sprout.uk.com
Project Editor: Laura Nickoll
Picture Researcher: Liz Moore
Proofreader: Lorraine Jerram

A CIP catalogue record for this book is available from the British Library

Hardback ISBN: 978-1-4711-8632-5

Ebook ISBN: 978-1-4711-8633-2

Printed in China